# Geography
# Yesterday
# and Tomorrow

# Geography Yesterday and Tomorrow

*Edited by*

## E. H. BROWN

FOR THE ROYAL
GEOGRAPHICAL SOCIETY

OXFORD UNIVERSITY PRESS
1980

*Oxford University Press, Walton Street, Oxford* OX2 6DP

OXFORD LONDON GLASGOW
NEW YORK TORONTO MELBOURNE WELLINGTON
KUALA LUMPUR SINGAPORE JAKARTA HONG KONG TOKYO
DELHI BOMBAY CALCUTTA MADRAS KARACHI
NAIROBI DAR ES SALAAM CAPE TOWN

*Published in the United States by*
*Oxford University Press, New York*

**British Library Cataloging in Publication Dates**

Geography yesterday and tomorrow.
1. Geography
I. Brown, E
910      G116      80–40193

ISBN 0-19-874096-4

*Printed in Great Britain by*
*Butler & Tanner Ltd., Frome and London*

# CONTENTS

# Introduction

PERHAPS the questions most commonly asked of geographers are, 'What is geography?' 'What do geographers do?', and 'What use is geography?' Quite frequently the questioners are mildly, but sometimes vigorously, critical and imply that geographers try to do everything but do nothing that is not better done by others. We are regarded as being neither fish nor flesh nor good red herring but trespassers into every academic discipline from physics to psychology. There is no doubt that a geographer's interests do range far and wide, but it is its breadth which makes geography so attractive to so many of us. Would-be university students of geography commonly give two types of response to the question, 'Why do you wish to read geography?' The first is, naturally enough, that in terms of examination performance they reckon they are good at it; the second that they like the breadth of the subject, the way its subject-matter spans the arts and sciences, that they are concerned about the future of the world and their place in it, and geography deals with such realities and that they do not wish at this stage in their lives to be tied down to a too narrowly defined discipline.

Human beings are territorial animals. We relate very strongly to the space around us, continuously mapping mentally the territory we occupy and traverse, defending it when it is attacked and enlarging and contracting it as need arises. This is equally true of the individual, the social group, and the nation. Most of us are naturally curious about what lies beyond our daily orbit and have a marked tendency to suppose that such places offer superior environments to our own. So we seek out and explore the resources not available to us in our own territory. In truth human beings, like all animals, are consumed with spatial curiosity, and that is the basis of an interest in geography. Sooner or later in the process of territorial enquiry the focus of interest descends from the general to the particular, from a concern for the whole continent, country, or region to a search for a specific resource, be it mineral, plant, animal, climate, or soil. So geography has on the one hand its generalities and on the other its particularities; its adherents to either a regional or a systematic approach; the geography of continents and countries or the pursuit of some branch of either physical or human geography.

It is not surprising, then, to find that geography is frequently defined as being concerned with the study of the areal differentiation of the earth's surface, particularly with the processes which determine the peculiarities of places. A more consciously humanistic definition, preferred by many, is that geography is concerned to investigate the relationships between human beings and their environments, both the influence of environment upon man and that of man upon the environment. But if the content of geography can be reduced to one word, as it might be said that botany is concerned with plants and biography with 'chaps', so we are concerned, not as Clerihew would have had it, just with 'maps', but certainly with places. People make places and places make people, but places, not people, are geography.

In 1978, in anticipation of the 150th anniversary of its foundation in 1830, the Royal Geographical Society set up a committee to determine how the event should be celebrated. Prompted by Mr Derek Diamond it made a recommendation subsequently accepted by the President and Council that the publication of a book concerned with the Society's academic interests should be included. A working group comprising Mr Diamond, Professor W. G. V. Balchin (Chairman Education Committee), Dr C. L. Bertram (Honorary Secretary, and Chairman Expeditions Committee), and Professor E. H. Brown (Honorary Secretary, and Chairman Research Committee) was asked to determine the form such a book might take. It thought the 150th anniversary of the RGS an opportune moment to review the growth of academic geography in the context of that of the RGS and to look into a crystal ball and seek out likely future trends. Professor Brown was asked to edit the book on behalf of the RGS.

The contents of the book reflect these aims in broad terms but individual contributors have been constrained and they have interpreted their briefs in ways which reflect the individual characters of their topics and their personal inclinations. No attempt has been made to cover every topic which might be included under the umbrella of geography but it is hoped that the reader will get an over-all, if not comprehensive, view of the subject. The nature of geography can only be understood in the light of the history of its study and Walter Freeman's contribution elucidates the origins of academic geography in Britain and the crucial role played by the RGS in encouraging its inclusion in the universities, first and abortively, in the 1830s, at University College London, later and successfully at Cambridge and Oxford, most notably in the financing of Mackinder's Readership at Oxford. The development of the teaching of geography in schools owed much to Francis Galton who, making amends for University College's earlier lapse, in the mid-nineteenth century was a staunch advocate of scientific geography and geographical education within the RGS. But the seminal work was that of J. Scott Keltie, whose famous report of 1886 was commissioned by the RGS. The foundation in 1893 by Mackinder of the Geographical Association, the association of teachers of geography, was a measure of the success of the RGS in fostering the teaching of geography in schools. On the other hand, the foundation of the Institute of British Geographers in 1933 was a response to what was then perceived to be a failure of the RGS to encourage and publish the products of research, particularly in the field of human geography. Happily relations between the RGS and its sister geographical organizations are today cordial. They jointly sponsor an annual lecture, their officers visit one another's conferences, the editors of their publications consult each other, the IBG has its offices in the house of the RGS, and there is much overlap in membership.

The theme of the development of geography in education is continued by Norman Graves, who outlines its growth in primary and secondary schools, recent curriculum developments, and the nature of current research in geographical education. The four major fields within physical geography—air, water, land, and life—are examined with a critical eye, and the relatively small contribution physical geographers are making to the environmental debate are evaluated. A strong move towards applied physical geography is noted and cautiously commended. Maps are models of the surface of the earth and the RGS has always had a strong link with survey and mapping, national and commercial. The making of maps has been revolutionized in recent years, new sources of data are available from orbiting satellites and computer-controlled mapping and map production are already with us. Recent developments in the whole field are reviewed by two authors themselves very much involved with research in cartography.

During the past twenty years human geography has been buffeted by winds of change blowing from a variety of directions: spatial analysis and the quantitative revolution, much more sophisticated approaches to economic geography, perception studies linking geography with psychology, concern for social welfare, consciously political approaches—all these and other trends are covered by Alan Wilson, Emrys Jones, and Hugh Prince in reviews of Human, Social, and Historical geographies.

By dint of the personalities of some of its professional geographical Fellows, the RGS has been at the centre of some major developments in the subject. Professor L. Dudley Stamp was an outstanding example, and three contributions, by Alice Coleman, Terry Coppock, and Melvyn Howe, are concerned with Land-Use Studies, Leisure and Recreation, and Medical geography, topics which he as an individual and in co-operation with others pioneered and encouraged. They are representative only of many themes within academic geography which the RGS seeks to foster. The book concludes with an enquiry by Bill Mead into what has happened to regional geography, for so long a focus of geographical teaching. It blossomed in Oxford under Mackinder, but never took root in Cambridge. Mackinder carried it to London with him, where it became firmly established, but in recent years, as geography has become more systematic and less regional, area studies, concerned not only with the geography of places but also with their history, language, and literature have come, in some respects, to replace it. It is perhaps dormant rather than dead; when pressed hard, many, perhaps most, geographers will say that systematic though their interests are, they still subscribe to the view that geography is a synthesizing subject and its proper concern is with the character of places, areas, countries, continents, and the world.

This book is designed to do many things. For Fellows of the RGS it seeks to record the events of 150 years of their society's concern for geography in education. For students of geography of any age and status it seeks to explain the nature of the subject, how it has developed and where it might be going. For the members of the public it is an attempt to answer two of the questions posed at the beginning of this introduction, what geography is and what geographers do. As for the third, 'What is the use of geography?', the reader may infer a great deal from the contents of the book itself, but a training in geography can be recommended at all levels in the educational system, primary, secondary, and tertiary, just because of its breadth and the bridges its builds between the physical and human sciences, between the theoretical and the applied, and between the literate and the numerate. In an increasingly complex but immediately interacting world in which most events in remote places become almost instantly matters of media reportage, it is even more necessary than in the past to be able to see the wood for the trees and to know the world outside our own cabbage patches.

In a very real sense our futures depend upon the need for us to appreciate how and why places and people differ.

ERIC H. BROWN
Honorary Secretary
Royal Geographical Society
*University College London*

# 1 The Royal Geographical Society and the Development of Geography

T. W. FREEMAN

### INTRODUCTION

THE Royal Geographical Society has a varied appeal to its members. Some care most for its meetings, especially those concerned with dramatic discoveries: thousands flocked to see and hear the heroes of nineteenth-century exploration as they did to welcome survivors of the Scott Polar expedition and the conquerors of Everest. Others prefer meetings of a more reflective character, such as those at which geographical papers are given, followed by discussions. Members rarely able to attend meetings of a society which did not until recently organize any outside London will be content to receive the Society's *Journal*, while a limited number of people will visit the splendid library and map collection at intervals.

Every society has to attract members and retain them: to do so it must have aims that seem worthy of support. In the nineteenth century the exploration of the world captured the imagination of a wide range of people and received the valued support of the royal family, not least of the Prince of Wales who gave a great deal of time to the work of the Society. The academic purpose must remain even if only a minority of the members of a geographical society care about it, though many more can be attracted to academic problems if these are presented in an interesting and jargon-free manner. Fortunately throughout its history the Royal Geographical Society has always tried to avoid encouraging esoteric academic cults though some would say that it has also been too conservative to welcome new developments in geography, too cautious to look beyond the safe and sound expression of familiar views, too obscurantist to care for new ideas and theories of which some—but only some—will prove crucial to academic growth.

Academically a society may aspire to lead or be content to follow developments; it may even appear indifferent to them. But if a society hopes to control academic work in universities, to set a scheme for research, to direct young researchers, it is hardly likely to be successful, at least in Britain. On the other hand, through its editorial policy a society may have considerable power, as shown, for example, in the emphasis on geomorphology and cartography that was characteristic of the RGS during the inter-war period of 1919–39. On pp. 33–6 the struggle to introduce more papers on human, economic, and regional geography at this time is told, but gradually, and especially after the War of 1939–45, a far wider range of papers was published in the *Journal*, covering most aspects of geography. This did not lessen the time-hallowed responsibility of encouraging exploration, not for its own sake but for its academic significance. Indeed, from the early days of RGS many of its wisest leaders have been rightly concerned with the quality of explorers and their work rather than with their thrilling adventures. Every effort has been made to help them, not only with funds but also through the publication of the famous *Hints to Travellers* from 1854 (Mill 1930; 244) and the award of a Diploma in Surveying from 1896 to 1948, when it was agreed that the Royal Institution of Chartered Surveyors, as well as several universities, provided courses and examinations which met this need (*Geogrl. J.* 112 (1948), 113–14).

Education has been a concern of RGS from its earliest years, and the association of geographers with students of classical and biblical antiquity developed in the 1830s, by which time the great figures of German geography, Alexander von Humboldt and Carl Ritter, were men of European fame. The appointment of Maconochie to the staff of University College London in 1834 was regrettably short-lived (see below) but Mackinder's appointment at Oxford in 1887 marked the beginning of modern geographical education in British universities, generously supported by RGS. Study of the earlier appeals made to Oxford and Cambridge in 1874 and 1879 shows that the idea of training potential explorers was dominant, as it was in the Medals Scheme for schools of 1869–84, but Mackinder had a much broader approach to geography, drawn from French and especially German scholars as well as from his own imaginative mind. He also saw the need to consider education at every level and gathered together a group of schoolmasters in 1893 to form the Geographical Association as a body serving teachers at all levels.

Although to academics RGS has often seemed obscurantist, traditional, establishment-minded, imperialist, as a body it has endeavoured to remain politically neutral. This does not mean that its members were—or are—politically neutral (how could they be?) and naturally each phase in the history of RGS has reflected the ethos of the time so that the emphasis on discovery and colonial expansion seems natural in the Victorian and Edwardian periods. This was part of a world outlook made possible by greater ease of travel than ever before and the advance of cartography from the reconnaissance sketches of explorers to the exact work of hydrographers and surveyors. Although the vision of RGS covered the entire world, Africa was always prominently represented in its publications, though even on Africa—despite the efforts of both a British Association committee during the inter-war period and of a few individuals—it was only after the War of 1939–45 that research papers began to appear except on rare occasions.

It would be easy to show that at times RGS missed opportunities of dealing adequately with current problems such as those of the new Europe created by the Treaty of Versailles in 1919; on the other hand a conspicuous contribution was made in its publications to the replanning of Britain after the War of 1939–45. A society does not exist, however, solely to provide material of contemporary interest that may attract the respect of future historians, for at all times there is the need to continue enquiry into such fundamentals as aspects of physical, human, historical, political geography, and even its own history, biographical and general. In recent years considerable contributions have been made to the history of cartography in a special cartographical section of the *Journal*. The possibilities are so vast that modern editors, confronted with a large number of papers on varied aspects of geography, may well wonder how to provide a selection that will give most if not all readers something that they will wish to read. This may lead, as it has at various times in the past, to a policy of caution based on the view that studies of exploration have a wide appeal. Sir Clements Markham never failed to remind audiences at the annual reviews given during his twelve-year presidency from 1893 that there was still much of the world to discover, much that was known only in general terms and in need of more detailed study and, particularly, that most of the world was inadequately mapped.

The present study is divided into four sections: 1830–80, 1880–1930, 1930–45, and the post-war years. The last section is divided into two parts, of which the first is concerned with the remarkable ten years that followed the war and the second with the period of increasing ferment and experiment in methodology that has marked the last twenty-five years. There is no ending, happy or otherwise, but rather a fascinating story of human activity in the field, the laboratory and the study which began in the reign of William IV and still continues as an expression of interest in men and environment.

## The First Fifty Years, 1830–1880

When the Royal Geographical Society was founded as the Geographical Society of London in 1830 it was the third in historical order (after Paris in 1821 and Berlin in 1828), though the African Association dated from 1788 and became part of RGS in 1831 (Mill 1930: 44). In theory the Royal Society had published geographical work from 1662, but in 1880 Clements Markham pointed out that, of the 5336 papers it had published, only 77 could be called geographical, and several of these had arisen from chance circumstances such as a visit in 1769 to Hudson's Bay to observe the transit of Venus (*JRGS* 50 (1880), 7–10). Many famous specialist societies were founded early in the nineteenth century, including the Geological in 1807, the Royal Astronomical in 1820, and the Royal Asiatic in 1823. The new Geographical Society was concerned with the whole world (Mill 1930: 7). Its aims included the publication of interesting and useful geographical facts and discoveries, the support of exploration, the instruction of explorers, and the accumulation of a library and map collection, the library enriched by periodicals received from all over the world in exchange for those of the Society and the latter strengthened by the gift of Ordnance maps and hydrographic charts from the government year by year.

Geographers at this time were looking forward to an understanding of the whole world, to be achieved both by accurate mapping and by studying relationships of climate, vegetation, soils, population, and natural resources. Colonel Julian Jackson, then the Society's Editor, gave a warm welcome to the *National Atlas of Historical, Commercial and Political Geography* by Alexander Keith Johnson in 1843 which included some maps and illustrations of the globe by Dr Heinrich Berghaus, the German cartographer. Among these was Berghaus's 'map-illustrations of Humboldt's system of isothermal curves, of the geographical distribution of the currents of air, of a survey of the culture of plants and of the mountain-chains of Europe and Asia'. The atlas also included 'explanatory memoirs' (*JRGS* 13 (1843), 156–60). This idea of relationship, later to be one of the fundamentals of Darwinian theory, was crucial to the modern development of geography. But adequate maps were still a hope of the future, even in Europe where, for example, little was known of Spain, 'one of the lacunae of European geography'.

Although exploration and map making were fundamental concerns of geographers when the Society was founded, some writers saw that these provided only an initial impetus for study. Among them was the same Colonel Julian Jackson who, said H. R. Mill, had a grasp of physical geography 'far in advance of his time' and was 'alert to classify and record the different phenomena of physical geography': he had, for example, while resident in St. Petersburg written memoirs in English and French on the freezing of the river Neva (Mill 1930: 53). As early as 1835 he showed his critical faculty (which Mill considered rather too well developed) by his acid comments on the 'travellers' tales' type of geography (*JRGS* 5 (1835), 381–7), suitable only for beginners. He appreciated the problem of subjectivity and said that twenty different observers would see twenty different landscapes. The traveller should not be content merely to enthuse but should also explain and re-create a landscape; and this involved more than physical geography, for the writer should so identify himself with the life of an area that he appeared to become part of it.

Colonel Jackson also understood the relation of a map to a landscape and in 1843 said: 'The ordinary map itself is not an exact transcript or fac-similie of the earth's surface as it really exists. The mathematical projection necessarily introduces something conventional into its composition; places are represented not in their actual relative position, but in a manner that is understood to indicate that position.' Clearly Jackson had three ideas generally conceived to be modern. Firstly he had an understanding of 'perception', of the subjective element in all geographical observation; secondly he thought that only by using this inherent quality could the geographer be effective; and thirdly, he realized that a map was only a model of reality. Too often a map was taken to be the reality itself (Goudie 1978).

German geography was advancing in prestige under the inspiration of Alexander von Humboldt and Carl Ritter, professor of geography in Berlin University, who, said an anonymous reviewer in 1843, 'by his occasional writing and still more by his academical lectures ... [had] given to geography a new impulse and a new form' (*JRGS* 13 (1843), 333–4). He worked on the geography of a particular district, discerning its varied features and the relationship between them, and so built up a world regional geography. Writing in 1893 Pyotr Kropotkin, the 'gentle anarchist', said 'If Oxford had had fifty years ago a Ritter occupying one of its chairs and gathering round him students from all over the world ... it would be this country, not Germany, which would now keep the lead in geographical education' (*Geogrl. J.* 2 (1893), 358).

Unfortunately the first modern entry into university teaching proved to be merely an incident. In 1833 University College London, asked the Society to provide a small endowment for a chair in geography, to which its secretary, Captain Alexander Maconochie RN was appointed. He continued to be secretary of the Society and editor of its journal and appears to have given regular teaching only from the later part of 1834. Few students were attracted and in 1836 Maconochie resigned from his posts with the Society and at University College on his appointment as Governor of van Diemen's Land (Tasmania), from where he sent occasional papers to the Society. No new appointment was made as no suitable candidate applied for the post and, regrettably, Maconochie with Captain (Admiral Sir) Francis Beaufort RN had refused the provision of an endowment on their own initiative, subsequently approved by the council (Ward 1960). According to Mill, this postponed 'for half a century a discussion of the duty of the Society towards the Universities' (Mill 1930: 40), but this generalization ignores the efforts made to establish university geography in the 1870s (p. 11).

Close association was made with the geographical society of Bombay, formed in 1833, and the Hakluyt Society in 1846. The Bombay society was a branch of RGS from 1833 to 1837 and maintained friendly contacts, with exchange of publications, until its eventual absorption into the Royal Asiatic Society in 1873. The founders of the Bombay society, nineteen volumes of whose journal were published between 1836 and 1873, fully appreciated the contribution to geographical knowledge that could be made from India. 'The difficulties which people residing in Europe experience in making any addition to geographical science can have no existence here, for a wide, and untrodden, and a profoundly interesting field of observation lies around; and however superficially we may turn up its soil, we may be sure to find a field for our labour.' (*JRGS* 3 (1833), x.) The contributions had come mainly from Army, Navy, and medical men. The Hakluyt Society was founded by a group of Fellows of RGS with Sir Roderick Murchison as president to publish standard works of travel in edited and annotated volumes, including translations. Possibly, as Mill suggested, RGS itself might have undertaken this work, but the two societies have remained in friendly contact and the distinguished record of the Hakluyt Society needs no emphasis. When Section E (Geography) of the British Association was formed in 1851, Sir Roderick Murchison became its first president and its meetings were arranged by RGS until 1914, with the Society's secretary, map curator, or librarian as its principal secretary. (Mill 1930: 62; Howarth 1951.)

To know the world and to map it were clear responsibilities of the RGS. The Ordnance Survey had some gifted officers, including Thomas Frederick Colby (1784–1852), whose great work was the survey of Ireland on the 6 inch to 1 mile scale, and Thomas Aiskew Larcom (1801–79), whose maps of Ireland on the 1:126720 scale and the Ordnance Survey memoirs of 1837 covering 20 square miles of County Londonderry received praise from W. R. Hamilton in his presidential address of 1839 (*JRGS* 9 (1839), lvi.). William Richard Hamilton (1777–1859), president in 1837–9 and 1841–3, gave the first presidential address in 1838 (these were to become lengthy and informative surveys as time went on) and in 1842 commended the work of 'Mr. Charles Darwin, Naturalist to the expedition of the "Beagle" under Captain Robert Fitzroy', 'on the structure and distribution of coral reefs', of which a long abstract with quotations was given in the journal (*JRGS* 12 (1842), lxxxiii, 115–20). Classical and biblical scholars had a natural interest in geography. In 1842, for example, Dr P. W. Forchhammer contributed an article on the topography of Troy and in the same year W. R. Hamilton included in his presidential address a reference to the work of his son, William John ('I must not be deterred on private considerations ... from inviting your attention to the Travels ...'), who had travelled through Turkey and 'brought home a large mass of Greek inscriptions, many of them indicative of ancient sites'. Shocking as this now sounds, it was frequently done (*JRGS* 12 (1842), lviii–lix).

Pragmatic in outlook the earliest presidents of RGS undoubtedly were, but this did not prevent two of them from giving an inspiring view of geography. In 1841 G. S. Greenough spoke of 'simple' and 'compound' geography. Simple geography had three essential qualities: a systematic classification of all the objects which belong to it, a precise and fixed terminology, and a 'good nomenclature'. It was a branch of physical science, having the sound basis of physical geography, concerned with the animal part of man. Compound geography was concerned with man in his social, civil, and political character and could develop many forms, political, civil, statistical, ethnographical, philosophical, classical, biblical ... (*JRGS* 10 (1840), lxix).

In 1842 W. R. Hamilton concluded his presidential address with a peroration that revealed both its author and the ethos of the early Victorian period. For the wealthy and influential, education was based on classics and the Bible, imperialism was beneficial to the population controlled, and great powers had military responsibilities. In Hamilton's view:

The study of geography is the most natural, as it is the most useful, of all human pursuits.... Where we are, whither we are going, whence we came, all enter into our daily actions, are all geographical and topographical interpretations; here we begin, and to these we must return.... though geography may be devoid of the charms of other systematic sciences, though it does not lend itself to brilliant theories, though it scarcely admits of the most innocent speculations, though it treats mainly of dry matters of fact, yet has geography other, and perhaps superior, claims.... It looks alone to truth as its object ... the absolute and relative positions of definite and fixed places ... It embraces the whole globe on and in which we live, and have our being; all the interests, all the occupations of men, are, more or less, dependent upon it. It is the mainspring of all the operations of war, and of all the negotiations of a state of peace; and in proportion as any one nation is the foremost to extend her acquaintance with the physical conformation of the earth, and the water which surrounds it, will ever be the opportunities she will possess, and the responsibilities she will incur, for extending her commerce, for enlarging her powers of civilizing the yet benighted portions of the globe, and for bearing her part in forwarding and directing the destinies of mankind. (*JRGS*, 12, (1842), lxxxviii–lxxxix.)

Eloquence could not solve the difficulties of a young society in the economic uncertainties of the 1840s, of which Mill says much of interest in his 1930 volume (Mill 1930: 51–63). Fortunately Roderick Impey Murchison gave his support to RGS and in 1844, during his first term as president, commented that 'no great European Kingdom, except England, is without some national educational establishment for general geographical purposes' (*JRGS* 14 (1844), cxxviii). Throughout the 1840s the main interests of the Society lay in exploration and cartography, though one sign of academic advance was the publication in 1848 of Mrs Mary Somerville's *Physical Geo-*

*graphy* (Baker 1948; Oughton 1978). There was a continuing interest in the work of European geographical societies and in 1848 the completion of Heinrich Berhaus's famous *Physikalischer Atlas* was noted (*JRGS* 18 (1848), xl). Ten years later Murchison showed his aspirations for the Society's journal by commenting with admiration on the success of the German *Geographische Mitteilungen*, edited by August Petermann and published by the Perthes firm at Gotha, with sales of 5000 copies a month (*JRGS* 28 (1858), clxvi). No part of the world was too remote to be of interest; indeed, the more remote the better, and in 1853 the Society's Gold Medal was presented to Francis Galton

for having, at his own cost, and in furtherance of the expressed desire of the Society, fitted out an expedition to explore the centre of Southern Africa, and for having so successfully conducted it through the countries of the Namaqua, the Damara, and the Ovampo (a journey of about 1700 miles), as to enable this Society to publish a valuable memoir and map ... relating to a country hitherto unknown; the Astronomical observations determining the latitude and longitude of places, having been accurately made by himself. (*JRGS* 23 (1853), ix)

In 1854 Captain Fitzroy RN and Lieut. Raper RN published the first edition of *Hints to Travellers* as a thin pamphlet on methods of surveying and in the same year the 'Journal' included several papers on this subject, among them one by Galton (*JRGS* 24 (1854), 345–58) which had several of the practical hints given in his fascinating *Art of Travel* (1855). This paper gave a long list of subjects on which questions were to be asked and answers recorded. They included aspect, surface, physical divisions, mountains, rivers, lakes, sea coast and ports, volcanoes and mineral springs, maps and charts, astronomical observations, instruments, meteorology, natural history, and ethnography. Under 'descriptive geography' came the name of the country, its boundaries, relief, hydrography, meteorology, zoology, botany, geology, and physical divisions; ethnography and statistics comprised population, habitations, communications, occupations, food, costumes, utensils, weapons, weights, measurements, divisions of time, language, music, religion and traditions, history, government, and foreign relations. In short, almost everything was included, with much practical information suitable for explorers; the famous *Hints* was republished in several later editions, of which the fourth (1880) was edited by Galton.

Murchison in his second and third terms of office (1856–8 and 1862–71) made the presidential addresses long and informative, covering progress in exploration and government cartographical work and giving excellent biographical material in the obituary sections. In 1858, for example, he observed that the opening of the Yangtze valley in China would 'bring Europeans into immediate commercial connection with the 100 million people who inhabit its fertile banks and those of its affluents'. The change in the course of the Hwang was 'one of the many proofs of the decline of vigorous government in China. In earlier periods the embankments of the rivers were carefully matched and repaired; but neglect has led to the breaking down of all artificial ramparts, and vast fertile tracts have consequently been sterilized' (*JRGS* 28 (1858), clxxxix–cxc). In 1870, Mr Ney Ellis's paper, 'Notes of a journey to the new course of the Yellow river in 1868', proposed that the river (Hwang) be returned to the channel it had occupied from 1194 to 1853 instead of flowing to the Gulf of Pohoi, but advised that English engineers would be needed: 'the whole works [should] be placed under the superinten-dence of Englishmen for organisation would seem to be absolutely necessary' (*JRGS* 40 (1870), 1–33).

The prospects of economic expansion were realized increasingly, for example in 1860 when Lord Ripon as President considered the possible future of Canada, where the areas drained by the Winnipeg, Red, and Saskatchewan rivers were then being explored:

With a vast area of fertile soil, and a climate favourable to the growth and cultivation of wheat; with lignite coal, iron ore, and common salt in abundance, a great future is in store for the Basin of Lake Winnipeg. Lying between the rich gold-fields of British Columbia and the powerful, populous and wealthy colony of Canada, it is only a ques-tion of time how soon its vast capabilities and resources will be developed, and that position assumed when, as a British colony, it will also become instrumental in carrying British institutions, associations and civilisation across the continent of America. (*JRGS* 30 (1860), cliv–clv.)

Australia attracted a number of travellers, including Captain Charles Sturt (1796–1869), who showed that the eastern interior was not one vast inland sea, as had been supposed. Although in 1870–2 a telegraph line had been constructed from Adelaide to Port Darwin, expeditions to the west of this link showed that the prospects of settlement were bad and that much of the interior was 'gently undulating desert, clothed with that plant of evil augury to Australian settlers, the spinifex grass' (*Proc. RGS* OS19 (1874–5), 446). Nevertheless the main areas promising for settlement had been discovered by the 1870s.

Africa was the prime interest in British exploration. This was due in part to Christian missionary enthusiasm, to concern about the slave trade, to the wish to penetrate a continent of which the margins had long been known, to the expectation of profitable commerce and even greater wealth, and to the desire to establish colonies. In addition, scientific curiosity was a powerful motive, for Africa was known to have a wide range of climatic and vege-tational regions. But the search for the unknown attracted far more attention than the production of useful papers such as R. J. Mann's study of the physi-cal geography and climate of the colony of Natal (*JRGS* 37 (1867), 48–67). Mann was Superintendent-General of Education in Natal and his article gives useful geographical information and meteorological data. Rivalry between Britain and Germany was clearly shown by Sir Henry C. Rawlinson who said in the 1872 presidential address that it would be wise to send 'an Expedition to ascend the Congo and pre-occupy the ground' before a rival German expedition did so (*Proc. RGS* OS17 (1872–3), 13).

Early in 1875 at a meeting on Africa, which was in fact entirely concerned with the Congo, Sir Rutherford Alcock as chairman stated that

any civilized nation that would take possession of the country would be in a position to confer upon that part of Africa one of the greatest blessings. The atrocities, tyranny, and continual massacres that take place were enough to make one's blood run cold. . . . Although the occupation of tracing African rivers to their sources led very rapidly to the end of life, yet such men as Livingstone and many others had not been deterred from their task, and he trusted such explorations would never cease until civilization had obtained a firm hold upon Africa, and had converted to useful and peaceful pursuits the people who at present were entirely given to frightful massacres and slavery. (*Proc. RGS* OS19 (1874–5), 133.)

Similar sentiments were frequently expressed at RGS meetings, and in 1877 the African Exploration Fund was founded to finance expeditions planned or approved by the Society: £4000 was raised and the expedition under A. Keith Jonston, of the Edinburgh map firm, landed at Dar-es-Salaam in May 1879. Unfortunately Keith Johnston died a month later, and his young colleague, Joseph Thomson, led the party through 2800 miles of country, 1300 of which was totally unknown territory.

As Mill has shown, expeditions to Africa were numerous at that time: the conference arranged by the King of the Belgians in 1876 had agreed that there should be an International Commission on African Exploration and that each country should have its national organization (*Proc. RGS* OS 21 (1876–7), 16), but RGS decided that 'African exploration would be more effectively prosecuted by England and the necessary funds more readily obtained through national enterprise than by international association' (ibid. 391–2). Unfortunately in Britain it was not easy to acquire government support. At a meeting of subscribers to the African Exploration Fund in 1878 it was explained that 'the Society cannot compete with governments or even with Missionary Societies in explorations on a large scale having neither the powers of the first, nor the pecuniary resources of the second'. German explorations were subsidized as trade was expected to follow from them; in Portugal a grant of £20 000 had been made by the Cortes for the exploration of western Africa; and in France a credit of 100 000 francs had been voted for a central African expedition (*Proc. RGS* OS 22 (1877–8), 463–8).

Meanwhile, in 1872–6, the great *Challenger* expedition was covering the world and accumulating data that were to lay the foundations of modern oceanography. It had been eagerly followed by RGS but full publication of the results was made elsewhere, though papers were read in 1886 by J. Y. Buchanan 'On the similarities in the physical geography of the Great Oceans' (*Proc. RGS* NS 8 (1886), 753–68) and in 1891 by A. Buchan on 'The meteorological results of the *Challenger* expedition in relation to physical geography' (*Proc. RGS* NS 13 (1891), 137–54).

In 1875 the Admiralty provided two ships, the *Alert* and the *Discovery*, which sailed to the Canadian Arctic under the command of Captain George S. Nares, who at a meeting in December 1876 advanced the theory that the polar sea might extend to the North Pole. One party reached 83°20′26″, the most northly position ever attained. Clements Markham, writing in 1879 on 'The Arctic expeditions of 1878', said that the next objectives of Arctic discovery 'should be the completion of work on Greenland', the examination of 'the supposed lands north of Siberia', and the further examination of Franz-Josef Land. He mentioned the variation of ice from year to year but did not comment on its significance in meteorological study (*Proc. RGS* NS

1 (1879), 16–36). In the discovery of new territory mapping was an obvious responsibility and a service to posterity. But observations, whether in steaming tropical or frozen polar areas, or indeed anywhere between such extremes, might prove far more significant later than anyone supposed at the time, and for this very reason RGS was anxious to improve techniques of observing and recording data which, even if apparently trivial, could be a contribution to a synthetic world study.

In 1877 RGS arranged a series of three annual lectures on physical geography as a contribution to world knowledge. The first, by General Richard Strachey, an 'Introductory lecture on scientific geography', suggested study of the mathematical and physical geography of the world with climate, vegetation, and the races of man and their dependence on geographical conditions (*Proc. RGS* OS 21 (1876–7), 179–203), and the second, by W. Carpenter, dealt with 'The temperature of the deep-sea bottom [*Proc. RGS* OS 21 (1876–7), 289–303] and the conditions by which it is determined' (clearly indicative of the interest raised by the *Challenger*'s researches). Then followed A. R. Wallace on 'The comparative antiquity of continents as indicated by the distribution of living and extinct animals'. Wallace's book *The Geographical Distribution of Animals* had appeared in 1876 and in this lecture he showed that though the distribution of land and sea had changed vastly through geological time, the human period was so short a final phase that it was reasonable to regard the continents and oceans as 'permanent features of the earth's surface' (*Proc. RGS* OS 21 (1876–7), 505–35).

The first two lectures of the following session were by P. Martin-Duncan on the formation of the main land masses (*Proc. RGS* OS 22 (1877–8), 188–216) and by Captain F. J. Evans, Hydrographer of the Admiralty, on the magnetism of the earth (*Proc. RGS* OS 22 (1877–8), 188–216). The third was by W. T. Thiselton-Dyer, then Assistant Director of the Royal Gardens at Kew, on 'Plant distribution as a field for geographical research' (*Proc. RGS* OS 22 (1877–8), 412–45). It was an excellent summary of the work done, with many comments on the world distribution of floras and the relationships between those already recognized. He showed that plant collections made by various travellers, including the Russian explorer Colonel Nikolai Prjevalski in central Asia, were valued contributions to the synthetic picture and he drew also on the fine work of Sir Joseph D. Hooker. The last three lectures given, in 1879, were by G. Rolleston on 'The modifications of the exterial aspects of organic nature produced by man's interference' (*JRGS* 49 (1879), 320–92), by A. Geikie on 'Geographical evolution' (*Proc. RGS* NS 1 (1879), 422–44), and by John Ball, 'On the orgins of the flora of the European Alps'. Although Geikie made a few comments on the relations between geography and geology, his paper was mainly on geological evolution, but Ball gave a fascinating account of the climatic influences on Alpine vegetation and showed the folly of using shade temperatures (*Proc. RGS* NS 1 (1879), 564–88).

In 1879 the RGS Council decided not to continue the course of lectures but to meet the same need, 'to give a more scientific direction to geography', by providing instruction and training for intending travellers, including the use of instruments for survey and astronomical observations and route-mapping (*Proc. RGS* NS 1 (1879), 453). This led to the diploma in surveying and practical astronomy of 1897 with instruction in photography, geology, natural history, and other subjects useful to explorers. With this, the famous *Hints to Travellers* was enlarged in size and scope in its various new editions.

Galton was a staunch advocate of geographical education and at the meeting of the British Association at Brighton in 1872 he gave a warm welcome to papers on geographical education and the scope of scientific geography 'although these topics do not possess the same immediate interest for the public as descriptions of travel and personal adventure, they are of far more enduring importance and, in fact, lie at the root of the science, for the cultivation of which this Society was instituted' (*Proc. RGS* OS 17 (1872–3), 20). In 1874 the Society made an appeal to the universities of Oxford and Cambridge for the recognition of geography 'in any future redistribution of Academical revenues'. This was to include University Fellowships, of which one might be called the 'Livingstone Fellowship' 'in memory of the great traveller who combined in himself the character of Missionary, Geographer and Scientific Observer' and whose example might be thereby kept before the eyes of English youths for generations to come (*Proc. RGS* OS 18 (1873–4), 451–2).

Nothing came of this somewhat emotional approach and in 1879 a second memorial was sent, explaining that

The Council wishes Geography to be understood in its most liberal sense, and not as an equivalent to topography. They mean by it, a compendious treatment of all the predominant conditions of a country, such as its climate, configuration, minerals, plants and animals, as well as its human inhabitants; the latter in respect not only to their race, but also to their present and past history, so far as it is intimately connected with the peculiarities of the land they inhabit. (*Proc. RGS* NS 1 (1879), 261–4)

The problems of man and his surroundings lay in assessing reciprocal influences, for, on the one hand, external nature influenced race, commercial development, and sociology, while, on the other, man influenced nature by clearing forests, introducing new plants and domestic animals, cultivating and draining the soil. Only through geography could the link between physical, historical, and political conditions be clearly seen and the wide contacts with the outer world of England through her 'colonies . . . , her commerce, her wars, her missionaries and her scientific explorers made such knowledge essential'.

A strong supporter of an approach to the universities was the Revd George Butler, headmaster of Liverpool College, who said that the systematic study of geography helped 'the intelligent appreciation of history, led naturally to the study of kindred branches of science and gave a student a comprehensive view of the greatness of the British Empire and of the undeveloped capabilities of the accessible world' (*Proc. RGS* NS 1 (1879), 469–70). Liverpool College had provided a number of candidates for the examinations organized by RGS from 1869 for the award of medals (p. 14). In 1871 the Society's President, Major-General Sir Henry Rawlinson, wrote to the Vice-Chancellors of the universities of Oxford and Cambridge seeking support for geography in their proposed examination scheme for schools, with some success, as in 1885 J. Scott Keltie noted that the inclusion of geography in the Oxford and Cambridge Schools Examination 'did much to improve the position of geography in middle-class schools' (Gilbert 1971).

Clements Markham in 1880 gave a splendid account of the first fifty years of RGS (*JRGS* 50 (1880), 1–126). Although the main emphasis had been on exploration and almost all the medals had been given to explorers, in 1869 Mrs Mary Somerville had received the Founder's Medal for her books, notably her *Physical Geography* (1848), and in 1880 the work of the classical geographer, E. H. Bunbury, best known for his great *History of Ancient Geography* (1879), was recognized by a special resolution of the RGS Council. In effect Markham gave a world review that still provides fascinating reading. Atlases revealed both the discoveries of fifty years and improved techniques of mapping. The Arctic in 1830 was shown only by unconnected stretches of the American coast with some speculative data on Greenland, Spitzbergen, and Novaya Zemlya, but by 1880 the whole coast of Arctic America was delineated and seven north-west passages had been revealed, though the interior of Greenland was still a mystery. Nothing more had been revealed of the Antarctic continent since the voyage of Sir James Ross in 1841–2. Africa was largely blank in 1830, with Ptolemy's 'Mountains of the Moon' still shown, but some details had been added by explorers of several countries, for it had become 'a glorious field of generous rivalry among civilised Europeans' though many areas were still unmapped or in need of systematic exploration and survey to replace reconnaissance data.

In Asia the main British work had been in India, but explorers in other countries, notably Russians, had made large areas known, though there were extensive gaps, especially the upper reaches of the Yangtze, Irrawadi, and Bhramaputra rivers, the northern part of Tibet, and large parts of Arabia. North America had been systematically explored and mapped, but in South America there was little knowledge of much of Patagonia and southern Chile. Almost the whole of the interior of Australia and also that of New Zealand had been discovered and explored since 1830, though considerable areas of the East Indies were still unknown. Knowledge of the ocean depths was negligible in 1830, but had extended considerably since, from the Ross voyages of the early 1840s to the great *Challenger* enterprise of 1872–6. The whole survey finishes with a statement that sounds like a roll of drums, and forms a fitting prelude to the next fifty years of the Society. In it, Markham said that 'When, in the far distant future, the whole surface of the earth has been surveyed and mapped, the study of physical geography may be recommenced on a sound basis and generalisations will become more accurate and will be founded on more correct and reliable data.'

### The Second Fifty Years, 1880–1930

Clements Markham obviously loved anniversaries, and in 1880 there were signs of encouragement although efforts to establish or re-establish geography teaching in the universities had failed and the Maconochie professorship at University College London, was a remote memory. The medals scheme for schools, from which Francis Galton and others hoped to see an educational advance, had failed, and the special course of scientific lectures at RGS had been abandoned in 1879 as so few came to hear them. Wisely the Society decided to substitute a programme of direct instruction for intending travellers, later to be markedly successful. Often the Society has been pilloried for its attention to explorers who became public heroes, but members were needed and the real question was the outcome of exploration rather than its drama. Scientifically the voyage of the *Beagle* in 1831–6 and of the *Challenger* in 1872–6 were of permanent value, as the former revolutionized thought on natural history and confirmed the concept of evolution while the latter laid the foundations of modern oceanography.

As Sir Joseph Hooker said in his presidential address of 1881 (*Proc. RGS* NS 3 (1881), 595–608) 'for a long time the origin of representative species, genera and families remained an enigma', but

Now, under the theory of modification of species after migration and isolation, their representation in distant localities is only a matter of time and changed physical conditions ... as Darwin well sums up, all the leading facts of distribution are clearly explicable ... such as the multiplication of new forms; the importance of barriers in forming and separating zoological and botanical provinces; the concentration of related species in the same area; the linking together under different latitudes of the inhabitants of the plains and mountains, of the forests, marshes and deserts, and the linking of these with the extinct beings which formerly inhabited the same areas; and the fact of different forms of life occurring in areas having very nearly the same physical conditions.

It was hoped that explorers would bring back specimens of plants, and so contribute to the definition of floral regions. Joseph Thomson had provided 'the most interesting herbarium ever brought from Central Africa', which included genera and even specimens known to exist at the Cape of Good Hope. Hooker wondered if there was a former land connection between continents of the southern hemisphere and observed that in general the south temperate floras were more intimately related to those of countries to the north rather than to one another.

Only by exploration could the world's mysteries be revealed. In 1883, for example, Baron A. E. Nordensköld penetrated Greenland for 250 miles and, having found ice-covered mountains 6000 to 7000 ft. high, disproved the idea that there were central valleys enjoying a 'comparatively warm and dry climate'. Progress of a more permanent scientific character was foreshadowed in the foundation of various observatories, by the British on Great Slave Lake, Canada, the Austro-Hungarian empire on Jan Mayen, the Norwegians in Lapland, the Swedes on Spitzbergen, the Russians on Novaya Zemlya. Explorers went everywhere: to New Guinea (but cautiously for even the London Missionary Society had only penetrated for twenty to thirty miles inland), Southern Patagonia (with Tierra del Fuego), North Borneo, as well as more usual fields of exploration in Africa, Asia, and Canada. Admiralty charts appeared month by month, and in 1883 Lord Aberdare, a president with political experience (he had been Home Secretary in a Liberal government), was able to say that 'the study of practical and scientific geography is being prosecuted with an ardour and energy never exceeded in any age in the world' (*Proc. RGS* NS 5 (1883), 387).

Two years later Lord Aberdare included a reference to the 'scramble for Africa' (and not only for Africa) in his presidential address, for he was well aware that scientific curiosity was not the only motive for exploration.

To the politicians of all the great European nations the period has been one of intense interest and anxiety, connected more or less with questions of vast territorial acquisitions. To the geographer the interest, although less painful, has hardly been less keen. The French in Asia and Africa and the Russians in Central Asia—the English on the Afghan frontier, on more than one border of India, on all sides of Africa, and in Oceania—and the Germans on the East and West African coasts and among the islands of the Pacific and Australasian seas—the Italians on the Red Sea—have, while pursuing measures of national policy, made large additions to our knowledge of the globe.... Never—and I need hardly even except that period of emigration which precipitated and followed the break-up of the Roman Empire—has the ferment among nations been so wide-spread or prophetic of such great consequences. The foundations of new empires, new civilisations, are being laid over vast portions of the earth. (*Proc. RGS* NS 7 (1885), 408.)

### GEOGRAPHY IN EDUCATION

From its earliest years, RGS was interested in the work of other geographical societies and it welcomed the foundation of the Scottish and Manchester societies in 1884, followed by those of Tyneside in 1887, Liverpool in 1891, and Southampton in 1897. The Manchester Society shared the concern of RGS for geographical education and from 1892 paid part of the salary of the first full-time teacher of geography in Manchester University, H. Yule Oldham, who was succeeded from 1894 to 1896 by A. J. Herbertson (*Geogrl. J.* 120 (1954), 118–19). The Manchester Society was interested also in the enquiry of J. S. Keltie, commissioned by RGS in 1884, into the teaching of geography in Britain and continental Europe. Scott Keltie's famous Report (*Supp. Pap.* RGS, 1 (1886), 443–594) showed that in Britain geography teaching was improving steadily in primary schools but that in secondary schools, including the great public schools, it was generally a 'poor relation' except in a few cases (notably Dulwich and Liverpool Colleges, which provided most of the medal winners), while in the universities it appeared only in rare instances, such as a paper in the Modern History Tripos at Oxford, of which the Revd H. B. George of New College said that in practice it carried so little weight that students paid it scant attention (*Proc. RGS* NS 8 (1886), 715).

The situation was far more favourable in Germany where twelve chairs of geography had been established and the subject was taken seriously in schools; several other European countries, notably France, had given recognition to geography. RGS decided to make representations to the universities of Oxford and Cambridge once more, and this time they were successful. The story of the modern growth of geography in these universities has been admirably told by D. R. Stoddart for Cambridge and by D. I. Scargill for Oxford, where geography had been taught at various earlier periods in the university's history. Readers of these two papers will realize that the early years of geography teaching in both universities were not easy and that chance misfortune might have ended geography's career in either. Both universities owed much to the financial support of RGS, which gave £7250 to

Cambridge from 1883 to 1923 and £11 000 to Oxford from 1887 to 1924 (Stoddart 1975; Scargill 1976).

From the new beginnings in Oxford and Cambridge came two different approaches to geography. At Cambridge the first lecturer, F. H. W. Guillemard (1852–1933), resigned only six months after his appointment in 1888 and was succeeded by J. Y. Buchanan who had been a member of the *Challenger* expedition. He planned and gave a wide range of courses, including general geography; the distribution of land and water on the globe; physical and chemical geography, with special reference to land surfaces and their development under climatic and other agencies; and oceanography. All these, it will be noted, were courses of a systematic character with a strong emphasis on physical geography, though in time human and historical geography was introduced. In Oxford, following Mackinder's appointment in 1887 the emphasis was on regional work with political geography.

Mackinder, having been invited to speak on the 'new geography' at RGS in 1887 when his time was spent mainly as a university extension lecturer, gave his famous lecture 'On the scope and methods of geography' (*Proc. RGS* NS 9 (1887), 141–60). He began by asking a fundamental question, 'Can it become a discipline instead of a mere body of information?' The 'roll of great discoveries' was near its end though there was still much to be done in New Guinea, in Africa, in Central Asia, and 'along the borders of frozen regions' and the tales of adventure were giving place to 'the details of Ordnance Maps' so that 'even Fellows of Geographical Societies will despondently ask "What is geography?"' Mackinder boldly defined geography as 'the science whose main function is to trace the interaction of man in society and so much of his environment as varies locally', in short, environment and man, with the emphasis on geographical environment, of which a fine example was *The Making of England* by J. R. Green (1882), who by deduction from geographical conditions had shown 'what must have been the course of history'. Geography was concerned with the configuration of the earth's surface, meteorology, and climate, and the products which a country offers to human industry. Expressed, he says, 'somewhat abruptly':

Physiography asks of a given feature 'Why is it?' Topography 'Where is it'? Physical geography 'Why is it there?' Political geography 'How does it act on man in society and how does he react on it?' Geology 'What riddle of the past does it help to solve?' And the first *four* are the realm of the geographer.

Trained in natural science as well as history Mackinder was determined to lay a sound scientific foundation, beginning with the idea of a landless globe, having 'three concentric spheroids, atmosphere, hydrosphere and lithosphere' on which climate is influenced by the irregularities of land, sea, plains, and mountains. Locally an 'environment' is a natural region, and the smaller the area, the greater tends to be the number of conditions uniform or nearly uniform throughout the area. Similarly a community is a group of men having certain characteristics in common, and the smaller the community, the greater will be the number of common characteristics. Man alters his environment, and new discoveries create differing circumstances, with the decline of the power of Venice following the opening of the Cape route to

India as a classic historical example. Modern communications by means of the electric telegraph had made possible the 'giant size of modern states' and now government, industry, trade, and commerce could have effective control and cohesion over a large area.

For Mackinder geography was a subject of vast scope, an approach to human problems such that 'to the practical man whether he aim at distinction in the State or at the amassing of wealth, it is a source of invaluable information; to the student, it is a stimulating basis from which to set out along a hundred special lines; to the teacher ... an implement for the calling out of the powers of the intellect.' Later, in 1893, it was reported that Mackinder had given a series of lectures on 'The relations of geography to history in Europe and Asia', of which the first showed the view of the world held at different periods from that of Homer, in which the 'inner geography' of the known and experienced was in contrast to the imaginative 'outer geography', to the enlargement of view in the Renaissance period and the revelation by Captain James Cook that there was no 'Antipodean oecumene' (*Geogrl. J.* 1 (1893), 157–8). This was followed by general lectures on subjects including the world belt of desert and steppe, the gates of India and China, the Alps in European history, divisions of Gaul, the geographical analysis of British history. Mackinder was highly successful as a lecturer and particularly enjoyed appearing before audiences of teachers, of whom hundreds were eager to learn more of the 'new' geography.

In 1895 he said in his presidential address to Section E of the British Association at Ipswich (*Geogrl. J.* 6 (1895), 367–79) that the British had no cause for dissatisfaction with their contribution to 'precise survey, to hydrography, to climatology, and to biogeography. It is rather on the synthetic and philosophical, and therefore on the educational side of our subject that we fall so markedly below the foreign and especially the German standard.' After giving a historical review of developments in various systematic aspects, he says that 'the treatment by regions is a more thorough test of the logic of the geographical argument than is the treatment by types of phenomena'. He uses the term 'chorography' as the nearest equivalent to *Landerkunde* but adds that it is 'a clumsy expression'—in fact it has never been used as freely in England as in America. He traces the regional method back to Humboldt's work *Essais politiques sur la Nouvelle Espagne* of 1809, in which for the first time relief, climate, vegetation, fauna, and human activities were related (incidentally fifty years before Darwin's *Origin of Species*), and pays tribute to the fine experimental work of nineteenth-century German cartographers, such as Berghaus, Perthes of the famous map firm, and August Petermann. But it was Peschel (1826–75) who 'restored physical geography to the geographer'.

Still the challenge lay in the study of environment and man, but environment is itself dynamic. Communities may move from one environment to another and two or more communities may fuse in any environment. Those who settle in a new environment carry genetic qualities with them so that, for example, the Normans were affected by the background of Norway and the Americans had developed a civilization in the Mississippi basin that could not possibly have arisen there without them. The facts of human geography are never independent of the natural environment and are 'resultant of the conflict between dynamic and genetic elements'. Human geography, properly studied, requires a knowledge of geomorphology, 'the half-artistic, half genetic consideration of the forms of the lithosphere: geophysiology, meaning oceanography and climatology, and biogeography, organic communities and their environments'. Mackinder also refers to what became known as 'geographical inertia', generally meaning the survival of an industry, or even a population, in an area after the initial attraction for its presence has passed. He recognizes the place of economic geography and, still broader, of political geography, which 'becomes reasonable when the facts are regarded as the resultant, in large measure, of genetic or historical elements, and of such dynamic elements as the economic and strategic'. In England geographers were 'good observers, poor cartographers, and teachers perhaps a shade worse than cartographers'.

The teachers did not seem to mind, for on 20 May 1893 Mackinder and ten schoolmasters founded the Geographical Association in his rooms in Christ Church, Oxford (Fleure 1943). Fears by the RGS that this new body might remove some of its support, especially when a new journal was planned, were quickly alleviated and the Geographical Association did fine work for geographical education in schools and preserved friendly relations with RGS which on occasion led to valuable co-operation. Mackinder's own textbooks were widely read, and revised with meticulous care when new editions were issued. He was a man of his time in his concern for education at all levels from the primary school to the university. In a letter to J. S. Keltie, dated 3 March 1891, he said, 'What I want to devote the remainder of my working life to is the modernisation of our English education. It appears to me that the whole future of Britain depends ultimately on this. And for this I required a combined basis of geography, administration, politics and writing' (RGS Archives, a). It did not prove an easy road, but in 1935 he wrote to another veteran of distinction, H. R. Mill, saying, 'It is indeed for both of us a glorious thing to have lived to see what is the beginning of the triumph of our youthful ideas' (RGS Archives, b). Twenty years earlier in a letter to Mill he had written that 'You and I have taken somewhat different paths in Geography—you rather the scientific and I the literary.... (RGS Archives, c).

### THE VISION AND THE REALITY

Mackinder and a few other prominent figures in academic geography shared a vision of its great future, but the progress towards its realization was hardly one grand sweet song. There were periods of doubt about the support for geography in both Oxford and Cambridge (p. 26-7), and at one stage Mackinder appeared to see the best hope of advancement in the creation of a London Institute of Geography similar to those he had visited in continental Europe, notably at Vienna, and also at Harvard (Cantor 1962). Ever practical, Mackinder negotiated with publishers interested in a series of regional texts which included his own *Britain and the British Seas* of 1902. Although RGS had welcomed Mackinder's academic enterprise, Sir Clements Markham was constantly insisting throughout his long presidency from 1893 to 1905 that the practical work of world exploration and mapping was not complete. Of course he was right: in 1896, for example, he noted that, of eighty-two papers published, the largest number were in surveying and oceanography, followed by physical geography and by questions relating to Polar and to African exploration.

This was safe ground and Markham regarded 'the instruction of intending travellers' as 'the most important and the most successful part of our educational efforts, e.g. in surveying, mapping, and fixing of positions by astronomical observations; in geology, botany and photography' with zoological observations and anthropological measurements. The real need was for the explorer to know how and what to observe (*Geogrl. J.* 8 (1896), 1–15). At the same time he welcomed new developments such as H. R. Mill's idea of a regional survey of Britain (p. 23) and the establishment in 1903 of a Research Committee. Perhaps the attitude of Markham and others explains why Mackinder found it necessary to climb Mount Kenya in 1899 and even why this episode was picked out as a highlight in his career in the speech given in presenting him for the Patron's medal in 1945 (p. 49). Geography of an academic character fought for space with the reports of various expeditions, not all of which were of much academic significance.

Definition of the various aspects of geography was a natural preoccupation of the time along with the relation of geography to other subjects and the placing of geomorphology in the academic structure of universities. In 1893 RGS arranged a discussion 'on the relation between physical geography and geology' at the British Association meeting in Dundee, with Sir Archibald Geikie as chairman (*Geogrl. J.* 2 (1893), 518–34). Clements Markham, the main speaker, quoted Sir Roderick Murchison's saying that 'the geologist was the geographer of former periods' and added that 'the geologist and the geographer use the same materials but for different purposes; the former to find interpretations of problems relating to the past, the latter to explain what is actually taking place in the present'. Of the latter he gave as examples the observed changes in the courses of the Ganges and other Indian rivers, as well as those from the medieval period in the Netherlands. Geology was the study of the condition of the 'Earth and of the changes on its surface during the cycle of ages before the dawn of history' and physical geography 'a knowledge of the earth as it is and of the changes which have taken place during historical times'. It is perhaps unfortunate that more geographers (and others) did not take to heart the comments of W. Topley, of the Geological Survey, who pointed out the fallacy of equating rock quality with geological age, for 'the soft clays, the limestones and sandstones of the Jurassic rocks of England are represented in the south and east of Europe by thick masses of limestone, forming prominent mountain ranges ....'

In 1895 W. M. Davis, then professor of physical geography at Harvard, expounded his views on 'the development of certain English rivers' at the RGS (*Geogrl. J.* 5 (1895), 127–46). He explained that the rivers of eastern England developed their present courses on an original gently inclined plain, composed of sedimentary strata of varying resistence. The land had been worn down at least once to a lowland of faint relief and afterwards broadly uplifted, so that another cycle of denudation developed in which the adjustment to structures was carried to a higher degree of perfection than in the earlier cycle. In this essay he defined some of the terms familiar to students of the cycle of erosion; 'consequent' was first used by J. W. Powell, 'subsequent' by J. B. Jukes in his much-quoted paper (Jukes 1862) on valleys in southern Ireland, and to these Davis added 'obsequent' for streams not showing any clear relation to initial slopes or to the strike or dip of strata, and also the term 'peneplain' for 'almost denuded surfaces'. The essay also included much discussion of marine denudation and subaerial erosion, the delight of argumentative geomorphologists for many decades.

Also in 1895 Charles Lapworth, professor of geology in Birmingham, reviewed Albrecht Penck's *Morphologie der Erdoberflache*, one of a series of books edited by F. Ratzel (*Geogrl. J.* 5 (1895), 575–81). The first of its three parts dealt with 'earth-morphology, an integral part of geography' on which progress had been made by W. M. Davis and others; the second dealt with the distribution of land and water and from it came the hypsometric curve familiar to generations of students, with constructional and erosive processes: the third, on the sea, dealt with coasts, waves and tides, deltas, coral reefs, and secular movements of the coastline. W. M. Davis and A. Penck were laying firm foundations for courses in geomorphology, already favoured in Britain. In 1894 an unsigned review of J. Cvijić's monograph on limestone regions published in Vienna and edited by Penck appeared (*Geogrl. J.* 3 (1894), 321–31). This was the earliest of a long series of papers on the subject by Cvijić through whom Serbo-Croat words such as *doline* and *polje* entered the literature, competing with French and German terms.

Many geomorphologists were concerned to give exact meaning to terms; in 1911 J. W. Gregory, for example, said that it would be unfortunate if important geographical terms had different meanings in Europe and America. He analysed the use of certain words by physical geographers working in English, French, and German and suggested the following usage (*Geogrl. J.* 37 (1911), 189–95):

*denudation*   wearing down of the land by any agency
*erosion*   excavation by rivers and glaciers of their beds (discarding 'corrosion' as an unnecessary synonym)
*solution*   for action of solvents.

Though it might seem easy to establish meanings for various terms, the attempts by RGS to compile a handbook on nomenclature ran into difficulties and the idea was eventually abandoned.

Oceanography was a favoured specialism of the late nineteenth century, and the *Challenger* voyages of 1872–6, reported by RGS at the time, were written up in fifty volumes issued from 1880 to 1895. H. R. Mill observed that the 29 500 pages, 3000 plates and maps, with innumerable blocks, provided 'the record of the greatest scientific voyage ever undertaken' (*Geogrl. J.* 5 (1895), 360–8). The aim was the scientific exploration, under the naval command of Captain George Nares and the scientific directorship of Professor Wyville Thomson, of the physical, chemical, geological, and biological conditions of the great ocean basins. As a result, the relief of three-fifths of the earth's surface was sufficiently known for general maps to be compiled and the circulation of air and water to be explained. With the help of observers on other ships, isotherms were given at intervals of 100 fathoms to 1000 fm; and then at 1500 fm., 2200 fm., and the sea bottom. Out of this came many novel facts, notably that beneath the surface the Atlantic was the warmest of the oceans and that, though ocean surfaces vary very much in temperature, water in the deeper layers is independent of surface temperature and apparently determined mainly by the cold Antarctic water moving northward into all the land-bordered oceans. Articles on oceanography were few in the *Geographical Journal* but in addition to those written by J. Y. Buchanan in 1886 and by A. Buchan in 1891 (p. 9), H. N. Dickson wrote on 'The movements of the surface waters of the North Sea' (*Geogrl. J.* 7 (1896), 255–67) and J. Y. Buchanan on 'The Guiana and Equatorial currents' (*Geogrl. J.* 7 (1896), 267–70).

In 1893 H. N. Dickson read a paper at the British Association meeting on an investigation of temperature and salinity on the north and west borders of the continental shelf (*Geogrl. J.* 3 (1893), 375–6). He began 54 miles due north of the Shetlands and ran a line for some 70 miles west to the Faroes, then north to the east of the Orkneys and Shetlands, finishing at Aberdeen. His observations confirmed and extended the results of H. R. Mill to the west of Lewis in 1888, and showed a warm layer, 53°F–56°F., 15–25 fm. thick, water that became increasingly cold with depth, reaching 30.9°F in the Faroe channel. Such investigations, as the *Challenger* voyage had shown, were of interest for fisheries, cable laying, physical geography, climate and weather, marine life, and much more. Their success was a source of great satisfaction to Clements Markham, who in his anniversary address for 1897 (on Queen Victoria's Diamond Jubilee) said that oceanography was 'founded by Major Rennell', made popular by Maury's 'charming' *Physical Geography of the Sea* (Leighley 1977), and had become of practical use through the need to examine the floor of the ocean for the laying of telegraphic cable (*Geogrl. J.* 9 (1897), 601).

Lake investigation, a fine contribution to geomorphology, was begun in England in the summer of 1893 by H. R. Mill, Mrs Mill, E. Heawood, and A. J. Herbertson (*Geogrl. J.* 2 (1893), 165–6, 362; 4 (1894), 237–46; 6 (1895), 46–73, 135–6), who worked on Derwentwater (800 soundings), Ullswater (maximum depth 208 ft.), Wastwater (258 ft.), Windermere, Ennerdale, Crummock Water, and Buttermere. According to Mill, previous surveys had been useless except for those of the Revd Clifton Ward in the 1870s (Mr Ward worked for the Geological Survey before becoming Vicar of Rydal). Haweswater was surveyed in 1894, and in the same year Mill published a summary account of the work done in which he showed that Wastwater, Windermere, and Coniston all had areas below sea-level (*Geogrl. J.* 3 (1894), 421). An estimate of the volume of water in each lake was made. In 1900 and 1901 Sir John Murray and F. P. Pullar contributed a double article on 'A bathymetric survey of the fresh-water lakes of Scotland', on which they also published several papers in the *Scottish Geographical Magazine* (*Geogrl. J.* 15 (1900), 309–53; 17, (1901), 273–95).

Vegetation survey too had its devotees, and in 1895 the work of Professor

Charles Flahault of the Botanical Institute at Montpellier was noted. This was based on the *Vegetation der Erde* by Engler and Drude (*Geogrl. J.* 5 (1895), 275). Flahault had found it best to base his mapping on 'a judicious selection of plants, which characterize a floral region as precisely as do certain geological strata'. He had also attempted to assess the possibilities of forestry, noting any experiments made during the previous century with archaeological and philological evidence. Several surveys were made in Scotland from 1903 by W. G. Smith with others and on the Pennines (Smith *et al.*, 1903, 1904). These surveys began with the geology, climate (especially rainfall), and plant associations. Farmlands with wheat were observed to an altitude of 600–700 ft and without wheat to 1000–1100 ft., birch woods in sheltered valleys to 1250 ft. and various types of moorland at higher altitudes.

In 1904, F. J. Lewis (then an assistant lecturer in Botany at Liverpool University but later to hold the title of lecturer in Geographical Botany) made a comparable study of the basins of the rivers Eden, Tees, Wear, and Tyne (*Geogrl. J.* 23 (1904), 313–31). Lewis said that plant associations were affected first by the chemical composition of the soil; second, by its physical characteristics such as its texture and the proportion of air and water retained; and third, by climatic factors influenced by slope of land, aspect, drainage, and altitude; and finally the influence of man and other animals. He found cultivation with oats to 800 ft., pasture and meadow to 1200 ft., and oak woods to 800 ft., above which planted conifers survived to 1300 ft., but birch only to 1000 ft. His classification of moorland was similar to that of Smith and Rankin. C. E. Moss, in an article also published in 1904 and obviously based on field-work, suggested that the peat moors of the Pennines had 'originated in morasses formed probably by the destruction of primitive Pennine woods' (*Geogrl. J.* 23 (1904), 660–71). Earlier authors had observed remains of trees in peat bogs, but theories of climatic change were only incipient. Much of this work became absorbed by plant ecology and later studies appeared in the *Journal of Ecology*, edited by Sir Arthur Tansley from 1916.

The world distribution of mammals was treated by W. L. Sclater in five papers published from 1894 to 1897. (*Geogrl. J.* 3 (1894), 95–105; 5 (1895), 471–83; 7 (1896), 282–96; 9 (1897), 67–76; 10 (1897), 84–91). He followed a scheme of regionalization suggested by his father, P. L. Sclater, in an essay on the distribution of birds published by the Linnean Society. At this time the Bartholomew map firm were producing their famous *Physical Atlas*, originally planned in five parts, of which only those on *Meteorology* (1899) and *Zoogeography* (1911) appeared. The atlas gives the Sclater schemes for land and sea areas with others. Six regions were demarcated: Australian, Neotropical, Ethiopian, Oriental, Nearctic, and Palaearctic. The Sclater scheme had been used by A. R. Wallace in his two-volume work, *The Geographical Distribution of Animals* (1876), which Bartholomew regarded as 'the standard work on Zoogeography'. Sclater warned readers against thinking in political terms of continents: Europe, for example, was only a peninsula of Asia, the African side of the Mediterranean had close links with southern Europe, and both central America and southern Mexico belonged zoologically to South America. Many of the nineteenth-century atlases, from Berghaus onwards, had given maps of faunal regions, though little is heard of this type of work today. Nevertheless floras generally have distribution maps for individual species, just as bird books do, and the changes observed from one year to another are widely studied, not least by students of climatic changes, notably in Norway, Sweden, and Finland.

The term 'applied geography' has been in common use since the War of 1939–45 but it is not known when it first appeared. It was used by H. R. Mill in 1892 in a short note on A. Haviland's *The Geographical Distribution of Disease in Great Britain* (1892), a second and much revised edition of a book first published in 1875. Such a book would now rank as medical geography (*Geogrl. J.* 2 (1893), 87). It is an interesting work, indeed a period piece, for its author had a wide knowledge of the natural sciences and related particular diseases to climate, soil, and many other natural conditions. In 1893 A. de Silva White, who had served as Secretary and Editor of the Scottish Geographical Society, used the term 'applied' in a sense generally covered later by 'human', which could in fact include almost anything.

White's geographical schema is best shown as a diagram; he uses the division between physical and political normally accepted in his day (*Geogrl. J.* 2 (1893), 178–9).

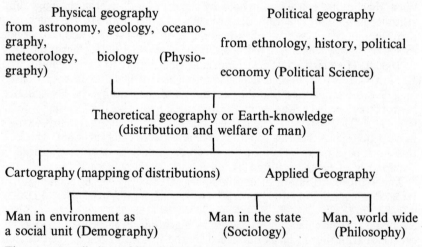

The term 'applied' could well be used for a splendid series of articles by J. F. Unstead on 'The climatic limits of wheat cultivation, with special reference to North America', published with the blessing of the RGS Research Committee in 1912 and 1913 (*Geogrl. J.* 39 (1912), 347–66, 422–46; 42 (1913), 165–81, 254–76).

Social geography became a favoured study after the War of 1939–45 (p. 60), but the term appeared in a paper by the American geographer, G. W. Hoke (*Geogrl. J.* 29 (1907), 64–8), in 1907. The term 'social' had been used earlier, for example by Paul de Roussier in 1884 and by Elisée Reclus in 1895 and also in his final work, *L'Homme et la Terre* (1905) (Dunbar 1978; 113). Hoke took a broad view, having much in common with continental geographers working on 'la géographie humaine', and stressed the need to study communities and groups, however defined. His paper opens with the statement that all geography deals with distributions in space, with the facts and products of human association as represented by group characteristics, industries, technology, customs, beliefs, and related phenomena; and he estimates the significance of the various factors which have influenced their distribution. The response of a social group to a given environment is determined not only by 'physical circumstance' but also by the status, technical and psychic, of that group as well. The three million people in the closed valleys of Kentucky were 'a relic of the England of Chaucer' (perhaps this statement is a little overdrawn) and those who went forward found a more prosperous way of life. Hoke assumed that groups normally do the best they can under their circumstances and all social geography must include 'in addition to the physiographical group of factors ... the sociological factors' which are 'no less fundamental'. To state an idea is easy, to work on it more difficult, but Hoke raised fundamental questions for geographers although his paper attracted little attention.

Regional geography on a local rather than a world scale was advocated by H. R. Mill in 1896, in his article on a 'Proposed geographical description of the British Islands based on the Ordnance Survey' (*Geogrl. J.* 7 (1896), 345–65). As always, Mill made his views abundantly clear, notably in saying

Geography as a science is the exact and organized knowledge of the distribution of phenomena on the surface of the earth. This involves the human race; and because the human race represents the culmination of organic evolution, the true understanding of man with his territorial environment is the final object of geography.... The attention of geographers has hitherto been directed mainly towards the collection of facts; we now require to discuss and arrange them.

He had in mind for each 1″ to 1 mile Ordnance Survey sheet (which in England then covered 216 sq. miles but in Ireland and Scotland 432 sq. miles) a survey based on material acquired from geological and hydrographic maps, meteorological data, census reports and local registrars' records, medical data, any governmental material available, local or national, county histories and photographs. Each memoir was to cover, in 8–32 or occasionally 48 pages, landforms, soils, climate, vegetation and agriculture, historical sites and political boundaries, and population distribution.

This sounds encyclopaedic, but Mill's article of 1900, 'A fragment of the geography of England—southwest Sussex', showed the possibilities. He noted particularly that when the cheap food from 'the fields of America, Russia and India' was no longer available, more attention would be given to British agriculture (*Geogrl. J.* 15 (1900), 205–27, 353–78). There would be regional memoirs, for the Weald, the Cornwall–Devon peninsula, the Lake District, the Pennine chain, East Yorkshire, with the Southern Uplands, the Central Plain (*sic*), and the Highlands in Scotland, and ultimately one great volume of perhaps 1000 pages. All this was in accord with the work of the Geological Survey, which issued memoirs for each 1″ sheet and regional monographs for larger areas. Mill was bitterly disappointed by the lack of support for his scheme, though in fact it was comparable with the work of the Land Utilisation Survey from 1929. That it would have been valuable is beyond doubt. Local regional work such as Mill envisaged became a normal student exercise in many universities, notably in Oxford, and in

1910 a study of 'The Reigate sheet of the 1 inch Ordnance Survey', by Ellen Smith was described by a reviewer as doing 'full justice to a region which for both scientific interest and picturesque beauty can scarcely be surpassed in England' (*Geogrl. J.* 38 (1911), 190–1).

Concepts of regionalism have been numerous and many of them were more inspirational than the somewhat pragmatic treatment advocated by Mill. In 1898 Patrick Geddes, in a paper on 'The influence of geographical conditions on social development', wrote that an ordered evolutionary unity was revealed in the world and correlations could be made but there was still the problem of 'how far nature can be shown to have determined man ... how far the given type of man has reacted, or may react, upon his environment' (*Geogrl. J.* 12 (1898), 580–7). Men were moulded by their occupations so that 'the salmon-fisher of Norway, the whaler of Dundee, the herring-fisher of Yarmouth, the cod-fisher of Newfoundland ... each is a distinct type'. Geddes went on to stress his famous triad of study, (derived from Frédéric Le Play), Place, Work, Family (so expressed though it was sometimes given as 'folk'), which could be modified and reversed as Society, Work, Place. Fully aware, like many of his contemporaries, of the growing interest in race, Geddes was aspiring at studying environment and people as totally as possible with a thorough appreciation of their social ethos, their religious outlook and practice, their tribal types or castes, their economic possibilities and aptitudes, their commercial and military history, as well as such basic concerns as relief, climate, and natural resources. This broad approach appealed to a large number of people.

By contrast, Herbertson, in his 'Major Natural Regions of the World' (*Geogrl. J.* 25 (1905), 300–12), based his classification mainly on seasonal rainfall and critical temperature (0 °C, 10 °C, 20 °C), though he also gave a map of the word's major structural divisions based chiefly on the work of Eduard Suess. Herbertson stated firmly that 'political divisions ... must be eliminated from any consideration of natural regions' though 'the factor of human development has to be taken into account as well as the possibilities of the natural environment. The density of population map is the most direct expression of the natural economic utilization of the natural region.' Nevertheless the regions of Herbertson were firmly based on physical factors, although he stressed that 'geography is not concerned with distribution of one element on the Earth's surface but with all'. Herbertson's paper was critically received: H. R. Mill said that it was 'quite impossible to get a description of the world that will satisfy the workers in several branches of sciences' though an orographical regionalization had some value. E. G. Ravenstein thought that, despite climatic differences, 'the great plain of India, extending from the Ganges to the Indus, had its analogue in the great plain of Scotland' which to some people might seem ridiculous, especially as the supposedly 'great plain of Scotland' is studded with hills. Ravenstein also pleaded for the use of the term 'chorography' instead of 'regional geography'.

Herbertson answered Mill's preference for relief by saying that it was not 'a complete guide'; for example in South Africa the relief was similar in the south-east and south-west but the rainfall 'as far apart as Spain and China'.

He also said it was not surprising that historians had done little 'to elucidate the influence of geographical conditions on history' for 'the geographer has not yet put his material into a form which the historian can use'. Finally Mackinder defended Herbertson by saying that 'regional geography' as a term was well understood, and that he hoped 'systematic geography' would not take root. Readers of the previous few pages, however, may think that systematic aspects of geography were already showing signs of vigorous life and that the definition of regional geography, at least as represented in the pages of the *Geographical Journal*, needed a great deal more thought, for so far they had been offered the empiricism of Mill, the inspirational but vague totality of Geddes, and the climatic delimitation of Herbertson. That Herbertson and others were looking towards a fuller treatment is clear.

In 1904 Mackinder's paper,' The geographical pivot of history', appeared (*Geogrl. J.* 23 (1904), 421–37). In it he argued that interior Asia and eastern Europe formed the 'heartland' of the 'World Island': in this northern and interior part of the Eurasian continent, where the rivers flowed to the Arctic or to salt seas and salt lakes, there were great possibilities of economic de-velopment with modern railways. Beyond the heartland there was an inner crescent of continental states accessible to sea power and also open to invasion by the land power of the pivot area; beyond this crescent of states lay an outer crescent of 'insular' states having sea power and inaccessible to the pivot area. Mackinder saw Russia as a modern equivalent of the Mongol empire and his work reflects the concern of nineteenth-century Britain with Russian expansion in areas such as the borders of Turkey, Iran, India, and China. The danger lay in the combination of Germany and Russia and he advocated the creation of a tier of independent states between them, on the lines of those defined after the War of 1914–18. In 1919 his *Democratic Ideals and Reality* was published but it attracted little notice in Britain, though more in America. Later his ideas of world strategy (including his famous jingle, 'Who rules East Europe commands the Heartland; Who rules the Heartland commands the World-Island; Who rules the World-Island commands the World') were taken over by Haushofer, the German geopolitician, and Mack-inder once more gained prominence as a political geographer. By 1943 Mack-inder was thinking of an international organization based on the North Atlantic (realized in NATO) partly because his original work was done before air power became so crucial in strategy.

The early years of the twentieth century were marked by polar exploration, the slow but fortunately sure progress towards the production of a 1 : 1 000 000 map of the world, the gradual acceptance of the regional idea, and a slow search for a broader understanding of the range of geography. Polar exploration is beyond the range of this study except to remark that its achievements grew in strength. An excellent account of the 1 : 1 000 000 map has been given by C. F. Arden-Close in his *Geographical By-ways* (1947), (pp. 108–18) in which he shows that though the scheme was originally sug-gested in 1891 by Albrecht Penck at the Bern International Geographical Congress, progress was slow until 1913, when the French Government called a conference in Paris attended by representatives from thirty-five countries. During the War of 1914–18 RGS compiled about a hundred sheets of a provisional 1 : 1 000 000 map covering practically the whole of Europe, with an extension through Asia Minor to Arabia and Iraq. After the war the American Geographical Society produced a set of 107 maps for Hispanic America, completed in 1946.

From 1901 to 1930 the *Geographical Journal* was primarily devoted to exploration and mapping and to many it seemed that the achievement forecast in eloquent words by Mackinder and others was as far off as ever. In 1912 J. L. Myres, already well known as a classical scholar attracted to geography, reviewed 'The work of the Research Committee' established in 1903 with an annual grant of £200 to finance new work (*Geogrl. J.* 40 (1912), 357–68). Fifty papers had been given at afternoon meetings (those in the evenings were reserved for 'first-hand recitals of striking geographical discoveries'). Of these papers twelve had dealt with 'new methods of surveying or mapping or computing'; three each with hydrology and climatology; four with regional and synthetic subjects; nine with geomorphology; three with crop and plant distributions; seven with evidences of physical changes in historical times, of which three included the human aspect; eight dealt with geographical exploration. There were no papers on historical cartography, on military geography, or on the special problems of the development and colonization of central and south America.

Nevertheless considerable progress was being made. Myres made a stimulating contribution to the questioning of the purpose of geographical study then (and since) prevalent. The aim was

to explain why anything which happens, happens just *where* it does; and under what circumstances it *does* happen *just there*. In this respect, geography will be recognized as the twin sister of history which, in the same largest and truest sense, is the branch of research which should discover *when* each thing happens, and tries to explain under what circumstances *this thing* happened or happens *just then*.... Historical geography ... is an attempt to answer the question, why did what happened just *then* also happen just *there*?

These are large questions, not susceptible to easy answers, but as another distinguished scholar (of biblical studies) George Adam Smith, said after a paper by J. W. Gregory (*Geogrl. J.* 56 (1920), 13–47) on rift valleys

... we are more and more coming to see that prior to any intelligent study or mastery of the subject of history lies a knowledge of the physical basis on which all history rests and is moved. Until you understand the lie and the lift of the Earth's surface you cannot understand either the political, or the military, or the economic movements of history.

That George Adam Smith made this view the basis of his famous *Historical Geography of the Holy Land*, first published in 1894 and still in print in its 1973 edition, needs no emphasis. Professor Myres contributed two articles on the Dodecanese in 1920 and wrote the Admiralty Handbook on the same islands during the War of 1939–45.

Inspiring visions often contrasted with disappointing results. The battle for the educational recognition of geography had not been won in the universities, for in 1904 A. J. Herbertson, somewhat enviously reviewing three French works (of which one was Vidal de la Blache's *Tableau de la géographie de la France*), said that in the great Victoria History 'there was no geography at all', that although French army officers were taught geomorphology it was not part of the British officer's training, nor had any attempt been made to deal systematically and fully with the agricultural geography of Britain (*Geogrl. J.* 23 (1904), 111–14). 'From the point of view of grasp of the value of geography by historians, soldiers and agriculturalists, we see that in France they are ahead of us.' In 1905 the military correspondent of *The Times* wrote an 'incisive' article on geography and war, supported by a leading article (*Geogrl. J.* 25 (1905), 17–22). Lack of geographical understanding was made clear in the Boer War as was the fact that 'the teaching of geography had not assumed its proper place in national education.'

In 1904 A. J. Herbertson looked at contemporary views of geography, German, French, American, and British, and reached the conclusion that its main educational value lay in its synthesis, involving 'the space relationships of groups of phenomena on the Earth's surface' (*Geogrl. J.* 24 (1904), 417–27). In his view 'no subject ... unless it be *sociology*, gives the comprehensive outlook on *existing* conditions which must be taken into account in all the larger problems of life.' The vision remained, but its realization came slowly. Students of the history of education will know that in these years there was a continuing struggle for the liberalization of teaching in schools, and those particularly concerned with geography will know that it was one of the main concerns of the Geographical Association. And perhaps the answer was to show, rather than say, what geographers could do. In 1911 Leonard Darwin as President of RGS noted that progress in exploration was still needed though the main discovery period was over and maps appeared 'with almost machine-like regularity and precision by Government officials' (*Geogrl. J.* 38 (1911), 1–7). The need was for

systematic and detailed examinations of comparatively small areas, and not merely to cover long distances with the result of doing little more than previous explorers. These surveys should be as good as is possible in the circumstances, and the information they collect should be extensive, varied, systematic, and recorded with reference to the needs of the students of science and history, as well as the man of commerce.

The last-named was considered only rarely by RGS, although at this time the Manchester Geographical Society's staff provided information helpful to exporters in the local textile and other industries.

Relationships between physical and human aspects of the environment were a constant concern of geographers, but some of the correlations made seem strange. In 1908 E. C. Semple's article (*Geogrl. J.* 31 (1908), 72–90) on 'Coast Peoples' (which in an extended form appeared as a chapter in her *Influences of Geographical Environment*, 1911) provided a fine example of her deterministic views. In it she said that

Inaccessibility from the land, a high degree of accessibility from the sea and a paucity of local resources unite to thrust the inhabitants of such coasts out upon the deep, to make of them fishermen, seamen and ocean carriers. The same result follows where no barrier on the land side exists, but where a granitic or glaciated soil in the interior discourages agriculture and landward expansion as in Brittany, Maine and Newfoundland. In all these the land repels and the sea attracts.

She went on to explain that one-fifth of France's sailors came from Brittany and that the Indians of British Columbia and South Alaska 'subsist chiefly by the generosity of the deep', and developed her theory by finding 'an historical development similar in not a few respects' on the fiord coasts of New England, Norway, Iceland, Greenland, Alaska, and southern Chile. This was a well-developed form of the 'environmentalism' favoured and then maligned at various times but it was not its only form: generally RGS showed no great enthusiasm for publishing papers with new hypothetical judgements on man and environment, and several authors of papers that could be described as 'speculative' had to seek other outlets.

H. J. Fleure, for example, did not find his work on the valleyward movement of population in prehistoric times or on human regions (RGS Archives, d, e) welcomed by the *Geographical Journal*. However, Griffith Taylor, having contributed papers on the 'Physiographic and glacial geology of East Antarctica' in 1914, also had his general paper 'The physiographic control of the Australian environment' published in the *Journal* for 1919 (*Geogrl. J.* 44 (1914), 365–82, 452–67, 553–71; 53 (1919), 172–92). In this paper he gave a summary account of the penetration and settlement of Australia and noted that 'the logical course is to consider the reliability of the rain in addition to the average annual rainfall ... only along the south is the inland rainfall at all reliable'. He added that 'economic climatology is a better guide to the future of Australia than the experience of the "man on the spot", which experience may have been limited to a series of unusually good seasons'. Taylor's later work on race and environment was not welcomed by RGS but received publication elsewhere, notably in the *Geographical Review*. One reason why RGS declined to publish some of his papers was that they required so many illustrations (RGS Archives, f). Or was that editorial tact?

Race was a possible guide to the understanding of the human environment in some geographers' outlook. Early in the War of 1914–18 L. W. Lyde gave a paper on 'Types of political frontiers in Europe' (*Geogrl. J.* 45 (1915), 126–45). He argued that as far as possible 'the racial unit should coincide with the political unit' especially if the 'racial unit' proved to be a people 'incapable of assimilation' such as the Albanians, 'the most ancient existing race in Europe, speaking a language older than classical Greek'. Secondly, in choosing 'a new political owner of any inhabited area, first consideration should be given to the capacity of the new owners to assimilate others'; this quality, he thought, was abundantly possessed by the French, conspicuously so in Alsace, but not by the Prussians. Thirdly, the feature used for a frontier should be one where people naturally meet, not a waterparting or desert likely to be sparsely inhabited. The Rhine, Lyde thought, was an obvious frontier. It was not a good military front, still less was it a political barrier, but it would be a good political frontier because

the whole population from Bavaria through Alsace and Luxemburg to Walloon Belgium is of a single general type—blessed with a rounded skull inside which the brain seems to have a freedom of development in all directions. Wherever that is found, you have a people who are naturally idealistic, artists and dreamers, willing to fight—and to die— for an idea such as honour or freedom.

Brave words perhaps, but H. J. Mackinder poured cold water on Lyde's eloquence and questioned the possibility of setting up a new Europe in accordance with supposed scientific ideals. Much more likely was a new Europe based on the 'old idea of the balance of power'. He also foresaw both the difficulty of giving Poland access to the sea without taking over some German territory and the possible conflict between Serbia and Italy on the Dalmatian coast. And as if that was not enough, the boundaries of Balkan states would continue to provide formidable problems. Mackinder finished by saying that although in physical geography it might be possible to establish general laws, 'if we try to obtain laws from one human geography, and especially laws which guide our action politically, we are attempting what I believe is doomed to failure. We shall cause both the scientific men and the historians to throw stones at geography.'

Several articles appeared at this time on the future prospects for Europe, among which was B. C. Wallis's 'Distribution of nationalities in Hungary' (*Geogrl. J.* 47 (1916), 177–88), partly based on work done at the RGS (p. 30). Surprisingly Wallis's later papers, of topical interest on post-war Europe, were published not by RGS but in the *Geographical Review*, which had greatly improved since Isaiah Bowman became Director of the American Geographical Society in 1915. The problems remained, and evocative words like 'race' and 'nation' were discussed endlessly. Marion Newbigin, for example, in 1917, in 'Race and nationality' hoped that 'it might be possible to find a solution for the problems of the disturbed belt of Europe without having recourse to the terrible solution of irreconcilable minorities'. But what was nationality? In her view it was 'not only race—whatever race may mean—not only religion nor language, nor history, nor tradition ... but, partially at least, a community of economic interests dependent upon geographic factors' (*Geogrl J.* 50 (1917), 313–28).

To what degree 'the precepts of geography ... were followed by the Congress of Paris' was a question asked in an unsigned communication of 1920 on 'Geography at the Congress of Paris 1919' (*Geogrl. J.* 55 (1920), 309–12). The Geographical section of the General Staff consisted of its chief, Col. Sir Coote Hedley, Major O. E. Wynne, who had extensive experience of boundary delimitation in Africa and elsewhere, and Captain A. G. Ogilvie. The 1920 note explained that while there had been much study of a geographical nature' by the Foreign Office, the Admiralty, the War Office, and the War Trade Intelligence Office, there had been little co-ordination between workers separated into departmental enclosures: there was far more co-ordination in the American Inquiry, with Isaiah Bowman in the geographical section as Chief Territorial Specialist.

Four main processes were involved in making the new map of Europe. The initial task was to decide on the major political divisions of the future, after which the best kind of frontier was sought. Then followed the definition and delimitation of the new boundaries, and finally the arrangements for its demarcation on the ground. In fact ethnic considerations were generally most important, though these could not always be followed for economic or strategic reasons, and it was clearly necessary to have some boundary that could easily be defined on the ground. The definition of boundaries involved legal as well as geographical matters and the line itself was drawn by survey experts. E. Romer from Poland and J. Cvijić for Yugoslavia were fine advocates for their people. Many detailed geographical points emerged, such as the problem of shifting rivers. The note, clearly written by someone with inside knowledge, was by A. G. Ogilvie (RGS Archives, g). Although RGS devoted far less of its space to the new boundaries than some other journals, notably the *Annales de géographie* and the *Geographical Review*, material on the new boundaries included a series of notes by A. R. Hinks. Ogilvie in 1920 published a long article on Macedonia, followed the then conventional order of regional study, geology and structure, landforms, vegetation, farms, human activity (*Geogrl. J.* 55 (1920), 1–34).

One wartime enterprise was the publication of *Handbooks* by the Geographical Section of the Naval Intelligence Division, under H. N. Dickson from 1915. By July, fifty books had been produced, all following the same pattern of physical features and boundaries, climate, history, administration, inhabitants, religions, trade and finance, hygiene and social conditions, economics, flora and fauna, communications, gazetteer, vocabularies and maps, with an index. Another enterprise much valued by later workers (including those working on the *Geographical Handbooks* of the War of 1939–45) was the Permanent Committee on Geographical Names, which was said by its first chairman, Major-General Lord Edward Gleichen, to have originated through the Admiralty, who wanted to know how to spell 'Walfisch' (Bay) as this name appeared to have English, German, Dutch, and even Hottentot forms. Representatives of the Admiralty, War Office, India Office, Colonial Office, and the RGS met in 1919 and appointed A. R. Hinks as secretary of PCGN. Later they conferred with representatives of the Foreign Office, the Board of Trade, the Board of Agriculture, and the Post Office and formed a sub-committee to draw up rules for the transliteration into English of thirty-two languages. Gleichen remarked laconically, 'It was not easy to draw up' (*Geogrl. J.* 57 (1921), 36–43).

After the war, there were hopes of a rich development of academic geography in Britain, for geography departments were established in most of its universities, and in schools—especially the new secondary grammar schools founded in the last pre-war years—geography was carefully taught, in many cases by people who had been eager students at Diploma courses or even at the summer schools in various universities (in a short article (*Geogrl. J.* 57 (1921), 76–84) H. J. Mackinder noted that there were at least ten of these in 1920). Unfortunately in the 1920s the *Geographical Journal*, though containing a number of interesting articles, appeared to be firmly entrenched in exploration and travel: even its work on colonial territories was conventional except when, for example, Clement Gillman in 1927 gave a fine study of southwest Tanganyika (*Geogrl. J.* 69 (1927), 97–131). As noted on p. 60, geographical research, supplementing the earlier pioneering and surveying, came to be published mainly after the War of 1939–45. These critical comments are general rather than specific, for there were occasional papers from Kingdon Ward on vegetation in Asia, a series of three papers by W. M. Davis in 1920 on 'the islands and coral reefs of Fiji' (*Geogrl. J.* 55 (1920), 34–45, 200–20, 377–88), and a splendid plea in 1926 by the revered botanist E. J. Salisbury for a more thorough and satisfying study of 'the geographical distribution of plants in relation to climatic factors', in which 'much remains to be done on the lines of primary survey both in the lesser-known parts of the Earth and even in well-explored lands like our own' (*Geogrl. J.* 67 (1926), 312–42). These and comparable articles were regrettably few, yet there were signs of new investigations to come later, including the 1927 study of 'The East Anglian coast' by J. A. Steers, whose work on Scolt Head Island had already begun. He was, said Professor F. W. Oliver, the first person to visit Blakeney Point for study other than as a bird man, a botanist or a biologist (*Geogrl. J.* 69 (1927), 24–48).

On the human side, a paper almost revolutionary in its day was C. B. Fawcett's of 1917, on 'The natural divisions of England', later called provinces and 'based on regional capitals such as Manchester, Birmingham, Oxford, Bristol, Southampton, Cambridge (or Norwich) and Nottingham'. These, Fawcett urged, must have some degree of self-government and would in time supersede counties (*Geogrl. J.* 49 (1917), 121–41). Fifty years later these ideas, modified in detail but not in essence by Fawcett and others, were much discussed. Another harbinger of later work came in a review by Fawcett (*Geogrl. J.* 63 (1924), 440–2) of the regional planning schemes for Doncaster and Deeside published in 1922 and 1923 and in his note of 1927 of the Leeds and Bradford Joint Town-Planning Committee's report (*Geogrl. J.* 70 (1927), 169–70), which, said Fawcett, would be more effective if it dealt with the whole of the West Yorkshire conurbation instead of merely with part of it. E. G. R. Taylor contributed papers on Tudor geographers from 1928, presaging her book *Tudor Geography* (1930) and confirming once again the friendly co-operation between RGS and the Hakluyt Society. This was traditional; this was safe. It was also fine scholarship. The *avant garde* of the time were not satisfied, for they wished to see as wide a range of papers published by RGS as by other journals, including the *Scottish Geographical Magazine*, the enterprising *Geographical Review*, and the *Annals of American Geographers* and the *Annales de géographie*. In 1929 a letter from A. R. Hinks to C. B. Fawcett (RRGS Archives, h), apparently following a conversation, explained current editorial problems: 'More than once the question has been raised at the Research Committee whether we could not have more papers on human and economic geography, and I have been obliged to tell the Committee the primary difficulty is that such papers are rarely submitted, and that when they are they are not always acceptable to independent referees.' As an example he enclosed a recent paper on economic geography in which 'many of the passages are platitudes and truisms decked out in rather grand language'. Fawcett agreed that the paper (unspecified) was not 'a good specimen of work in either economic or historical geography'.

Several academic geographers were looking forward to an advance in the study of its human aspects, and RGS asked—quite reasonably—whether the articles really existed. Some were accepted by the *Geographical Review* and a number of authors looked to two helpful editors, Marion Newbigin in Edinburgh and H. J. Fleure in Aberystwyth (later Manchester); the *Scottish Geographical Magazine* was approaching its finest years and the journal of the Geographical Association, formerly the *Geographical Teacher* but renamed

*Geography* in 1927, was also gaining strength. There were already moves to establish the Institute of British Geographers, finally achieved in 1933, though the initial purpose was to publish monographs rather than to establish another journal. Much of the hostility generated in those years now seems ridiculous, for as the following pages show RGS in time gave a welcome to a wide range of geographical studies, including many of a controversial character on such subjects as the replanning of Britain after the War of 1939–45. This did not preclude interest in exploration that was scientific rather than merely adventurous and not unusually both. To some extent the centenary celebrations of 1930 came as a truce in a time of reasonably gentlemanly hostilities, to be resumed later and merging into the tolerant understanding that the RGS, the Geographical Association, other geographical societies, and the Institute of British Geographers all had their value, their opportunity of extending the knowledge of geography, of encouraging research in a changing world.

## From the centenary to the War 1930–1945

At the celebration of the centenary of RGS in 1930, Sir Charles Close as president said, 'One source of our strength is that we have no narrow, restricted view as to the qualifications of a geographer, nor as to the content of our subject, which is no less wide than the study of the Earth itself, but particularly includes the study of the Earth as the home of man.' In the same year he had introduced a paper by R. O. Buchanan on hydroelectric power in New Zealand by commenting that the Society had been anxious for some little time to increase the number of contributions to its proceedings that dealt with various aspects of human geography (*Geogrl. J.* 75 (1930), 444–57). There had been few of these in comparison with the number in the *Geographical Review* or the *Scottish Geographical Magazine* of the post-1918 years. Volume 75 for the first half of 1930, for example, deals mainly with exploration, of which some was presented in a bright journalistic fashion, while the rest was of academic significance such as 'Forests and plants of the Anglo-Egyptian Sudan' by T. F. Chipp, Assistant Director at Kew Gardens (*Geogrl. J.* 75 (1930), 123–45), which showed that 'every type of vegetation of northern tropical Africa is represented'. Other papers of value include two by E. Trinkler on the Ice Age on the Tibetan plateau and on explorations in the eastern Karakorum and the western Kunlun (*Geogrl. J.* 75 (1930), 225–32, 505–17), and two papers on the 1929 Italian expedition to the Karakorum (*Geogrl. J.* 75 (1930), 385–411). J. M. Wordie gave an account of the Cambridge East Greenland expedition of 1929 (*Geogrl. J.* 76 (1930), 126–37), which had been preceded by others in 1923 and 1926 all designed to reveal more of the then little-known east coast of Greenland (*Geogrl. J.* 75 (1930), 481–504). The 1929 paper included a section on meteorology by G. Manley.

All these papers, and others, not only provide fascinating reading but also contain data useful to students of physical geography in its broadest definition. Nevertheless the overall impression of the volume is that its main concern was the remote and rare; people are seen mainly as possibly unfriendly natives in tropical valleys or, in the case of Albanians and others, as peasants somewhat hostile to the work of the boundary commission in the Balkans described by Col. Frank L. Giles, the British Commissioner on the Yugoslav–Bulgarian and the Yugoslav–Albanian Commissions (*Geogrl. J.* 75 (1930), 300–12). Sir Charles Close, at the beginning of the discussion, expressed once again the official policy of RGS in saying, 'Clearly this Society does not deal with politics. Though some questions in the Balkans are of political interest [this was an understatement] I hope no speaker will touch on political matters.'

Douglas Freshfield, who was president of RGS from 1914 to 1917 and, in his own words, 'the oldest ex-president available', spoke at the centenary meeting of the long presidencies of Sir Roderick Murchison to his death in 1881 (*Geogrl. J.* 76 (1930), 462–5). 'During the next twenty years,' he observed, 'his successors were mainly peers or proconsuls' none of them expert in geography with the exception of Sir Henry Rawlinson and Sir Richard Strachey. They worked in co-operation with the honorary secretaries and the permanent staff and were grateful for 'the silent but impressive sagacity of Bates and the persuasive good sense of Keltie'. The Council was 'essentially conservative', concerned with exploration and naturally eager, necessarily for financial stability, to win and hold popular support and therefore 'disposed to concentrate ... on discovery ... by taking advantage of the national passion for hero-worship'. There was the appeal of Arctic exploration, including the search for Franklin and his lost ships, and of equatorial Africa, with the Nile sources, the great mountains, glaciers, and forests, indeed of many unknown and fantastic areas of the globe.

To the Victorian Councils it seemed that 'geography in the wider sense in which we have learnt to pursue it was in this country hardly yet in the air. Our universities did not find time to trouble their heads about so humble a branch of knowledge. There was no voice to remind Cambridge of the eloquent appeal made to it by Milton—to turn from its abstruse pursuits to the study of the world we live in.' Freshfield went on to mention the educational enterprises of the RGS, the medals scheme (p. 14) and the series of advanced lectures (p. 10)—both failures—and also the Scott Keltie investigation of 1884 (p. 14), more influential than most people realized even in 1930. He also referred to Clements Markham's long-held wish that the Society should publish a journal comparable to the revered *Petermanns Mitteilungen*, a wish that was realized from 1879 to 1893 in the *Proceedings* edited successfully by Bates and Keltie and transformed into the new *Geographical Journal* from 1893. The redoubtable Sir Francis Younghusband, also speaking at the centenary meeting, mentioned that Freshfield had hoped to foster the climbing of Everest during his presidency but that some Council members regarded this—and also expeditions to the South Pole—as 'verging on the sensational'. 'And I entirely disagreed,' he adds. 'I held that there was not a square foot of the planet's surface to which Fellows of this Society should not at least try to go. That is our business. That is what we are out for.' Sir Francis said that during his presidency (1919–22) he had sought to stimulate 'the spirit of adventure and the interest in natural beauty' (*Geogrl. J.* 76 (1930), 467).

At the centenary celebrations papers on the 'habitable globe' were given by A. Penck, J. W. Gregory, and L. S. B. Leakey (*Geogrl. J.* 76 (1930), 477–98). Penck reached the conclusion that over the 7000 years of the post-glacial period, the surroundings of man have undergone no important changes in Central Asia. But in the glacial period the state of things was different from what it is now. J. W. Gregory reached a similar conclusion for Palestine, having considered the work of previous writers, including Ellsworth Huntington, whose views were markedly different. L. S. B. Leakey was concerned with East Africa's habitability from the Pleistocene period and argued that Africa had been inhabited much longer than Europe. Observation suggested that there was a pluvial period about 1000 BC 'so that big areas which are today practically waterless during a large part of the year had an abundant rainfall and a big and flourishing population'. He also considered the suitability of East Africa for permanent white occupation (a question closely argued at the International Geographical Union Congress in 1938) but concluded that 'the combined effect of high altitude and a latitude practically equatorial upon the health, and especially the nervous health, and mental capacity of the white races has not yet been properly studied ... until there have been several generations of whites in East Africa none of whom have come back to Europe the question cannot be settled.'

A. M. Carr-Saunders dealt with population pressure, actual and potential, even in some cases propagandist as in Germany before the War of 1914–18, with immigration laws that 'may block the most promising outlet for the relief of pressure', and with possible 'optimum' populations (*Geogrl. J.* 76 (1930), 500–4). His paper was a welcome treatment of an issue much discussed at the time, for the World Economic Crisis had revealed fears that unemployment was politically dangerous, both nationally and internationally. Emigration, the nineteenth-century remedy, no longer offered a solution for population pressure except on a limited scale. C. B. Fawcett (*Geogrl. J.* 76 (1930), 504–9) gave a crisp summary account of 'the extent of cultivable land' in the world and showed that '40% was desert, 30% cultivated land and 30% poor grazing, forest waste and high mountain'. The symposium on the habitable earth also included an excellent study of soil erosion in South Africa by F. E. Kanthack (*Geogrl. J.* 75 (1930), 516–21), who showed that in the areas having a rainfall of less than 20″ per annum on the average soil erosion was due to unwise farming practices. The report of the Drought Investigation Commission, published in 1923, had shown that there was no proof that the mean annual rainfall in the Union had altered appreciably in recent historic times. Soil erosion was becoming a subject of increasing interest in the 1930s.

## THE CLAIMS OF HUMAN AND REGIONAL GEOGRAPHY

The papers on the 'habitable globe' had been commissioned by the RGS committee on human and regional geography, formed in 1929. This committee (RGS Archives, i) was formed in response to representations from university geographers, notably C. B. Fawcett who in 1928 succeeded L. W. Lyde as professor of geography at University College London, that human and regional geography did not figure prominently enough among the

Society's lectures or in its journals. P. M. Roxby argued that at least one-quarter of its space should be reserved for such material but modestly did not mention that his own (much-admired) papers on China had appeared in *Geography*, the *Geographical Review*, and the *Scottish Geographical Magazine* (then in its finest years under the editorship of Marion I. Newbigin). Some papers rejected by RGS had aroused wide interest when published elsewhere, including H. J. Fleure's on human regions which appeared in *Geography* and, in French, in the admirable *Annales de géographie*.

Many young authors working on human and regional geography assumed that there was no point in approaching RGS at all. The outlook did not seem to be encouraging when, for example, F. Debenham of Cambridge said, 'Without being an opponent of human geography in any way, I cannot regard much of the descriptive stuff they favour with any enthusiasm, but I suppose it can be raised to a plane corresponding with the physical side of geographical inquiries. At all events, unless it is to give us something more than enlightened guide-books, I do not think the Society can give it very strong support.' A. R. Hinks had asked C. B. Fawcett for a list of papers that might have been offered to RGS but had in fact been published elsewhere, but no convincing response followed. However, at a meeting in December 1929, A. R. Hinks said that papers must have originality (exercises on a well-worn theme would not be acceptable) and readability (must not be in jargon) (RGS Archives). He described Fleure's human regions (Fleure 1919) as 'vague' and criticized it for using terms like 'parts of the Mediterranean' of which the only essential information appeared to be that 'this place is fertile'. He added that Fawcett's paper on Edale (Fawcett 1917) reached conclusions much the same as for all country places.

The university geographers gave some revealing comments on their work and their relations with RGS (RGS Archives). C. B. Fawcett said that university courses were largely human and regional, and in this he was supported by H. J. Fleure, who added that divergences between individual departments of geography were of historical origin and that RGS must look to a future when exploration lessened, and a considerable expansion in human geography might be expected: geography was currently in an experimental phase (a view he put forward to the end of his life). P. M. Roxby, like Fleure an admirer of French geography, stressed the need for synthesis in geography teaching ('few are able to take in details of unknown regions') and in his own work found most help in the *Geographical Review*, *Annales de géographie*, and *La géographie*, all of which published papers in his view not likely to be welcomed by RGS. F. Debenham, echoing his view expressed above, saw no problem in the lack of human and regional papers, but J. F. Unstead found little help in university work from RGS publications, and much more in French and Germans journals which had 'better papers'. With this L. Rodwell Jones was in agreement. A. G. Ogilvie, often more tactful than some of his brethren, regretted that RGS published few papers on the British Isles, while H. J. Mackinder, by now drenched in political and university experience, expressed a somewhat lofty view in saying that he was 'not impressed by any definition of geography'. Tradition was the only basis on which a subject might be defined and division was 'a concession to human weakness, especially of Professors', university geography in Britain was 'rather philosophical than scientific', and the way to improve the *Geographical Journal* was to have better specialist meetings: anything 'below the best becomes futile and dangerous'. Disappointment with the work of the committee led to its closure in February 1934, and G. R. Crone commented that the idea that 'a large number of such papers' (on human and regional geography) existed 'was probably erroneous' (Crone 1955: 5).

Meanwhile, in January 1933, a group of geographers, concerned with publication problems, founded the Institute of British Geographers, which published six monographs before the War of 1939–45, after which papers also were published. Crone observed (1955: 5) that 'the Institute did not find first class material plentiful, having accepted only one monograph for publication by September 1934.' This experience is also reflected in the rule adopted later that 'it was not necessary that work accepted for publication should be fully mature or finally authoritative'. To many experienced workers this would seem realistic for any publication in any journal. An absurd attitude of antipathy between RGS and IBG was fortunately abandoned eventually by all save a few people on either side, especially after the War of 1939–45 when the number of geographers in academic and other occupations sharply increased, the policy of RGS became more liberal, and authors eagerly provided material for publication in all the journals, with a growing number of books but only a few monographs.

Despite all the controversy on the scope and merits of human and regional geography, its growth was reflected in the pages of the *Geographical Journal* during the 1930s but slowly at first. In 1932, C. B. Fawcett's paper on 'The urban population of Great Britain' included a review of the 'conurbations', areas of major urbanization with interlocking towns, originally defined by Patrick Geddes in 1915, and other large towns (*Geogrl. J.* 79 (1932), 100–13). The paper was widely read and discussed, and provided a basis for the definition of conurbations for the 1951 Census and for the 'metropolitan counties' of the 1974 administrative reorganization of Britain. In the same year J. F. Unstead gave a paper on 'The Lotschenthal: a regional study', which he described as 'apparently descriptive' but bringing out those geographical elements (e.g. structure, relief, vegetation and human activities) which mutually interact and thus may be regarded as causal factors (*Geogrl. J.* 79 (1932), 298–311). In a footnote he explained that the study of the evolution of each of the phenomena, for example, the landforms, the plant coverings, or the economic systems, requires a specialist training and must be left to specialists.

To Unstead, as to other regional geographers of the time, 'the essential characteristics of the regional method in geography is to take as the units of study definite areas of the Earth's surface; it regards these as wholes, and examines the way in which their physical and human elements are interrelated and the influence of the unit-areas upon one another.' The demarcation of regions rested not on single factors such as 'relief, structure, climate, natural vegetation, land utilization ...' but on 'the combined effect of all the factors'. This was a view widely held at the time, but some geographers, while not condemning such aspirations, saw that in some areas towns dominated the landscape: the area on which an ever-expanding London was built, for example, had interesting physical features and—as later research was to show—an 'urban climate' of special character, but the experienced reality as well as its visual or mapped expression was London itself. And in rural areas, with which in the 1930s, geographers were largely concerned, agricultural life made the human landscape. Therefore an approach to the problem of regional classification might profitably include both the conurbation work of C. B. Fawcett and the Land Utilisation Survey of L. D. Stamp, on which a paper was read to RGS in 1931 (*Geogrl. J.* 78 (1931), 40–7).

Stamp's presentation of the work begun in 1929 opens with a summary of comparable enterprises already in being, including the 'now famous paper' of H. R. Mill in 1896 (p. 23), the work of individuals in universities and elsewhere, of groups linked with the Institute of Sociology, the Regional Survey Committee of the Geographical Association under the chairmanship of Sir John Russell, and other organizations. Land use survey was one part of a wider conspectus of local survey which drew inspiration from the work of Frédéric Le Play and Patrick Geddes; and work in Northamptonshire had shown that schools could provide workers able to map the use of the land on 6″ to 1-mile sheets. Stamp parried the question, 'What will be the value of such a survey when completed?' by asking another one, 'What is the value of the Survey that was carried out for the Domesday Book?'; it is perhaps from this that the survey was often referred to as a 'modern Domesday'. The abundant literature of county reports gave a fascinating study of Britain during the 1930s, a time of agricultural depression, splendidly summarized in L. D. Stamp's *The Land of Britain: its use and misuse* (1947). The reports were of mixed quality, but the best could be regarded as fine regional studies though their authors worked within county boundaries. Explanation of the land use led authors on the one hand to study the physical background and on the other to view the complexities of economic and social geography. Put differently, this means that land use provided a focus for regional study of a pragmatic character, to be valued later for the replanning of Britain. In 1933 E. C. Willatts gave a paper on 'Land Utilisation in the London basin, 1840–1932', which used the Tithe maps of 1840 and later material to show the recent historical development of the landscape (*Geogrl. J.* 82 (1933), 515–28). This was to be a feature of many of the county reports of the Survey and notably of Willatts' *Middlesex and the London Region* published in 1937.

An approach to regional geography from the historical angle came in H. C. Darby's 'Human geography of the Fenland before the drainage' in 1932 (*Geogrl. J.* 80 (1932), 400–35). This was followed in 1934 by Gordon Fowler's 'Extinct waterways of the Fenlands', which was warmly welcomed (*Geogrl. J.* 83 (1934), 30–6). Major Fowler and others had formed a Fenland Research Committee which included archaeologists, botanists, geologists, and others; although the committee survived only to the war years, the work continued and led to the publication, in 1971, of the fifth RGS Research Memoir, *The Fenland in Roman Times*, edited by C. W. Phillips. The subtitle describes this work as 'a study of a major area of peasant colonization with a gazetteer covering all known sites and finds'. Gordon Fowler had shown that a vast system of natural and artificial waterways laced the Fenland landscape, and aerial survey revealed an intense density of Romano-British settlement on the silt marshland and the peat margins of the fens. Darby's own researches received fuller treatment in two books published later. From this time also there were articles on the geography of Domesday Book, for example H. C. Darby on Norfolk and Suffolk in 1935, forerunners of the series of books edited by Darby and others covering the whole of England (*Geogrl. J.* 85 (1935), 432–47).

While there were abundant signs of advance in historical geography, human geography was variously taught in the universities and those who opposed it—or at least wanted to see its claims to academic respectability justified—remained critical. C. B. Fawcett, reporting on the meetings of the International Congress of 1933 in Warsaw, said that 'we are as yet in the early stages of investigation of the many problems of human geography, and have not reached well-established generalizations' (*Geogrl. J.* 84 (1934), 427). In 1934 E. G. R. Taylor welcomed Daryll Forde's *Habitat, Economy and Society* and said that the author, armed with a training in both geography and anthropology, could 'point out the fallacy underlying the attempt to link man's activities directly and causally to his geographical environment. Geographical determinism cannot succeed, for as Febvre has said long since "wherever 'man' and natural products are concerned, the 'idea' intervenes".' Forde had himself said that between the physical environment there was always 'a middle term, a collection of specific objectives and values, a body of knowledge and belief; in short a cultural pattern' (*Geogrl. J.* 84 (1934), 537–8).

The search for 'relationship' went on, and in 1939 C. B. Fawcett, writing on 'the distribution of rural settlements', studied by the International Geographical Union Commission on the rural habitat from 1925, said that more than a thousand papers had been written on the subject (*Geogrl. J.* 93 (1939), 132–5). Pioneer generalizations proved dubious, for dispersed settlement was found among people of varying racial types (this was a refutation of the hypothesis of Meitzen in 1895). Though physical facts, such as those of climate, relief, and soil, set limits to the extension and character of the rural landscapes, 'those limits are usually wide; and within them the differences are determined by man himself . . . on the type of rural economy and the stage of its development'. Old correlations between villages with certain characteristics of water supply and dispersion with others were questioned, and while in some areas it seemed that people built villages for security against natural hazards or invasions, in others they did not. One may observe in the labours of workers on rural settlement the same desire for correlation and synthesis that marks many model building enterprises, exercises in quantification, and enthusiasm for paradigms, that has marked the 1960s and 1970s.

Gradually through the 1930s interest in population problems was reflected in the *Geographical Journal*; for example in 1934 C. B. Fawcett reviewed the Ordnance Survey map 'Population of Great Britain 1931', of which the Scottish area was mapped by A. C. O'Dell, but commented that it was not easy to distinguish the twelve tints from one another (*Geogrl. J.* 84 (1934), 258–60). In 1935 Fawcett and others held a discussion session on population maps and dealt with the use of dots or shading (*Geogrl. J.* 85 (1935), 142–59). Some speakers favoured contours, and A. R. Hinks mentioned the Hungarian map drawn by B. C. Wallis at the close of the War of 1914–18 which 'showed by interlacing coloured contours the percentages of different nationalities in a mixed population'. E. G. R. Taylor and Brigadier Winterbotham agreed that dot-maps had a 'pseudo-accuracy', but Fawcett mentioned the fine work of Sten de Geer in Sweden and J. G. Granö in Finland on dot-maps though in general he favoured contours or hatches for population studies. A contour technique was used by A. C. O'Dell in a population map of London published in 1935.

Map presentation became more sophisticated as research workers developed an increasing concern with population study and human welfare. In 1936, for example, Arthur Geddes presented a paper on the population of Bengal, of which at the 1931 Census 93 per cent was rural, though some two million people lived in the 'Hooglyside conurbation' focussed on Calcutta (*Geogrl. J.* 89 (1937), 344–61). Geddes dealt with the fluctuations of population from 1854 in relation to land colonization schemes and health conditions as well as to the agricultural life of the people year by year and season by season; his paper gives a close view of the environment and led him to some interesting conclusions, for example that 'the calamitous state of the whole of the western and central Delta calls for urgent action . . . in this land, which less than eighty years ago was an abode of health and of high civilization, the water and silt supply for agriculture can be increased, and with this malaria can be largely controlled'. In a paper (*Geogrl. J.* 97 (1941), 248–53), 'Half a century of population trends in India: a regional study of net change and variability 1881–1931', Geddes came to the conclusion that 'the recognition of stagnation where both increase and variability are virtually nil' was 'particularly necessary', for in such area misery was not occasional or catastrophic but perennial: for example, 'among the densely packed but stagnant population of the least prosperous of the rice tracts of the Indian plains which extend from central Bengal to the eastern United Provinces or southwards to the delta of the Cauvery'. Both these Geddes papers show a close concern with human welfare and a fine knowledge of Indian environments.

In the later 1930s industrial areas of Britain made their first appearance in the *Geographical Journal*, though the plight of decayed coalfields and other depressed areas had for several years been a matter of intense public interest. In 1938 A. E. Smailes dealt with 'Population changes in the colliery districts of Northumberland' (*Geogrl. J.* 91 (1938), 230–9) and J. T. Gleave with the Tees-side iron and steel industry (*Geogrl. J.* 91 (1938), 454–67); a discussion on the 'geographical distribution of industry' in 1938 (*Geogrl. J.* 92 (1938), 122–39) was introduced by K. Mason's kindly comment that 'a great number of young geographers in the country . . . take much interest in it'. The main speaker was E. G. R. Taylor, ever young in heart and mind, who said that the pattern of English industrial development had taken shape in the late eighteenth century on a basis of free trade and *laissez-faire*, but had resulted in distressed areas, markedly so in areas such as Tyneside, though there were flourishing service industries at nodal points, for example Manchester, Leeds, and, above all, London. The country was dominantly rural in land use, and ports like Hull and Southampton were separated from other industrial areas by some forty miles of rural countryside.

England's main industrial belt lay in a SE–NW belt from the London area to Liverpool and Manchester and extending into West Yorkshire. Only part of this was suited to industrial development, as Professor Taylor showed by her famous 'sieve' map which marked off areas unsuitable owing to surface relief, scanty population, or poor accessibility, with some other unpromising areas such as 'narrow thronged valleys near the northern rim' of the South Wales coalfield. It was, she commented, 'remarkable how extensive is the area to which industry could not wisely be transferred without a migration of population (including provision of housing, schools, and so on) or an improvement in access, or both'. Many industries were tied to the location of raw materials, notably iron and steel, heavy chemicals, bricks, tiles, and cement, and some depended on local water supplies. Accessibility was closely studied, with an emphasis on railways that would be less marked forty years later.

Several problems remained, of which not the least was the spread of London and the demand for spacious sites for the horizontal layout of modern factories. Even greater difficulties lay in the decline of industries in such areas as the valleys of South Wales or Clydeside which, D. L. Linton (who contributed the section on Scotland to the memorandum sent to the (Barlow) Commission on the Royal Commission on the distribution of the industrial population) noted, 'found itself over-specialized and over-equipped with plant (not infrequently antiquated and badly sited) and with labour'.

Even in the darkest period of the War of 1939–45 there was concern for the future planning of Britain. This was shown in a vigorously written article by J. N. L. Baker and E. W. Gilbert in 1943 on 'The doctrine of an axial belt of industry in England', in which the authors viewed with dismay the prospect of widespread industrialization within what had become unfortunately named the 'coffin-shaped belt'. They called attention to the intersection of this by a belt of rural country marked by the road and rail routes between Oxford and Cambridge (*Geogrl. J.* 103 (1944), 49–72). Inevitably some of the findings of E. G. R. Taylor, L. D. Stamp, and others concerned with planning had been generalized and some of the maps produced had led to alarm, especially those based on rural districts which, though including industrial areas, were still primarily rural and agricultural in character. Published articles dealing with planning included useful pragmatic studies such as those by W. D. Evans on the opencast mining of ironstone and coal and by S. H. Beaver on minerals and planning in 1944 (*Geogrl. J.* 104 (1944, 102–19, 166–98). At the meeting when Beaver's paper was read the then Minister of Town and Country Planning (W. S. Morrison), the Minister for Agriculture (R. S. Hudson), and the Director of the Geological Survey (Dr E. B. Bailey) gave appreciative speeches.

Greater knowledge of Britain was required, not only for the satisfaction of the academically minded, but for the welfare of the country and its people. In 1938, at the British Association meeting in Cambridge, a committee had been formed to inquire into the possibility of publishing a national atlas. Of this E. G. R. Taylor was an enthusiastic supporter, and to some extent the need was met by the maps published by the Ministry of Town and Country Planning. The problem of regional units in England was raised in 1939 by E. W. Gilbert (*Geogrl. J.* 94 (1939), 29–44), who showed the wide range of regional divisions made for statistical purposes, special governmental or other public services, private bodies for their own needs, and finally for military and civil defence. As it happened, regional units were used to a considerable extent during the war and, with modifications at various times, for the increasing variety of public services afterwards, and on these developments Gilbert and others made various comments and suggestions.

In 1942 RGS held a discussion on 'geographical aspects of regional planning' with E. G. R. Taylor as chairman (*Geogrl. J.* 99 (1942), 61–80). Five speakers represented bodies concerned with the subject, the Association for Planning and Regional Reconstruction (Lord Forrester), the Nuffield College Social Reconstruction Survey (G. D. H. Cole), the Oxford Preservation Trust on Planning and Reconstruction (G. Montagu Harris), the Reconstruction Parliamentary Group (Henry Strauss MP), and the West Midland Group on post-war Planning and Reconstruction (G. W. Cadbury). There were also contributions from L. D. Stamp and E. W. Gilbert. Stamp noted the assumption that central government would delegate its powers to regional authorities in local or provincial centres. Both he and E. W. Gilbert agreed with some reluctance that in any replanning of Britain existing county boundaries would have to be retained except in a few cases, though Gilbert suggested that all the Tees valley should be in a new 'Northumbria' county, that a new Upper Thames planning region should include Buckinghamshire, Berkshire, and most of Oxfordshire, that a Fenland region should be defined, and that the area around Stoke-on-Trent should be part of Lancastria. As a whole the 1942 meeting showed the interest of a variety of people and though, as several speakers noted, compromise was essential, the eventual if much delayed outcome of 1974 was achieved only after long and careful thought.

During the war years reports on the replanning of cities and larger areas appeared, and in 1943 Patrick Abercrombie addressed the RGS on 'Some aspects of the County of London Plan', issued in the same year and widely read (*Geogrl. J.* 102 (1943), 227–40). Some geographers thought that he underplayed the quality of its physical site, except for his emphasis on the Thames, when he mentioned almost casually 'some low hills on the north and some low hills on the south'. But there was much in the London study that was to stimulate the study of urban geography after the war, including the emphasis on the two main central areas, one commercial and the other political and administrative, with perhaps a third, less cohesive, entertainment area. The history of London has been well covered, especially by the London Society. For planning purposes crucial considerations were traffic, housing, occupations, communities, open spaces, and public services. For all of these much of the basic provision was underground and consequently road alteration was difficult. But it was obviously needed, and Abercrombie wanted to keep through traffic from various recognizable quarters of London, including the City, Whitehall, the professional area south of Regent's Park, the university area, and the legal quarter. He also favoured ring roads, an issue to become controversial later, with a tunnel under Hyde Park. In his view a great city must remain closely settled (perhaps he foresaw the problems that were to arise from the movement of scores, even hundreds, of thousands from major cities within the next forty years), for although 'a great deal of industry can be moved ... a certain density, even a somewhat high density of population, is essential in the centre of a great city such as London because much of the work is tied to the centre; it is the great financial centre of the Empire; there are also the docks, and the industries tied to these docks.' For future housing development, he favoured the idea of the neighbourhood unit developed first in the USA with populations of 6000 to 10 000 grouped round schools and other facilities in constant demand, and he also advocated the 'shopping precinct' now familiar in towns.

Rural life was considered less frequently than urban problems, though the visual preservation of the countryside was always a popular cause. L. D. Stamp in 1943 commended the Report of the (Scott) Committee set up in 1941 to consider the conditions which should govern building and other constructional development in country areas, of which he was vice-chairman (*Geogrl. J.* 101 (1943), 16–30). The main recommendations were an overdue improvement of rural housing and village life; the preservation of the countryside with its footpaths, national parks, nature reserves, and coasts; the location of industries on vacant or derelict sites in towns or in country areas in existing or new small towns; and the location of new towns and villages away from the best farm land. For the townspeople, who numbered over four-fifths of the entire population of Britain, the countryside was a place to visit, an amenity area, but the threat to rural areas had been one impetus that led to the Town and Country Planning Act of 1932.

In 1935 Vaughan Cornish (*Geogrl. J.* 86 (1935), 505–11), writing on the scenery of England and the preservation of its amenities, observed that opinion was moving towards the formation of national parks. He wished to see the coastlines preserved and suggested that a belt 110 yards wide would be suitable for 'the proposed pleasance of the cliffs'. In his book *The Preservation of our Scenery* (1937) he regretted that 'few councillors have made a special study of the aesthetics of scenery, of the pictorial unity of architecture and nature in a rural scene, or, when rates and the schemes of local builders are on the agenda, have vision and enthusiasm enough to see the future.' Much later, in 1944, A. J. Steers followed his varied coastal studies with a paper on 'Coastal preservation and planning' (*Geogrl. J.* 104 (1944), 7–27). He began by saying that coasts had a variety of scene (in earlier years he had argued that low lying coasts were just as worthy of preservation as those of a more dramatic and cliffed character) and that 'no assessment of quality' was possible. (Not all would agree with this—though the present author does—for in some modern books scenery is graded on a numerical scale.) Some coasts had been defaced by industrial development, mining and quarrying, and 'nearly every sandy cove in Wales and southwestern England where bathing is reasonably safe is partly, or wholly, spoiled'. There were areas of shacks and huts on the south-east coast, and also the coasts of Norfolk, Lincolnshire, Holderness, and Wales, except where coastal access had been restricted for bird sanctuaries and other purposes. Careful replanning and control was needed, especially in a coastal zone with cliff and beach paths.

Although at the time of the centenary the attitude of RGS to studies in human and regional geography was cautiously friendly, by 1945 several notable papers had appeared on contemporary problems and planning prospects. But during the depressed years of the early 1930s RGS had no published contributions on the declining industrial areas, nor was there any mention of the special legislation passed from 1934 to assist the Depressed (Development) Areas by establishing trading estates and other measures. On such areas reports were written primarily by economists, some with a firm understanding of the local economic geography. That geographers had much of relevance to say was abundantly revealed by their contribution to the work of the Barlow commission and the Scott committee (with the Uthwatt committee of 1941 on land values) and by the work of the Land Utilisation Survey. The study of current and future distribution of population was one to which geographers could make valuable contributions, and through it the amenity of the landscape became a concern not only of the aesthetically minded such as Vaughan Cornish but also of the general public. Geographers who worked on these lines found their views welcomed in various papers such as the intellectual weeklies and the quality Sunday papers. The success of the *Geographical Magazine* from its first publication in 1935 was notable, and among its varied contents were attractively presented articles on current problems of human and regional geography.

If some of the more traditionally minded people of the RGS thought that concern with current problems was unworthy of the Society's heritage, they could draw comfort from the continued interest in exploration, in cartography, in place-names (especially the work of the Permanent Committee on Geographical Names established in 1919), in physical geography, and political geography. Perhaps it was fortunate that human geography came into its journal through its social relevance rather than through a demonstration of the remote and rare or through the association of racial characteristics with prehistoric and historical migrations across the habitable world (a favoured study in the 1930s) or methodological debate (always attractive to academics, especially younger ones).

PHYSICAL GEOGRAPHY

Whether physical geography could be regarded as a specialism in its own right was to be argued later (cf. p. 80); many of the papers that fell into this category from 1930 to 1945 had obvious significance for regional geography. One such was 'The glacial drifts of Essex and Hertfordshire and their bearing upon the agricultural and historical geography of the region' by S. W. Wooldridge and D. J. Smetham of 1931 (*Geogrl. J.* 78 (1931), 243–69). It was based on a close field-work survey of the various drifts and other superficial coverings of rocks with a study of agricultural soils, improved by farmers and in some areas used for crops or pasture according to current economic circumstances. This was followed by a study of the land use, including forests in Roman, Saxon, and Norman times. Although the paper could be regarded as largely geomorphological, Wooldridge presented it as a contribution to regional geography for which, said the RGS president Sir William Goodenough, 'there is always a welcome.... We do not spend all our time chasing about after new places in the world.' The paper is a fine example of the type of regional geography, having a strong foundation in physical geography, that Wooldridge was to advocate later (p. 70).

Coasts remained a fascinating subject, for on them the evolution of land-forms could be studied by continued observation and in some cases map evidence was available for earlier periods. This was one theme investigated by C. Diver (a clerk in the House of Commons), who in 1933 gave a paper on 'The physiography of the South Haven peninsula, Studland Heath, Dorset', warmly welcomed by Professor E. J. Salisbury, the plant ecologist, for the light shed on the plant life of an environment subject to constant change (*Geogrl. J.* 81 (1933), 404–27). A similar welcome from botanists came for A.J. Steers's paper on Scolt Head Island (*Geogrl. J.* 83 (1934), 479–84) of 1934, and the book he edited, *Scolt Head Island: the story of its origin: the plant and animal life of its dunes and marshes* (1934), a series of papers published for the Norfolk and Norwich Naturalists Society in 1934, was praised by a reviewer for the co-operation between the Society and Cambridge University.

Another paper in 1936 on the north Norfolk coast from Hunstanton to Brancaster was a detailed study of an area close to Scolt Head Island (*Geogrl. J.* 87 (1936), 36–46), and in 1939 Steers discussed recent coastal changes in south-eastern England, including some treatment of floods though, not surprisingly, the Wash and the north Norfolk coast figured prominently (*Geogrl. J.* 93 (1939), 399–418). Also in 1939 Steers gave a paper on 'Sand and shingle formations in Cardigan bay' in which he showed that the most conspicuous feature was the movement of beach material towards Tremadoc Bay (*Geogrl. J.* 93 (1939), 209–27). In 1931 W. V. Lewis wrote the first of a number of papers on coasts (*Geogrl. J.* 78 (1931), 129–48) based on a study of various shingle beaches on the south coast and in Glamorgan and entitled 'The effect of wave incidence on the configuration of a shingle beach'; this was followed in 1932 by 'The formation of Dungeness foreland' (*Geogrl. J.* 80 (1932), 309–24), a subject of fascinating complexity as there had been changes in the level of land and sea from Roman times and various efforts to drain marshes and control rivers. A further study of this area, with W. G. V. Balchin, was 'Past sea levels at Dungeness' in 1940, in which little appeared to be capable of proof, either on how the curious landscape was formed or the processes involved. As in comparable studies there was much of interest in the study of vegetation, past and present (*Geogrl. J.* 96 (1940), 258–85).

Glaciation was discussed in numerous journals and the contributions made in the *Geographical Journal* came partly from the Cambridge University Vatnajökull expedition of 1932, of which W. V. Lewis was a member (*Geogrl. J.* 81 (1933), 289–313). He also wrote a paper on his observations, 'Nivation, river grading and shoreline development', published in 1936 (*Geogrl. J.* 88 (1936), 431–47). This was followed in 1939 by a study of 'Snowpatch erosion in Iceland', based on fieldwork during a visit in 1937 (*Geogrl. J.* 94 (1939), 153–61); these features, though familiar to climbers, had received little study, and Lewis, with the other members of the Cambridge East Iceland expedition, carefully analysed the evidence available. Similar work was done during the Cambridge University Spitzbergen expedition of 1938, reported in 'Nivation and corrie erosion in west Spitzbergen' by I. A. McCabe (*Geogrl. J.* 94 (1939), 447–65). Expeditions have inspired many people to carry on research in the field and have received encouragement and assistance from the RGS though equal—perhaps greater—value lies in long-continued work in some chosen area. One man who needed no such support was Thomas Hay, who from 1928 to 1944 wrote twelve papers on the topography of the Lake District. Hay was headmaster of a school in East Anglia who spent his holidays studying mountains, valleys, lakes, and rivers.

In his final article on 'Mountain forms in Lakeland' Hay reached four conclusions. First, the result of ice action depends greatly on the nature of the rock: a heterogeneous rock or one possessing widely spaced big joints gives rise to a rugged landscape, while a homogeneous rock with small closely spaced divisional planes gives rise to much more regular outlines. Secondly, Dr Marr's interpretation of the subdued summit land as a relic of the pre-glacial surface is supported by an examination of the erratics on the high-lying tracts. Thirdly, there seems to have been a continuous icecap near the central and western parts of the area with a thinning-out of the ice on the more easterly elevated ridges. Fourth, an Ullswater glacier can be recognized as separated in parts from the icecap but still joining it on the south and west (*Geogrl. J.* 103 (1944), 263–9).

Possibly the work of Hay might be regarded by modern quantitative geomorphologists as characteristic of its time: observational, inductive, even hypothetical. They could hardly lay such charges at the tent door of Brigadier R. A. Bagnold (*Geogrl. J.* 85 (1935), 342–65), whose 'Movement of desert sand' in 1935 and 'Transport of sand by wind' in 1937 (*Geogrl. J.* 89 (1937), 409–48) were based on mathematical methods. They might, however, look askance at the dramatic presentation given by E. B. Stebbing (*Geogrl. J.* 85 (1935), 342–65) in 'The encroaching Sahara; the threat to the West African colonies'. This had a mixed reception, and in 1938 Francis Rodd argued that deforestation had been arrested. In this he agreed with J. D. Falconer (*Geogrl. J.* 91 (1938), 354–9), who had reviewed Stebbing's book *The Forests of West Africa and the Sahara*, that the spread of desert was far less than Stebbing suggested; in fact, there had been some spread of desert but in other areas trees had become more abundant, suggesting a fluctuation of climate. There was undoubted danger in the 'uncontrolled expansion of shifting cultivation'. Stebbing was concerned with 'increasing erosion ... due to man's upsetting Nature's balance by an ignorant over-utilization of vegetation and soils'. He suggested that 'once Nature's balance has been interfered with to the point at which the rainfall becomes irregular and unreliable, a stage has been reached ... the Intermittent Rainfall Stage ... at which man can no longer rely either on the amounts or the times at which the rain will come.'

All this was contested by Brynmor Jones (*Geogrl. J.* 91 (1938), 401–23) in a long article on 'Desiccation and the West African Colonies'. He argued that West Africa had experienced several climatic changes since the beginning of the Quaternary era of which the last was towards more humid conditions;

much could be done to prevent the loss of headwaters and to control floods, and to guard against soil erosion. Some forms of farming were unwise, and trees should be preserved to break the force of the wind; in particular, shifting farming threatened the economic future of the drier parts of the West African colonies. Nevertheless the nearest masses of live desert sand were 150 miles north of Nigeria; they were not encroaching, nor was a vast shelter belt needed, as in any case a thorn forest existed.

The need for more research on Africa was stressed by various writers including Daryll Forde (who himself made a fine contribution to African studies). In an extended review of *An African Survey* by Lord Hailey, Forde said that he (*Geogrl. J.* 93 (1939), 430–6) 'had never allowed himself or his collaborators to forget the simple but all-important fact that it is not an area but a body of some 145 million human beings and their future descendants that are the subject of their survey'. The essential need was the application of the natural sciences to the land and people of Africa with a co-ordinated research policy for British territories supported by the Treasury. Devoted workers had been attracted, including G. Milne whose work on soils was reviewed with that of Clement Gillman on Tanganyika in 1936 (*Geogrl. J.* 88 (1936), 465–6), but workers were few and the problems great. A paper that showed how man could make the earth and also destroy it was E. J. Wayland's 'Desert versus forest in Eastern Africa' (*Geogrl. J.* 95 (1940), 329–40). The climatic history of Africa was controversial, with L. S. B. Leakey and others locked engagingly in argument. Wayland held that in 1940 Africa might be 'in a dry period' and on the past as well as the present man's influence

almost from the first, has been profound. For many millennia man has been felling trees, burning forest, and (later) adopting other practices detrimental to the vegetal cover, so that very much, perhaps all, of the counter-forest has been destroyed.... The zone of man-made desert is wide and much of it is still cultivable. What will happen if, through unwise cultivation, destruction of forests, overstocking of cattle, the soil deteriorates and erosion carries it away? The area will become a dreary wilderness.

He advocated—and in this was warmly supported by E. B. Stebbing—counter-forests as a buffer against erosion and the direction of research towards 'determining how far reclamation of now waste marginal ground may be possible, desirable and economically practicable'.

Research in physical geography, it is clear, had many practical applications and therefore could rank as 'applied'. This raised issues of world significance, as Stamp noted in a review of *The Rape of the Earth* by G. V. Jacks and R. O. Whyte, 1939. Stamp (*Geogrl. J.* 94 (1939), 345–6) hoped that readers would not be put off by such an unfortunate title (who would cavil at it forty-odd years later?) as despite some errors in geology and geomorphology it 'rightly stressed that disastrous flooding of such rivers as the Mississippi ... cannot be controlled by engineering works but only by removing the local causes, destruction of natural vegetation and misuse of land ...' The real problem was to make a scientific study of land utilization. All this research led back to climate and weather as well as to soils and vegetation. In countries where abundant data (at least relatively) were available, such as Britain, the hope was for more understanding of microclimatology, as G. Manley (*Geogrl. J.* 103 (1944), 241–63) showed in his 'Topographical features and the climate of Britain: a review of some outstanding effects'. In Africa, as in much of the world, far fewer data were available and broad generalizations were based on unavoidable ignorance except for a few local studies.

As will be seen, papers on physical 'geography' covered a wide spectrum of interest in the fifteen years to 1945. Many were research enquiries that could profitably have been repeated elsewhere. Some were concerned with fundamentals of regional geography, of which Wooldridge and Smetham's paper (p. 43) is a fine example. Others, particularly the African studies, had obvious human applications, but this was equally true of the paper by Brysson Cunningham (*Geogrl. J.* 85 (1935), 531–44) on the water supply of Britain, 'National Inland Water Supply'. There were other articles, perhaps meriting classification as geology or geomorphology or a mixture of both, such as those of George B. Barbour on the loesslands of China in 1935 and the physiographic history of the Yangtze in 1936, for of these vast areas little was known (*Geogrl. J.* 86 (1936), 54–65; 87, 17–32). There was still interest in the whole world even though the homeland appeared to have been rediscovered. Perhaps even physical geographers were sufficiently interested in literature to appreciate T. S. Eliot's lines in 'Little Gidding'

> We shall not cease from exploration
> And the end of all our exploring
> Will be to arrive where we started
> And know the place for the first time.

## THE WIDER WORLD

Papers on political geography were not numerous in the *Geographical Journal* between 1930 and 1945, though some were notable, particularly a study of 'The geographical factor in Mongol history' in 1938 by Owen Lattimore who so identified himself with the Mongol people that he wrote from inside rather than outside his subject (*Geogrl. J.* 91 (1938), 1–16). Toynbee's *Study of History* had already aroused great interest (cf. p. 73), but Lattimore, having accepted climatic variation as part of the Mongol environment, did not accept the view that adverse climatic periods necessarily led to invasions, for 'the relation of history is much more complicated ... steppe society has been modified by both evolution and devolution, and also by shifts between extremely extensive forms of economy and relatively intensive forms'. Also in Asia, in 1943 O. H. K. Spate (*Geogrl. J.* 102 (1943), 125–36) wrote on 'Geographical aspects of the Pakistan scheme', on which he was to say much of interest later; this article deals not only with obvious economic and social questions but also with the attitudes of populations, crucial in such matters however unreasonable they may appear to be. 'Perception' has been a concern of geographers much longer than people sometimes think.

Despite all the problems of Africa or Asia before the War of 1939–45, the real threat to peace lay in Europe and three articles by H. G. Wanklyn discussed some of the issues in eastern Europe. The first makes sad reading for those who have visited Auschwitz, for it deals with the Jewish settlement east of Germany, in the Jewish Pole between the Dvina and Dnieper rivers on the east and the Vistula and Dniester on the west, with an extension west of Warsaw (*Geogrl. J.* 95 (1940), 175–88). The author comments that 'when the standard of living dropped [in the world economic crises of 1931–4] and trading activity shrank in Poland, Hungary, Romania and Yugoslavia, the victims of general ill-will were again the Jewish groups, just as they had been in the difficult years before 1914'. For a time after 1936 industrial revival alleviated this problem, but hope vanished for the Jewish population with the German invasion of 1939. The other two articles dealt with peasant Hungary and the artisan element in the Slav countries (*Geogrl. J.* 96 (1940), 18–34; 103 (1944), 101–19). Whatever happened after the War of 1939–45 it was clear that a new world order must emerge. Much had been written in the *Geographical Journal* on the replanning of Britain, and in the wider world there was a general realization that Mackinder's comment, when he received the Charles P. Daly medal of the American Geographical Society from the American Ambassador in London, was true. 'Henceforth', he said, 'nothing significant can happen in human affairs but what its effects will spread, like the Krakatoan dust, over the whole global surface. We have now to adapt our minds to living in a closed environment' (*Geogrl. J.* 103 (1944), 132–3).

## After the Second World War

As the War of 1939–45 ended, everyone looked forward with questioning uncertainty about the future but with the hope that the rebuilding of Europe might be accompanied by the achievement of world peace and prosperity. Although the Royal Geographical Society's house had been damaged by bombing and shared in the generally depressing shabbiness of London and, indeed, of the whole country at that time, its reduced publications over the war years had shown a spirit of optimism on the future replanning of Britain. As it happened, much of the work done before the war, discussed above, was to prove helpful in the difficult years that were to come, and the value of work that began as detailed and local studies, such as those of the Land Utilisation Survey, was revealed in due time.

An interesting sidelight on the Society was given at the annual general meeting on 25 June 1945, after the end of war in Europe but before the end in the Far East. The President, Sir George Clerk, noted that it was still not 'possible to assess the relative value of the geographical work done by officers of His Majesty's Forces' but they were anxious to honour 'the work of distinguished senior British geographers' (*Geogrl. J.* 105 (1945), 230). Regrettably much of the work of geographers during the war was to remain unrecorded, for in the uneasy years of peace that followed it was wiser to avoid publication, especially if one had read the small print in signing the Official Secrets Act. Even the general article by L. S. Wilson on 'Some observations on wartime geography in England' which appeared in the *Geographical Review* (36 (1946), 597–612), though written with obvious military discretion, revealed more than it would have been possible to publish in England at that time. Significantly the paper's last few pages consist of an appendix by L. D. Stamp listing the various 'British town planning schemes' known to exist. There were sixty of these as well as eleven 'regional planning schemes'.

The senior geographer chosen for the Patron's Medal in 1945 was Sir Halford Mackinder, who as always matched his oratory to the hour (*Geogrl. J.* 105 (1945), 230–2). Sir George Clerk said that for Mackinder geography had been 'not merely the concern of the schools; he has striven to secure the recognition of its fundamental role in national life, and its value as a bridge between science and philosophy. To many his *Britain and the British Seas* has revealed for the first time the fascinating interplay of environment and human society. As a political geographer his reputation is indeed world wide.' Earlier the President had noted that Mackinder had established his name in the history of African exploration by 'his first ascent and survey of Mount Kenya in 1899' and went on to say that at Oxford and later in the University of London he had 'inspired and guided successive generations'. In his reply Mackinder said that he was a survivor (with H. R. Mill) of 'a small group of men who a generation and rather more ago, here in London and a few of them in Scotland, achieved a revolution in the conception of geography in this country'. Later he says of his 'Kenya year 1899' that 'at that time most people would have no use for a geographer who was not also an adventurer and explorer'. Twelve years had elapsed since he had given the famous lecture (resented by some traditionalists) at the RGS in February 1887 (p. 15), but 1899 also saw the opening of the Oxford School of Geography 'with an adequate staff and the necessary apparatus'. This was a more permanent contribution than the Kenya episode, especially as there were others only waiting their chance to climb the mountain and the expedition had therefore to be planned with the utmost secrecy.

Mackinder described how Bates, the Amazon naturalist who was then Assistant Secretary of RGS, asked him to call in 1886 and write down what he meant by 'the new geography', on which he had given University Extension lectures. (At that time Keltie's famous report had just appeared (p. 14).) The resultant paper was read at an evening meeting in February 1887 (p. 15) and

was followed by an important debate in a full house, due not only to Keltie's and my activities, but also to the fact that, unknown to me at the time, the Council of the Society was divided into embattled parties with rival programmes. The progressives were led by Douglas Freshfield, Francis Galton, and Clements Markham, and in the end they won. As a result I was sent back to Oxford in June 1887 to inaugurate the teaching of Geography there on a pittance subscribed partly by the Society and partly by the University.

Mackinder goes on to explain how, hearing that 'a young Scottish geographer, Herbertson, a disciple of Patrick Geddes, on whom I had my eye for my principal assistant', had been offered a professorship in New York which he was about to accept, he travelled at once to Edinburgh, breakfasted with the currently reigning John Bartholomew, from whom he gained a favourable report, and drove to Herbertson's home and arranged that he should come to Oxford. Finally Mackinder explains that he 'passed gradually into spheres of wider experience' than academic life normally provides but that 'now and again, in essays and addresses, and in a book, it became my privilege to write on the bridge function of geography in spanning the interval between scientific and political facts and philosophical values.'

A source of fascinated interest to many geographers, Mackinder is a figure who cannot be ignored in the history of modern geography. All through the changes of the thirty-odd years that have elapsed since 1945 his name recurs, on occasion in terms of warm praise that is almost adulation but also with less favourable judgements for his former embroilment in the political affairs of Russia and for his imperialistic outlook. That his views of world political strategy had been of interest to German geopoliticians is hardly remarkable and in no way to his discredit. On his death two years later in 1947, the obituary by E. W. Gilbert (*Geogrl. J.* 110 (1947), 94–9) was both a contribution to the history of modern geographical thought and a study of a man who through his papers originally published by the RGS stimulated others to further research in regional and political geography and who also had a marked interest in physical geography that is less well remembered.

In 1945 the President of RGS clearly hoped that much of geographical interest might emerge from the experience of serving soldiers. Not all the articles that appeared on this aspect could be regarded as contributions to academic geography. There is for example a paper by Brigadier Bernard Ferguson (*Geogrl. J.* 107 (1946), 1–9) on 'Upper Burma 1943–44' in 1946 which provides thrilling reading, but the author begins by saying wryly that he and his companions were 'somewhat preoccupied' with fighting the Japanese and therefore scarcely able to make full use of the opportunity of making geographical observations. More serious contributions came from Brigadier Bagnold and others having military experience of deserts (p. 45), among which J. K. Wright's 'War time exploration with the Sudan defence forces in the Libyan desert, 1941–43' (*Geogrl. J.* 105 (1945), 100–11) concludes by recording that 'As a result of the war a fair-sized area of the Libyan desert has been explored in considerable detail; it is doubtful whether maps of greater accuracy than these we produced will ever be worth while in the central desert.' Another paper that drew strength from wartime experiences was on dust storms in Egypt from 1939 to 1945 by F. W. Oliver (*Geogrl. J.* 106 (1945), 26–49). This was based on six years of observation of an area of 'semi-desert' west of Alexandria and, like other studies of the desert, owed much to the book by Brigadier A. R. Bagnold on *The Physics of Blown Sand and Desert Dunes* published in 1941.

When Lord Rennell of Rodd became president in 1946, he too said that the journal should include some papers on the 'work and exploits of personnel serving in the armed forces of the Crown ... in strange and difficult places and circumstances' (*Geogrl. J.* 107 (1946), 83). Some such contributions were of academic significance even though the appeal for many readers may have been primarily the adventures. A paper of 1946, by P. Joubert and A. R. Glen (*Geogrl. J.* 108 (1946), 1–21), 'High latitude flying by Coastal Command, in support of convoys to North Russia; is in effect a fascinating adventure story with valuable information on the limits of ice at various times and excellent aerial photographs. In December 1945 and February 1946 papers were given by Wing Commanders D. F. Mckinley and K. C. Maclure on Arctic flights which, apart from photographic reconnaissance, were concerned with climatic observation and the location of the earth's magnetic fields (*Geogrl. J.* 107 (1946), 90–126). The second paper concludes with the comment that (p. 123) 'before many years have passed flying across the Arctic will be as much a matter of routine as crossing the Atlantic by air is to-day'. In volume 109 (1947), 147–8, it was reported that aircraft from the Arctic Institute at Leningrad had made a number of flights in the western sector of the Russian Arctic 'with the object of making autumn ice observations', many of which penetrated to 82 °–85 °N. Now at last aviation was able to solve the mysteries of the north-west passage of the American continent and of the north-east passage of Asia that had been so great an impetus to adventurous discovery for centuries. The significance of such observations for climatology, meteorology, oceanography, geomorphology, glaciology and cognate studies, however defined and subdivided, needs no emphasis.

Every society has to consider the tastes of its members and, following the appointment of L. R. Kirwan as Director and Secretary in November 1945, the RGS sent a questionnaire to its Fellows asking for preferences in subjects for evening lectures, many of which would later be published (*Geogrl. J.* 107 (1946), 265). As always in such enterprises the definition of categories offered difficulties, but the 1133 cards returned give a clear indication of interest. First in order of preference was exploration and travel (16.9%), followed by geography and international affairs (12.2%), human geography (11.1%), and history of exploration (9.7%). These four categories account for almost half the points scored, and the appeal of exploration, in two of the categories named above, is apparent. The remaining six categories score in the following order: geographical aspects of other sciences (9.3%), physical geography (9.2%), economic geography (9.1%), historical geography and cartography (7.8%), maps and survey (7.7%), and finally town and country planning (7.0%).

Younger readers may not realize that in 1946 town and country planning was regarded with suspicion by many people, but in fact geographers were well equipped to make a substantial contribution to the replanning of Britain, in part because, as shown earlier, several of them, including C. B. Fawcett, E. W. Gilbert, L. D. Stamp, and Eva G. R. Taylor, had already contributed to the *Journal* papers that had obvious implications for planning. Among the efforts of geographers, the contribution made to the Barlow report on industrial location, published in 1940, is noted on pp. 39–40; naturally the implementation of this report, like that of the Scott Committee on Land Utilisation in Rural Areas of 1942, had to wait until after the war. Opposition to the idea of planning, never powerful, declined and it was never necessary, in British (as in some American) universities, to hide an interest in redevelopment behind such vague euphemisms as 'environmental studies and development'. By the end of the war the term 'applied geography' (which incidentally was welcomed by some prominent planners such as Sir William Holford) was in general circulation, though geographers were firmly reminded by Eva G. R. Taylor that the success of applied geography depended on the underpinning of an adequate pure geography. Many of the planning reports, which were eagerly read at the time, seemed visionary and uplifting but appeared to have been produced in offices with efficient staffs of draughtsmen and photographers rather than from actual knowledge of the places and the people who would be affected by the redevelopment. 'Uplift' was fashionable.

## HOME PROBLEMS

Now, when peace slowly emerged, it seemed, as S. W. Wooldridge said in a lecture given to the RGS and the Geographical Association in 1947 on 'Geographical Science in Education' (*Geogrl. J.* 109 (1947), 198–206), that:

This country, in common with a large part of the world, is moving in the direction of a conscious control of its environment. Though opinions in the political sphere may legitimately differ as to the desirable mode and rate of such control, no informed mind can reject the conclusion that the state of the world or the stage of man's development calls for some measure of deliberate calculation and adjustment. Since Humboldt and Ritter founded modern geography more than a century ago, the population of the world has more than doubled and the whole vast range of the Earth's productive resources has been brought into use by long-continued trial and error.... It is at least certain that there is no room for the care-free drift of *laissez-faire*. It is not the whim of tyrants or the passions of parties which have brought present looming issues to their crux, but rather tides of the spirit and surges of social process, while lying behind them both are the physical facts of man's expansion to the limits of his habitat. If this verdict seems to ignore the intimations of an approaching atomic age, it can hardly be doubted that any such new major factor will complicate rather than simplify earth patterns and human problems.

In 1948 RGS, in co-operation with the Geographical Association and the Institute of British Geographers, formed the Education Committee, with E. D. Laborde as its first chairman, to deal with teaching in universities, institutes of education, colleges, and schools (Crone 1955: 35). Wooldridge, like many of his contemporaries, was deeply concerned with the future of Britain and with the contribution to national welfare that geographers might make.

He argued that man 'had taken his physical and social world to pieces with exemplary zeal' but seemed to view a synthetic, regional approach sceptically ('it cannot be done or is not worth doing'). He wanted to see a synthetic, regional approach and commended the work of H. R. Mill in 1900 on local memoirs covering defined areas (in that case a 1:63, 360 sheet of the Ordnance Survey) from which a major regional synthesis could be made (p. 23). He viewed sympathetically the statement of A. L. Rowse that a sense of history was part of the 'self-awareness of our environment' but added that part of the environment is geographical. An opportunity for geographers lay in the current need for planning, and in fact this was particularly true in colonial studies: of Britain's colonial empire people were 'calamitously and shamefully ignorant'. (As will shortly appear, efforts were made to remedy this situation, at least by pilot studies.) In this, as in much else, Wooldridge was an enemy of complacency. He was not denigrating analytical and specialist studies (he was known mainly as a geomorphologist) but regarding them as contributory to a broader, regional synthesis. The day was yet to come when a large number of geographers would view the human environment solely in a social, economic, or political context.

Infinitely slow as many physical processes are, physical hazards still remain, some of them unheeded. Probably few people paid much attention to a paper by W. H. Ward in 1945 (*Geogrl. J.* 105 (1945), 170–91) on 'The stability of natural slopes', based on research done at the Building Research Station of the Department of Scientific and Industrial Research, which impelled W. D. Evans, in discussion, to observe that there were great dangers of disaster in 'the central and eastern parts of the South Wales Coalfield': in 1966 the Aberfan disaster revealed this to the whole world. The catastrophic element always exists in geomorphology, and some of the challenges actual events provided were taken up by geomorphologists with good results. Regrettably, the floods in the east of England that followed the rapid thaw in March 1947 did not attract such attention in the pages of the *Geographical Journal*, though fortunately the East Coast floods of 31 January–1 February 1953 were treated in an article by J. A. Steers based on his own observations and those of other workers (*Geogrl. J.* 119 (1953), 280–95).

Planning in Britain is concerned with the whole environment, rural and urban, and in some areas particularly with holiday and tourist provision. This was clearly appreciated by J. A. Steers, who in 1946 (*Geogrl. J.* 107 (1946), 57–60) wrote a supplementary section to his major paper of 1944 on 'Coastal preservation and planning' (p. 42). This was allied to his work for *The Coastlands of England and Wales* (1946), a substantial book welcomed in an extended review by S. W. Wooldridge (*Geogrl. J.* 109 (1947), 107–9), who comments that 'it is in no sense a mere appraisal of our endowment of coastal scenery' but more fundamental; it seeks to examine the significance of our coasts in terms not only of present form but of genesis, leaving aside the secondary or 'applied' problem for the 'landscapist'. Rather sadly, Wooldridge completes the review by saying, 'It must be left to others, at what is commonly known as a "high level", though too often informed by a much lower level of knowledge and scholarship, to decide the fate of our coastal heritage.' Nevertheless Steers had given a clear indication of the major problems of planning on coasts, made notorious by much undesirable building and camp-site development permitted during the inter-war period. That an increasing population would look to the coasts for recreation was all too clear, but fortunately in later years some of the fears of wholesale despoliation were not realized. In time, coastal paths were laid out and the need to preserve coastal scenery became generally accepted. People like Vaughan Cornish, whose 'aesthetic geography' was regarded with some amusement during the 1930s, later saw their ideas respected.

Town expansion was known to be inevitable after the War and the problems of heavy losses of population in many of Britain's older industrial and commercial areas were complemented by the attraction of other areas, particularly around London and in towns of the south-east, for new industries. This had been clearly seen during the inter-war period when the contrast between depressed areas, including some of the coalfields, and the expanding industrial belts such as north-west and west London, with much of Middlesex, became marked. Inevitably the question was how contemporary and future growth could be controlled, and what measures would be likely, perhaps through legislation, to prevent the decay to extinction of old mining and industrial areas that had lost their *raison d'être* through the working-out of mines or dependence on declining industries such as cotton in Lancashire.

In 1946 G. H. J. Daysh and A. C. O'Dell (*Geogrl. J.* 109 (1947), 103–6) spoke on some of these problems and notably of the central challenge for 'planning should mean the integration of hitherto separate lines of thought and action in respect of industry, and other forms of development'. They pleaded for co-ordination, which was one purpose of the Town and Country Planning Bill of 1947, and argued that the geographer

is not handicapped by being too much of a specialist; claiming no particular pheno-menon as his own, he is trained to study the relationships between phenomena, to view a particular problem in its setting, to consider trends and the interplay of factors. He is moreover accustomed to taking evidence from specialists and applying it, and to mak-ing surveys and depicting his findings on maps.

Daysh and O'Dell were looking forward to a type of regional research based on local studies in which not only geographers but also a wide range of other specialists would be involved. Their conception of the geographer as a co-ordinator was widely shared and similar ideas were discussed in the United States some twenty years later at meetings of the International Geographical Union's Commission on Applied Geography, of which one merit was the sharing of experience between different countries. But their views of a geo-grapher's work would not be universally approved, especially by differently-trained specialists who might think that they would be effective co-ordinators of the work of others.

Rural Britain, with its farming, forestry, and scenic amenity, was threatened by urban expansion after the War of 1939–45. Following the pre-sentation of L. D. Stamp's paper on 'Wartime changes in British agriculture' in 1946 (*Geogrl. J.* 109 (1947), 39–57), the Minister for Town and Country Planning, Lewis Silkin, assured the RGS that the government was well aware of the need to preserve the balance between the farmed countryside and grow-ing towns. Stamp had shown that agricultural production had been greatly increased during the War and that the maintenance of this prosperity depended on the wise allocation of land for agricultural and other uses in the future. All too easily the inevitable urban expansion, including suburban-ization and new industrial growth, the construction of new roads, airports, and especially New Towns, could consume much of the richest agricultural land of Britain. Stamp argued particularly at this time for security of tenure by farmers, so that they might plan improvements likely to be profitable on a long-term basis (*Geogrl. J.* 109 (1947), 39–54). He also noted that

we should realize how precious to us as a nation are our very limited areas of really good land—the level, well-drained land suitable for intensive cultivation for fruit and vegetables. Its very character renders it the most vulnerable to factories and housing development, though how dull is a housing estate on such a site! To use land which to the farmer is poorer gives far greater scope to the planner as well as economizing in the use of the nation's one great asset—the land.

There was the challenge, the basis for endless argument to be experienced from the minutely local to the national and even international plane. The location and relocation of industry was no longer a matter on which only management was to have a deciding vote; suburban residents could (and did on many occasions) weld themselves into societies and groups to express their views and public opinion at times and in places became more vocal than ever before. Academic studies dealt with practical issues of concern to millions of people, and research, far from being a withdrawal from the real and living world, became increasingly an entry into practical affairs.

Town study, later to become so widespread and, in some workers' hands, decidedly esoteric, had developed only to a limited extent in Britain by the late 1940s, though it was more advanced in continental Europe and the United States. An article which attracted attention at this time was E. W. Gilbert's 'Industrialization of Oxford', read in 1946 (*Geogrl. J.* 109 (1947), 1–22). Gilbert makes some comparisons with England's other ancient university city, Cambridge, 'tucked away on the edge of distant East Anglia and the even remoter Fenland'. Oxford stands in a central position in southern England, 'at a sufficient distance from London to be able to dominate the region which surrounds it, a region which, like the city, was entirely non-industrial until a very recent date'. The layout of Oxford has a close relation to its site, for the flood plains had never been suited for building and so were still agricultural and recreational enclaves. The main story told by Gilbert is of the vast industrial growth associated with the production of cars at Cowley and of radiators in a residential district of north Oxford. All this happened almost by chance, and of the various consequences lamented by Gilbert one was the crowding of the central city area and the extreme pressure on the roads because thousands of the workers lived several miles from their employment, many of them in surrounding villages and towns. Naturally, this raised the question of future development.

Some prominent planners of the 1940s favoured the prohibition of further industrial growth in particular areas: for example, Gilbert noted Sir Patrick Abercrombie's stricture that in Oxford, Reading, and Luton 'further industrial growth should be banned'. This judgement some would contest, especially for towns where one industry employs a very large proportion of the workforce and in which rationalization or changes in demand might create severe unemployment at a later time. Looking somewhat enviously at Cambridge, devoid of heavy industry, Gilbert worte:

There has been an almost complete lack of control over the modern development of Oxford. It has been stated that within 6 miles of Carfax every possible mistake has been committed. In the years between the wars, especially in the twenties, there was no clear and far-sighted policy about the future plan of Oxford. The half of Oxford which lies east of the Cherwell was never considered as a whole. If Cowley had been planned as a separate entity, then towns like Warwick and Leamington might have come into being. This was not attempted, nor was there any real attempt to integrate Oxford into one city (p. 19).

Gilbert concludes his paper by making three crucial points. Firstly, though materially prosperous, Oxford, like many other towns had 'urban sprawl, ribbon development, overcrowded streets, danger on the roads, destruction of beauty'. Secondly, Oxford—not unique in this—could have 'benefited by more urban research, with appropriate land use maps'. Lastly Oxford, like Cambridge, was 'a unique part of Britain's cultural heritage ... it is the duty of the whole nation to restrain industry from laying hands too heavily upon them, and to spare them, as far as may be, from the tumult of modern traffic'. Clearly Gilbert dealt with a city of quite special character, but despite all the efforts of urban geographers to assess the common qualities and problems of towns, it remains true that each has some unique quality, or more aptly combination of qualities.

Land-use studies of towns were only the beginning of a far wider, and still expanding, study of economic and social aspects of urban life. The outward spread of towns into rural areas was a process known from Roman times and probably regarded as inevitable in most phases of human history, but in the immediate post-war period it was questioned, for one of the first effects of planning control was to set limits to this growth while at the same time providing the opportunity for expanding villages by commuter settlement. The aim was to prevent the spread of a 'rururban' belt with ribbon develop-ment of housing along roads, a notorious feature of the inter-war period though never as prevalent as in American cities. Gilbert merely refers to this in his Oxford article, though he was fully aware of its prevalence in the Oxford area. Nevertheless, rural areas depended increasingly on towns for employment and conspicuously also for a wide range of commercial and pro-fessional services. That the market towns had their tributary rural areas, their hinterlands, their urban fields, or whatever phrase might be used, was clear enough but these needed definition, especially as by 1945 it was unquestion-ably true that some redefinition of administrative areas would be needed.

From the nineteenth century (and earlier in some places) industrialization had stimulated town growth, and presumably this could happen again: one speaker in the discussion feared that Oxford could grow to the size of Coventry (with three times its population) and just as rapidly. The heritage of economic history was plain. Though there as a basic town pattern of mar-ket and trading centres, the country towns at intervals of ten miles or so from one another, there were also the greater towns, of which the largest—apart from London, at all times unique—had become the country's major regional capitals; these included Bristol, Birmingham, Liverpool, Man-chester, and Leeds in England; Cardiff in South Wales; Glasgow, Edinburgh, Dundee, and Aberdeen in Scotland. Several other towns, such as Norwich, Nottingham, Leicester, Hull, Sheffield, Portsmouth, Southampton, Ply-mouth, even Exeter, also served large areas by their commercial and pro-fessional facilities. As Gilbert was to show in his work on seaside towns, Brighton, though only slightly industrial, was by its size able to attract a wide range of commercial and professional services for its own population and for people who lived beyond its boundaries.

Walter Christaller had drawn attention to this feature of modern urban growth in an industrialized society having a basic pattern of country market towns. His work was appreciated by R. E. Dickinson and A. E. Smailes among others, though their findings, based on a close study of English life, were published in books and articles in journals other than that of the RGS. In 1950 F. H. W. Green's paper on 'Urban hinterlands in England and Wales:

an analysis of bus services' was widely studied and stimulated a great deal of local work in various parts of the country (*Geogrl. J.* 116 (1950), 64–81). It was well timed, for by 1950 bus services had thoroughly penetrated the villages and countryside and the number of private cars was small in comparison with what it was to become in the following decades. Some of the bus services were available only on market days, but the use of towns as market centres was what Green had in mind. In 1957 Ian Carruthers' paper, 'A definition of service centres in England and Wales', also used bus services as one indication of the administrative and trading activity of towns (*Geogrl. J.* 123 (1957), 371–85). The significance of such work was to be seen later, as the inevitable redefinition of administrative districts was studied, endlessly if not always patiently, until a final redefinition was made in 1974. For the geographer, especially one with an interest in planning, part of the value of such work lay in its contribution to a new regionalization of Britain, based essentially on the distribution of population and the constant movement of people for work, trading, and recreation (this last to receive increasing attention in time). Many British geographers had an acute appreciation of the dynamism and change of life and the constant remodelling of the environment, however this most dangerous of words might be defined.

Gilbert came forward again in 1948 with a paper on local government boundaries, based on the Report of the Local Government Boundary Commission for 1947 (*Geogrl. J.* 111 (1948), 172–98). Few people at that time would disagree with his comment that

the present system of local government areas in England and Wales is in many respects out of date: the needs of local administration have long outgrown the existing framework of boundaries. In the words of *The Times* (9 April 1948) 'the petrol engine, the telephone, vast shifts of industry and population have transformed the human map of England while leaving the local government map essentially unchanged.' It would be difficult to make radical changes in the boundaries of local government areas without reorganizing the system as a whole. The problem is not merely of revising the existing boundaries; it is a question of creating new types of unit more consonant with the geography of the country as it is at the present day. Unfortunately, while there is almost complete agreement on the necessity for change of some kind, there is extreme diversity of opinion as to the design of any alternative structure.

The last sentence proved to be abundantly true in the following thirty years. Gilbert's first task in compiling the paper was to draw a map of the projected new administrative units, all of which were based on existing boundaries. Theoretically it would be desirable to ignore these and begin *de novo*, but this would multiply the difficulties. The Commission was obviously concerned with the appropriate population for various services provided, and for these they favoured a minimum of 200 000, with a maximum of 1 000 000.

Experts on local government, in the universities and outside them, were to argue endlessly about the appropriate population for particular services, reluctantly admitting that any standard figure was a compromise. The complexity of the problem was shown in the guidelines for the Commission, who were asked to consider the following aspects (given in alphabetical order): (*a*) community of interest; (*b*) development or anticipated development; (*c*) economic and industrial characteristics; (*d*) financial resources; (*e*) physical features, including in particular but not exclusively, suitable boundaries, means of communication, and accessibility to administrative centres and centres of business and social life; (*f*) population-size, distribution and characteristics; (*g*) record of administration by local authorities concerned; (*h*) size and shape of the area concerned; (*i*) wishes of the inhabitants. It was only veneration for the alphabet that put the wishes of the people last but possibly the range of considerations involved in administrative definition explains the difficulty of achieving it. It also illustrates the vast possibilities open for the study of social geography in the post-1945 world.

All the papers discussed above show the extension of interest, especially in practical problems for the people, of geographers from the immediate post-war period. Although it was a time of austerity and of grave international problems, academically the scene was bright. Higher education was expanding and geographers hoped that their academic studies might be fruitful in a period of reconstruction and development. In education, however, the position of geography seemed to be threatened by the enthusiasm for 'social studies' and 'environmental studies' (*Geogrl. J.* 116 (1950), 221–41). The former appeared to be, at least in some schools, 'an amorphous hotchpotch of geography, history and civics' and the latter a substitute for ordinary courses in geography, history, biology, and the physical sciences, while the emphasis sometimes given to 'local studies' could be uninspiring, even boring, and might mean the exclusion of 'the broad geographical features of the principal countries of the glove'. Sir Cyril Norwood, always a strong advocate of geography in education, perhaps went further than some geographers in stating that 'Geography is ... more important than a foreign language or a science, highly important as these are, for the simple reason ... that the intelligent person must understand something about the country and the district in which he is to spend his life.'

The RGS Education Committee (p. 52), in a report on 'Geography and Social Studies, (1950), said that the essentials of a school geography course should include, first a training in the use of maps; secondly, an appreciation of the orderly grouping of scientific facts, physical and human, leading to a balanced conception of the neighbourhood and its setting in the home country, and of the major varieties of natural environment and human activity in the world; and thirdly, 'some understanding of the relation between man's activities and his natural environment'. The Committee also published a report on careers for geographers (Crone 1955:35) in 1949 (revised in 1950, 1954, 1960). Having recognized that geography did not as a study lead directly to a job, like law, engineering, or medicine, and having given the view that it was transitional between the naturalistic and humanistic sciences and in essence correlative, the Committee regarded its 'synoptic and philosophical view of the world' as helpful for a career is public administration, industrial and commercial management, politics, transport, journalism, the armed services, banking, and insurance. In fact at this time geography graduates entered a whole range of careers, but planning was to become a major source of employment later.

In 1951 the Schuster Committee on the training of planners, having received evidence from Eva G. R. Taylor, L. D. Stamp, S. W. Wooldridge, and G. H. J. Daysh, included geography as a basic discipline suitable for post-graduate study in Town and Country Planning. Other degrees mentioned included Greats at Oxford, history, and economics; the Committee suggested that it was not essential for a chief planning officer to possess a technical qualification in architecture, engineering, or surveying. In the universities of Manchester and Durham there were five-year first-degree courses in planning but others, for example London, Liverpool, Nottingham, Reading, and Sheffield, favoured post-graduate diplomas. Cambridge, it was noted, was against any participation by the university in either a full-time degree course or a post-graduate diploma. Since 1951 many geographers have become successful planners, having followed post-graduate courses in universities (*Geogrl. J.* 117 (1951), 248–9).

From 1945 a new literature of geography and cognate subjects appeared, for publishers were eager to welcome authors, and books were relatively cheap. Many of the latter were on planning and related problems, and a number were noted in the review pages of the *Geographical Journal*. On four pages of volume 113, 1949 (98–102), three works of particular interest were reviewed consecutively, the first two by C. B. Fawcett. L. D. Stamp's *The Land of Britain: its use and misuse* (1947) was welcomed as 'probably the greatest contribution to our knowledge of Britain made by the present generation as it summarized all the work, including the comprehensive local reports, of the Land Utilisation Survey'. The second book, *Conurbation*, published by the West Midlands Group in 1948, was a study of Birmingham and the Black Country by a team of local workers; it was welcomed as 'a great contribution to our knowledge of the geography and economics of an industrial conurbation'. The problems of living conditions and dereliction were vividly shown by a series of photographs taken on a railway journey from Birmingham to Wolverhampton. Fawcett notes that the authors wanted no new industries to enter the area and also suggested that no immigration should be allowed. 'How is this to be enforced?' he asks. That in time it was not, and could not be, needs no emphasis. The third review, by E. W. Gilbert, was of J. A. R. Pimlott's *The Englishman's Holiday* (1947). This book, described as 'lively and amusing' and 'primarily historical in outlook', covered all kinds of holidays from the Grand Tour to the modern holiday camp. The reviewer observes that it could 'suggest many new lines of thought to social geographers'. In the following year Gilbert published 'The growth of Brighton', a fascinating analysis of the town which, like numerous other papers published at various times in the journal, was followed by a fuller treatment in book form (*Geogrl. J.* 114 (1949), 30–52). This was work on the social aspects of geography, including what was later to become fashionable as the 'geography of recreation'. In the case of Brighton, however, and other places which attracted large numbers of visitors, initial popularity depended partly on the beneficial effects of sea bathing and in some cases also of spa waters; Brighton had both, and its appeal was enhanced by royal patronage until 1845, when Queen Victoria and Prince Albert found the lack of privacy intolerable and moved their seaside home to Osborne on the Isle of Wight.

COLONIAL GEOGRAPHY

Papers on overseas subjects, particularly on the Commonwealth, continued to be prominent in the *Journal* after the War of 1939–45. In 1945 a paper on 'Social geography and its place in colonial studies' was published by E.W. Gilbert and R. W. Steel (*Geogrl. J.* 106 (1945), 118–31). This use of the term 'social' was, at least in Britain, relatively new at the time and was discussed in the *Journal's* correspondence pages. Earlier the word 'human' had been widely used as a contrast to 'physical' or as a complement to it, but the range of work, and therefore its increasing variety, made a wider choice of categories desirable and, some would say, necessary. Gilbert and Steel included four main topics under the term 'social': the distribution of population over the earth's surface, the distribution and form of rural settlements, urban geography, and finally the distribution of social groups and their 'way of life'. The last category included housing, health, and the conditions of labour of different communities. They also claimed that 'economic' geography had obscured the 'social' aspects of geography, which had been 'sometimes neglected', and added that in the case of Africa the initial colonial assumption that 'social welfare would automatically follow economic development' had proved questionable. Experience had led to a broader view of life, in which agriculture, education, health, and housing were all of concern. The 1926 British Association Committee set up to investigate the geography of Africa had accumulated some local information. Gilbert and Steel note that the need for direct population study in Tanganyika had been stressed by Clement Gillman (p. 46), who thought that if such a study had been made earlier the railways might have been more wisely aligned. Regrettably, Gilbert and Steel observe, 'the colonies are far too full of examples of schemes without adequate geographical and sociological investigation'; for example, 'the study of population in British colonial territories in relation to geographical environment has not ... been pursued very far', and even less attention had been given to the study of cities and towns. Similar problems existed in the West Indies. On rather a high note the two authors quote Strabo's definition of a geographer as 'a man who busies himself with the investigation of the art of life, that is of happiness'.

Work on Africa continued. In 1946, for example, S. J. K. Baker and R. T. White (*Geogrl. J.* 108 (1946), 198–210) wrote a study of 'The distribution of native population' covering what was then the two Rhodesias, Nyasaland with parts of Portuguese East Africa, Tanganyika, and the Belgian Congo. Except in Nyasaland the census statistics were not reliable; the authors, however, reach the conclusion that 'the population is below its optimum level' and that a far more considerable population would be needed 'to operate the agrarian revolution implied in the introduction of scientific methods of irrigation, manuring and crop rotation; and there is certainly demand for increased manpower in connection with the industrial revolution which has come to certain areas.' In 1947 M. Fortes, R. W. Steel, and P. Ady (*Geogrl. J.* 110 (1947), 149–77) gave an example of an interdisciplinary survey carried out by Fortes as an anthropologist, Steel as a geographer, and Ady as an economist. It was presented as 'Ashanti Survey, 1945–46: an experiment in social research' at a joint meeting of the RGS with the Royal Anthropological Institute, and was supplemented in 1948 by a further paper by Steel on 'The population of Ashanti: a geographical analysis' (*Geogrl. J.* 112 (1948), 64–77). At this time studies of distribution and density of population were favoured, for the explanation led naturally to study of a regional character without the rigid progression from the physical to the human aspects that had become traditional (and could be cramping) and could lead to social and economic problems connected with overpopulation and underpopulation, though the latter was generally assumed to be less prevalent.

In 1948 F. Debenham's interest in Africa, which continued until his death in 1965, was shown in his paper on 'The water resources of Central Africa' (*Geogrl. J.* 111 (1948), 222–33). He said; that 'One of Central Africa's needs [is] ... reconnaissance surveys to establish the facts, ... a field party to do some levelling, measure discharges, set up gauges and map tributaries. ... Many of these surveys could be done by junior geographers in vacation periods under direction by a senior person.' Also in 1948 Debenham produced the 'Report on the water resources of the Bechuanaland Protectorate, Northern Rhodesia, the Nyasaland Protectorate, Tanganyika Territory, Kenya and the Uganda Protectorate'. This was published by HM Stationery Office as no. 2 of the *Colonial Research Publications*. It was reviewed by R. W. Steel (*Geogrl. J.* 115 (1950), 98–100), who observed:

Geological surveys in the African colonies have too often been exclusively concerned with the search for economic minerals, and water resources have been largely neglected. Only rarely are there Hydrological or Water Supply Departments, and where they exist they are often poorly staffed and inadequately equipped. It is significant that one of the first and most successful, that in Tanganyika Territory, owed its success more to the late Clement Gillman ... than to any other individual ... it was Gillman who established that two-thirds of Tanganyika's population are concentrated on the one-tenth of its area which possesses enough water to support such a concentration. Whether new developments in the British African territories are meant to increase the African contribution to world supplies of food or to maintain the existing indigenous standard of living where populations are increasing, often at an alarming rate, little can be done unless water supplies are adequate.

Later Steel noticed that Debenham was 'essentially practical and realistic ... at pains to push the small, generally inexpensive, local scheme rather than the grandiose, costly plan, whether it be for water supply, drainage, or even hydro-electricity'. In 1950 E. B. Worthington (*Geogrl. J.* 116 (1950), 29–43), Scientific Secretary to the East Africa Commission, gave a lecture on 'Geography and the development of East Africa', the first of a series to be given on East Africa normally biennially under a benefaction from Mrs Will Gordon. Here too, as Debenham pointed out in the discussion, the problem of water supply was noted. Indeed, the lecturer went further and frankly admitted that those interested were 'still beset by ignorance about the natural resources of East Africa and how best to use them'. As he showed from a series of local studies, the challenge was for a varied team of scientists to recognize the environmental problems and to plan the land use of the future. Virtually no mention was made of the political problems that were to arise all too soon in Africa and to attract more attention than problems such as the use of natural resources and land, which were directly related to the welfare of the people and on which too little was done too late.

WORLD PROBLEMS

Population problems are of perennial interest, though many of them appear so intractable that politicians and others find it convenient to forget them. But they will not go away, as G. C. L. Bertram (*Geogrl. J.* 107 (1946), 191–205), a Cambridge biologist, showed in a lecture given in 1946. He commented that 'We have begun in recent years to appreciate just how large is the gap between a diet sufficient to maintain life and one adequate for full health and vigour.' He used the material brought to the Hot Spirngs Conference of 1943 which ultimately led to the foundation of the United Nations Food and Agricultural Organization. At the conference it was clearly stated that there had 'never been enough food for the health of all people' but this view is not universally held. One by one Bertram dealt with the problems, even with the narrow distinction 'between virtuous thrift and avaricious hoarding'. He also studied population trends, with the help of a book then much revered, A. M. Carr-Saunders's *World Population* (1936), and showed the futility of regarding emigration as a remedy universally applicable. He stated that, as health standards improved, so too would the expectancy of life and the infant mortality rate.

Progress had been made, for example in India, where the area irrigated was three times as great as that in the USA, and in Egypt, where for several months of the year no water from the Nile entered the sea until it had been used for irrigation. Crops giving a greater yield per acre, or capable of surviving in marginal climates, had been developed, and fisheries offered substantial contributions to the world's diet though, as Bertram noted, before the War there had been severe overfishing with depletion of stocks and this might happen again. (That it has done so is well known.) The paper concluded with the view that

the plea in this matter of population trends relative to the world shortage of biological resources [is] ... let there be appreciation that ... eventually active and voluntary adjustment of population will be as inevitable as it is already desirable. In our attempts to make supplies match needs more adequately than now, let policies be based on real assessment of quantitative data rather than on half-knowledge and guesswork. Let there be appreciation that these matters are desperately urgent.

Concern for world needs and problems, as the various papers show, was abundant in the years immediately following the War. As always in the *Geographical Journal*, speakers tried to conceal their own political views and to achieve objectivity. Nevertheless political geography was not neglected and in 1948 one of the world's most famous political geographers, perhaps the most famous of them all, appeared. Isaiah Bowman, then President of Johns-Hopkins University, Baltimore, gave a lecture that even in print is a *tour de force* (*Geogrl. J.* 112 (1948), 129–42). It was given on 21 June 1948, a time when the difficulties of achieving world peace and even the fear of renewed hostilities strengthened the mood of reaction that inevitably followed the end of the War of 1939–45. Bowman points out that the biological principle that nature abhors an equilibrium had its analogy in human affairs and therefore

the tendency towards disequilibrium is the root of the difficulty in devising any system of power control in the United Nations. As geological exploration and scientific discovery proceed, and new resources of astounding potential are discovered, power disequilibrium comes into play concurrently.... With vast oil deposits, two otherwise weak countries, Saudi Arabia and Venezuela suddenly rank high in the power field. It is often said that geography does not change. In truth geography changes as rapidly as ideas and technologies change: that is, the *meaning* of geographical conditions changes.... Every major technical and power development must be followed by a resurvey of world geography, region by region and indeed point by point.

Bowman argues that it is 'the great merit of the United Nations that it attempts, not the unattainable but an attainable goal'. Searches for a utopia of tension-free peace are meaningless, for

The processes of adjustment of men's claims and rights must be as continuous as life itself. Confronted by the eternal difficulties of world politics, minds that wish to escape from reality prefer an heroic solution, hence the popularity and appeal of 'world government' and peace by force. Heroic solutions solve few problems and always create a crop of new ones, for all rest essentially on force without prior agreement on how to use it in unpredictable circumstances.

Looking to Russia, Bowman saw the dangers of political theory as a solution for human problems: 'Conceit leads them to assert that they have solved at last the complex art of governing men.' The USSR had a vast territory and resources but for thirty years had been governed by leaders 'unqualified by a free and popular referendum'. Similar danger lay in the confidence possessed by the USA, which was imbued with 'a mystic quality' and expressed as its 'manifest destiny'. Russia had its 'urge to the sea' historically and currently, just as America had its urge to the Pacific, and the two powers therefore met in the West and in the East as 'two totally unlike stocks with different historical and cultural backgrounds, Slav on the one hand, and on the other that mixture known as American with basically Anglo-Saxon institutions'. Political ideals in USA were 'neither materialistic nor imperial in origin'. But it was wrong to assume that democracy on the American pattern was wanted everywhere, and just as fallacious to assume that responsibility for the rest of mankind could be avoided, for 'if isolationism did not die at Pearl Harbour, the virtual destruction of our Pacific Fleet at least permanently disabled this mistaken policy.' And the dangers of over-confidence were seen in the depression from the late 1920s; indeed, 'the economic crash of 1929 ended the era of over-confidence' and taught America that 'we had to be sensitive to the realities of the whole globe if our American system was to have any meaning or stability'.

Distant countries could no longer be thought of as far away, merely as sources of trade, or fields of Christian missionary endeavour. They were part of the same world, to be studied along with the American homeland, where forests were cut at a rate that exceeded natural growth, where soil erosion operated on a continental scale, where water conservation was a major concern in an area covering more than two-thirds of the country, and where the mineral resources were so depleted that imports were vital to industry. No longer was hemisphere defence practicable, for it was impossible to maintain 'a political curtain drawn down over the Atlantic'. And, as Europeans so often forget, the USA had a vital interest in the political alignments as well as in the economic resources of countries in Central and South America (and even Cuba as events turned out, but Bowman did not foresee this). Bowman looked at the world and said that 'the great design for peace had yet to be forged'. He finished his remarkable paper with a quotation:

It is a gloomy moment in history. Not for many years ... has there been so much grave and deep apprehension; never has the future seemed so incalculable as at this time. ... In France the political cauldron seethes and bubbles with uncertainty; Russia hangs, as usual, like a cloud, dark and silent upon the horizon of Europe; while all the energies, resources and influences of the British Empire are sorely tried, and are yet to be tried more sorely....

These words were written not in 1948 but in 1857, in *Harper's Magazine*. It perhaps gave comfort to some in the 1948 audience. Read now, with hindsight, Bowman's fears have proved to be overdrawn. Possibly those present did not agree with the statement by the President, Lord Rennell of Rodd, that 'There is only one in the Hall tonight who can adequately comment on Dr. Isaiah Bowman's paper, and that is Lord Halifax' (British Ambassador to USA). They may have disagreed even more after Lord Halifax had spoken.

Political geography may be a subjective study in which the sympathies of the author are apparent. Characteristically O. H. K. Spate, in his article of 1947 on 'The partition of the Punjab and of Bengal' (*Geogrl. J.* 110 (1947), 201–18), disarmingly admits at the outset that he was a technical adviser for a Muslim group and 'in effect became an "expert witness" for the Muslim League ... I was not above the battle but in the thick of it, and being a political animal I thoroughly enjoyed it.' For all that, he gives an objective treatment and shows that a new boundary made on a social and religious basis could produce severe economic problems for the people on either side, apart from such obvious details as cutting across a vital railway line. Spate had the local knowledge essential for sound judgement in so complex a problem. In 1948 he gave an account of the readjustments made by the Indian states, (*Geogrl. J.* 111 (1948), 288–92), of which there were 584, many of them very small though in all they covered 716 000 square miles with a population of 93 000 000.

Working on a broader view, C. A. Fisher, in two consecutive papers on 'The expansion of Japan: a study in oriental geopolitics', gives a fascinating account of the Greater Asia Co-prosperity sphere, based on the Japanese view that the East Indies should be a source of new materials (*Geogrl. J.* 115 (1950), 1–9, 179–93). The Dutch aim had been to build prosperity around the industrialized and densely populated island of Java. Fisher was not afraid to make some broad generalizations and prophetically looked forward to the problem of the relations between the USSR and China. He was clearly an admirer of the work of Mackinder in political geography and contributed a number of papers on political geography to the journal in later years.

Contemporary studies of various countries had a special appeal after the War of 1939–45. One such, which now, thirty years after publication, has historic interest, is the paper by Harriet G. Steers on 'The middle people: resettlement in Czechoslovakia', based on a journey made in the autumn of 1947 and on earlier study (*Geogrl. J.* 112 (1948), 28–40). It shows vividly, in photographs and text, the drastic solution of the Sudeten German problem which tore many thousands from their homes and sent them as migrants to Germany. Another article with an element of tragedy, 'The revival of northern Norway' (*Geogrl. J.* 109 (1947), 185–97) by D. H. Lund, Director of the Finnmark Reconstruction Office at Harstad, shows the catastrophic devastation and certainly helps visitors to Arctic Norway to understand why almost every building dates from the post-war period. W. R. Mead dealt with the political geography of Finland in articles on 'The Finnish outlook: east and west' in 1949 and on Karelia in 1952; both studies were enlivened by obvious local knowledge (*Geogrl. J.* 113 (1949), 9–20; 118 (1952), 40–54).

There was an immense and only partially satisfied curiosity about other countries after the War, and especially those in which military campaigns had been fought. This was not new; in 1946, for example, A. Perpillou (*Geogrl. J.* 107 (1946), 50–7), in an article on 'Geography and geographical studies in France during the War and the Occupation', having provided an analysis of what had been done under very difficult circumstances, commented:

During the war the majority of Frenchmen heard of distant countries as theatres of operations which they recognized as critical of the course of the war and for the future of their country, of which even the names were scarcely familiar to them. From the realization of this lack and the surprises which the war has revealed, has now sprung without doubt the desire to know the world better, which is certainly a good augury for the future progress of geographical science.

This extract may appear to be merely a demonstration of curiosity, but knowledge of the world is education. It is certainly a confirmation of the view that geographical knowledge acquires significance and prestige in wartime.

Knowledge of the USSR was slight for most of the inter-war period, though the achievements of its cartographers were admired. In 1945 the Swedish geographer H. W:son Ahlmann visited Moscow and Leningrad to celebrate the 220th anniversary of the foundation of the Academy of Science, founded by Peter I (*Geogrl. J.* 106 (1945), 217–21). In 1943 the Academy had been moved from Leningrad to Moscow and its work greatly developed. Ahlmann paid special attention to the wide scope of research in physical geography in Russia. A. A. Grigoryev, in a paper on 'Soviet plans for irrigation and power' published in 1952, said that, for the Stalin plan of agricultural expansion with irrigation, power, and canal development:

Specialist in every branch of science and technology have been drawn into the work of solving the numerous urgent problems connected with the construction works, and the best methods of using the huge new areas reclaimed.... Soviet geographers have taken a very active part in the work of the big integrated expeditions in 1949 and 1950, organized by the Academy of Sciences of the USSR to study the areas through which the future state shelter belts and other forests will pass. (*Geogrl. J.* 118 (1952), 168–79.)

The work was in microclimatology, soils and geomorphology; geographers were 'mostly concerned with the detailed study of the physical geography of the districts to be irrigated and supplied with water' and were also concerned with the economic geography of the newly irrigated lands. In 1953 Grigoryev provided a second paper on plans for 'the reclamation of the forest belt of the USSR in Europe' (*Geogrl. J.* 119 (1953), 411–19). This deals with the problems of the glaciated, boulder-strewn belt with its water-logged soils 'particularly under the conditions of the flat relief prevailing in the European part of the USSR'. Contributions from Russian geographers were read with interest but with the realization that they were generalizations lacking any examples of local sample studies. In later years material on the USSR proliferated and a considerable number of textbooks appeared, as readers will no doubt be aware. From 1960 Russian geographers came in some strength to Congresses of the International Geographical Union, and in 1976 the Congress was held in Moscow.

## THE PEACEFUL GROVES OF ACADEME

Throughout the post-1945 period, a crucial time in the modern history of geography, individual authors considered matters of special interest not obviously or in some cases, conceivably, connected with the new world of planning, reconstruction, and continuing tension. It would be difficult to imagine a more peaceful study than Edward Heawood's *Watermarks: mainly of the 17th and 18th centuries*, edited by J. Labarre, published posthumously at Hilversum in 1950 and reviewed by R. A. Skelton (*Geogrl. J.* 117 (1951), 221–2). Heawood's first paper on watermarks was published in 1924, and what Labarre called his 'patient and expert labour' continued to the end of his life. For this volume G. R. Crone had completed Heawood's index and contributed a memoir. The history of cartography is an absorbing subject to a number of people, perhaps increasing in number, and in the pages of the *Geographical Journal* there are various articles on this theme; for example, in 1951 Helen M. Wallis wrote on 'The first English globe: a recent discovery', in fact a globe of 1592 by Emery Molyneux discovered in July 1949 in Lord Leconfield's Library at Petworth (*Geogrl. J.* 117 (1951), 275–90). Globes were made in considerable numbers during the Age of Discovery and Dr Wallis notes that Shakespeare in *The Comedy of Errors* assumes that his audience would be familiar with them. Of the vast additions to the literature of geographical discovery made by the Hakluyt Society from its foundation in 1846 it is not possible to speak here, but it was appropriate that its centenary was celebrated by a joint meeting with the Royal Geographical Society at which Eva G. R. Taylor gave an excellent lecture on Hakluyt (*Geogrl. J.* 109 (1947), 165–74), on which E. W. Gilbert commented:

Richard Hakluyt concerned himself with the politics of his day and proved that the study of Geography can benefit the state.... Professor Taylor ... not only is a leading authority on the geographers of the Elizabethan period but ... also one of the British geographers who have not been afraid to teach that modern political planning must be linked with geographical learning; like Richard Hakluyt she too has given counsel to the State.

Some English geographers and historians have a deep and rich appreciation of the English landscape as a living revelation of many centuries of effort by farmers, craftsmen, road builders, and others over thousands of years. The first volume of the Domesday Survey, H. C. Darby's *Domesday Geography of Eastern England*, appeared in 1952 and was reviewed in volume 119 (1953), 337–40. The reviewer welcomed the book as a useful contribution to the study of England in the eleventh century while praising the author's caution; he also said that 'the problems connected with the making of the Survey are probably much further from solution than the introductory chapter would seem to suggest'. For some years the aim of historical geographers had been to reconstruct the regional geography of the past, choosing certain epochs for detailed investigation. This avoided the broad sweeps through historical and even prehistoric time for which some writers (never favoured by the perhaps overcautious editors of the *Journal*) had become known, but while the rigid acceptance of periods was a challenge to accuracy of scholarship, there was also the dynamic element in history and in the evolution of the human landscape. This was elegantly shown in H. C. Darby's paper of 1951 on the 'Changing English Landscape' (*Geogrl. J.* 117 (1951), 377–94) in which he deals with the Anglo-Saxon settlement and clearing of forests, the medieval pattern of settlement and agriculture, the agricultural revolution from the eighteenth century, the embellishment of parks by landscape gardeners, and the final massive impact of the Industrial Revolution. How much remains visible on the ground despite all the changes is a question every field-worker asks. The answer is probably 'far more than people realize', especially when aerial survey is employed, though in the end the field-worker, like the patient infantry, has to walk.

Archival sources, including maps and plans, provide part of the answer to the mysteries of the past. In 1951 M. Beresford's paper on 'The lost villages of medieval England' showed both sources of wisdom (*Geogrl. J.* 117 (1951), 129–47). At the time his work, to become greatly respected later, was in its early stages, though he had gone far enough to state:

In terms of the distribution of the rural population this movement meant a retreat from a number of sites, some with a history of settlement as old as any in England. Others of these sites had arisen in the outward push of colonization in the twelfth and thirteenth centuries; the high-water mark of medieval settlement lay further into some marginal land than even the ploughing-up campaign of 1940–44. A map of settlement in the early fourteenth century would include most modern villages *plus* many of the sites whose fates we have been describing. But in lonely places, like the ridge and furrow of medieval ploughing in Northumberland and Durham, there are visible signs of how far the high-water mark of ploughing once stretched. ... There were forests, heaths and marshes untouched by the plough in medieval times but colonized subsequently. This later colonization of the sixteenth, seventeenth and eighteenth centuries was not a return to the sheep-runs; these were left to their productive and profitable alternative use in grass. It turned to the open fields of the villages, seeking to use them more economically by enclosure, and to the still plentiful areas of forest, heath and fen. ... More villages probably existed about 1320 than at any time before the industrial revolution. To plot the living villages and the lost villages on the same map is to discover how far there has been a retreat of village settlement.

M. Beresford is a professor of economic history, and in welcoming his paper H. C. Darby quoted the comments of Sir John Clapham, 'an economic historian with a geographical leaning', that 'he who is not a tolerable geographer is a very imperfect economic historian and that he could not picture a useful historical geographer who had not got a working knowledge of economic history'.

That many mysteries remain is shown, for example, by an article by W. R. Mead on 'Ridge and furrow in Buckinghamshire' of 1954 (*Geogrl. J.* 120 (1954), 34–42). When these were formed, Mead points out, had led to 'lively thrust and parry among economic historians'. He lists some of the numerous theories but is concerned mainly with the distribution of each phenomena, though he notes that heights vary from a yard to 'scarcely discernible inches revealed in the shadow of the setting sun'. Sometimes they are clearly discernible, for example when a snow cover has almost melted and the last of it remains as a shallow cover in the furrows. The present writer remembers reading a remarkable description of ridge and furrow in a thesis on a rural area near Cheltenham, of which the student author explained that he had acquired his knowledge by cross-country running. Mead shows that aerial photography picks out—at least in favourable weather conditions— examples of ridge and furrow not normally visible on the ground. Assisted by children from several schools the author found that ridge and furrow is associated with the clay lands and virtually absent—though not completely— from the Chiltern chalklands. There is a frontier zone, separating Bucks into northern and southern halves', which 'follows the line of the Chiltern scarp and extends over a distance of one or two miles on the Vale side', that is the Vale of Aylesbury. For limited areas, maps of historical value are available: one such, for 1769, showing ridge and furrow as it then was, is complemented by a map showing its far wider distribution in *c*.1950. Mead concludes his paper by saying that some new facts had arisen but they apply only to the areas in which they occur. Generally

questions are raised rather than answers given. The only indisputable fact is that all ridge and furrow land has at some time been under the plough. Most of it today persists under grass. . . . Ridge and furrow land in water meadow, reverted scrub and plantation hint at how much land marginal to present-day arable husbandry was under the plough. The plough produced the pattern and the pattern has raised the problem.

Mead points out that there are comparable patterns of ridge and furrow in France and Germany 'in a form familiar to the English eye', as well as in Denmark and Sweden. Hopes of further light rest on detailed studies, especially of areas for which map evidence is available, of mathematical measurement, and of friendly co-operation and discussion with workers elsewhere.

From what has been said of the work of Darby, Beresford, and Mead, the clearest of cases could be made for fieldwork in historical geography. And in all such work the name of W. G. Hoskins comes up as the advocate of local history using all possible evidence on the ground as well as archival sources (Hoskins 1959). Such work will attract more scholars at some times than others, but the continuing interest may be seen for example in the steady increase in the number of museums devoted to past periods, including the quite recent past. In the 1960s a campaign to rescue many industrial premises from destruction won wide support, and 'industrial archaeology' became popular. With this there came a realization that ordinary small houses, used by workers in town and country, should also be preserved as examples of domestic architecture. Despite all the interest shown by geographers, however, the number who were attracted to historical geography remained small.

## PHYSICAL GEOGRAPHY

If the people interested in various aspects of historical geography remained peaceful and contented, the same could hardly be said of physical geographers. After the War of 1939–45 there was a fear that geomorphology might be absorbed into geology, or that it might be almost entirely neglected and even ignored by some university departments of geography. This perhaps partly accounts for the somewhat aggressive attitude shown by some of its exponents. In 1948 a Norwegian geomophologist, K. M. Strøm gave a paper (*Geogrl. J.* 112 (1948), 19–27) in which he showed that the present morphology of Norway is conditioned by the interplay of at least two erosion surfaces elevated during Tertiary times, with the forms resulting from Pleistocene erosion and deposition, both mainly glacial. He added that 'the importance of structure is also everywhere apparent'. In fact the article is a splendid study of the main physical features of Norway, including its strandflat and the highly varied country around Oslo Fiord, which the present writer at least found to be an excellent introduction to a course on Norway. A letter of vigorous opposition, signed D. E. Sewell, said that it was 'of little interest to geographers' and that it was 'about time that a stop was put to much muddled thinking and practice (particularly the latter) on the relationship of geomorphology to the field of geography' (*Geogrl. J.* 111 (1948), 147). He also stated that through the influence of W. M. Davis 'academic geography had—apart from certain notable exceptions—stood still' in work on regional and economic geography 'the most important fields'.

This drew a very sharp rejoinder from S. W. Wooldridge (*Geogr. J.* 112 (1948), 125–6), who said that a pre-Cambrian eorsion surface was an element in an *existing* landscape, that the South Downs was undeniably a 'cuesta', and that such genetic concepts were relevant in studying the earth as 'the home of man'. After some stronger comments Wooldridge ended by quoting W. M. Davis's comment that 'To look upon a landscape without any recognition of the labour expended upon producing it, or of the extraordinary adjustment of streams to structure and waste to weather is like visiting Rome in the ignorant belief that the Romans of today had no ancestors.' Not all of us care much for argument by analogy but everyone appreciated Wooldrige's determination, in association with a not inconsiderable group of colleagues and friends, 'that the links between geography and geomophology should not die'. Among the friends was D. L. Linton, who also attacked another correspondent, W. L. Handyside, who had said that 'River action as such is not Geography; it becames so when it is studied in its relation to man.' Having declared that geography is concerned with the study of areas, Linton added that 'there can be no reasoned, explanatory description of areas that does not draw on geomorphology for basic facts .... The mid-Tertiary plantation of the Weald is a fact of the present landscape and a factor to be reckoned with in any serious attempt to write the geography of the area.' Finally he said that 'regional description is the geographer's central task'. In that he was in accord with the general view of the time (*Geogrl. J.* 112 (1948), 126–7).

As the correspondence continued, and extended into other aspects of what J. F. Unstead noted as geography's 'connection with humane studies and its application to human activities', the simpler assumptions of former years were rejected.   In 1950 Unstead, whose work on regional geography was perhaps more revered in Germany than in Britain, said that the character-istics and the origins of particular landforms are matters of geomorphology; 'they enter the sphere of geography only when it is incorporated into an in-vestigation of the complex in which they occur, such as the nature of the soils or of the vegetation, the facilities or the hindrances which they present to human activities and settlements' (*Geogrl. J.* 116 (1950), 138–40). This reveals to some extent the view that any geographer, and particularly a regional geographer, should be acquainted with the entire range of the sub-ject. The traditional association of physical and human distributions had been prevalent not merely during the nineteenth century but far back into history and geomorphologists of the time felt obliged to defend this work by stressing its ultimate value to regional analysis. The age of unashamed specialization was to come later in Britain.

It was not so everywhere. The Swedish geographer, H. W: son Ahlmann gave two admirable papers on his research into snow and ice between 1916 and 1940 and on the current climatic fluctuations in 1946 and 1948 (*Geogrl. J.* 107 (1946), 11–25; 112 (1948), 165–93). These studies, with other material, were also developed in the first publication of the RGS Research Series, no. 1, 1948, *Glaciological research on the North Atlantic coasts.* (The series was inaugurated in 1945 'to facilitate the publication of scientific memoirs on any aspect of Geography, too long or too specialized for publication in the *Geographical Journal'*.) The fascination of glacial study lies in the possibility of observing current changes in climate and perhaps of gaining insight within a lifetime of processes that over thousands of years have moulded the land-scapes of much of the world. This also applies to coastal study, of which one published in 1949 by C. A. M. King and W. W. Williams on 'The forma-tion and movement of sand bars by wave action' cast light on the problems of landing craft on enemy beaches during the War. Some useful research had been done on the beach at Blackpool as well as on beaches in North Africa (*Geogrl. J.* 113 (1949), 70–85).

Landscape analysis on a far greater scale was given by the famous New Zealand geomorphologist, C. A. Cotton, in 'The erosional grading of convex and concave slopes' (*Geogrl. J.* 118 (1952), 197–204), published in 1952. A student and disciple of W. M. Davis, Cotton attacks the view of A. N. Strahler, who had said that 'the common illusion of the concavity of slopes comes from viewing profiles of lateral spurs or other lines which are not true orthogonals' (right-angled). This attack was one expression of conviction that geomorphology had become, or perhaps always was, oblivious of mathematics, prone to accept much from merely visual impressions, likely to make startling assumptions, even to make the hypothesis into the reality. The denigration of W. M. Davis was not surprising as the quantifying movement gained force. Cotton, then in his late sixties, finished his paper by saying 'It is impossible to believe that a long succession of competent field observers from G. K. Gilbert onwards through Davis, Fennemann and Penck, to Baulig and Birot, have been deceived by a concavity of profile which Strahler claims to be commonly (but mistakenly?) shown in elementary textbooks.' His appeal to established authority such as this is unlikely to deter researchers young in spirit and nimble of mind, but they still had to meet Cotton's assertion that

Broad convexity of upper slopes and hilltops is obvious even to untrained eyes, and concavity of valley sides is found commonly enough to merit serious attention and thought. It will be held very generally that straight slopes on the sides of minor, submature valleys such as have been the subject of Strahler's very interesting statistical study, cannot be accepted as reproducing in miniature the profiles of fully mature landscapes.

Cotton's paper of 1953 (*Geogr.. J.* 119 (1953), 213–19) dealt with tectonic relief 'resulting primarily from earth movements' of geologically recent origin and was splendidly illustrated by block diagrams and photographs. In the following year he contributed a paper on coasts, again using New Zealand examples, to illustrate a new theory by a German worker, H. Valentin, which had focused attention on advance and retreat (*Geogrl. J.* 120 (1954), 353–61). Classification of coasts, he admitted, demands 'intensive examination of coastal details, a task entailing much field work in many regions'. So many factors are involved.

It is obvious that students of descriptive and systematic geomorphology must recognize and develop analytically the viewpoint of non-cyclic erosion; but it would at the same time be suicidal to abandon rather than supplement the helpful cyclic concept, without which description tends to become chaotic at the present stage of scientific progress. Few geomorphologists would agree to cut adrift from the 'sheet-anchor' of base-level, whether in the study of coasts or landscapes.

Even the last sentence might be regarded as question-begging, but on that geomorphologists will have their own views.

Papers on the physical landscape of England included two by S. W. Wooldridge and W. G. V. Balchin. Wooldridge's paper, 'The changing physical landscape of Britain', begins by noting that in 1895 W. M. Davis had 'sketched the physiographic evolution of south-east England' (*Geogrl. J.* 118 (1952), 297–305). The central feature of his 'interpretation was his recognition of the widespread hill-top plain of the area regarded by him as an example of an uplifted and dissected peneplain, his own term for that undulating surface which marked the termination of a long continued cycle of erosion'. Wooldridge, in a comprehensive and searching review of the work done since Davis, noted that since his time 'his lucid pen and graphic pencil had bred a lineage of workers now in its second generation'. Not surprisingly 'the record of the changing physical landscape of Britain had proved to be more complex and subtle' than Davis suggested. In 1895, and for some time after-wards, the emphasis in British studies was on marine denudation but this proved to be 'in part ... demonstrably in error'. The paper by Balchin dealt with Exmoor and adjacent areas and opens with the comment that during the previous fifty years the evidence in South Wales and south-west England of a polycyclic landscape with erosion surfaces had been largely accepted (*Geogrl. J.* 118 (1952), 453–76). From laboratory and field-work and with a fine group of maps illustrating points made in the article, Balchin deals with the various erosion surfaces that he recognized in and around Exmoor. Much of the argument rests on changes of sea-level from post-Alpine times.

In the discussion the geologist Professor O. T. Jones raised a point which besets all field-workers: 'Looking at a surface, there might appear to be a definite cliff in the background: when looked at from another direction the cliff was not a cliff at all but a long sloping curved surface ... the only satisfac-tory way of showing the true form of the surface was to take a series of pro-files.' Balchin had himself drawn attention to the perils of field-work, particu-larly in viewing sloping surfaces of such gentle gradient that they could appear to be flat. But Jones went further and deplored the apparent eagerness to recognize surfaces as 'a series of platforms due to marine erosion cutting cliffs'. He thought that the idea had come from Henri Baulig, author of *The Changing Sea Level* published by the Institute of British Geographers in 1935, a work of which he did not approve. Support came for Balchin's views, how-ever, from other speakers, including some whose detailed research apparently confirmed his findings; the views of Jones were indicative of a questioning attitude to work then much in vogue not only in Britain but throughout the world.

Similar, indeed greater, difficulties were encountered by J. N. Jennings, whose *Origin of the Broads* was published as the second of the RGS Research Memoirs in 1952. It included work on the natural evolution of the river val-leys, as distinct from the origin of the Broads themselves, which were thought to have originated as discontinuous lakes of natural origin. Dr J. M. Lambert, however, between 1950 and 1952, proved that they were artificial, formed in basins down to 3 metres and more below the present level left by deep excavation of peat in historical times (*Geogrl. J.* 119 (1953), 91). In 1960, the third RGS Memoir appeared under the names of five contributors: a botanist (J. M. Lambert), a geomorphologist (J. N. Jennings), a historical geographer (C. T. Smith), an archaeologist (C. Green), and a civil engineer (J. N. Hutchinson). The Broads were 'not as was generally believed up to now formed naturally at the meeting-place of marine and riverine influences by the river-water being ponded back against deposits laid down by the encroaching sea. They are rather the result of the cutting of turf from about the eleventh to the thirteenth century, during a period when the land was relatively much higher than it is today and the bogs were above water.' Few records of this period had survived though the specialist authors were able to collect material on the human settlement, including sites and names of villages and towns, and also on fluctuations in the level of land and sea. Initial difficulties in this case proved to be only the prelude to a valuable publication, warmly welcomed by a Dutch expert who in his review suggested that it could stimulate research on both sides of the North Sea (*Geogrl. J.* 126 (1960), 341–2).

## A PAUSE FOR MEDITATION

It may seem that in the 1950s geographers were so concerned with their various pragmatic studies that few had time for wider philosophical speculation. O. H. K. Spate in 1952 brought back the problem of determinism, possibilism, and probabilism that had fascinated geographers for years and he was not, like some of his contemporaries, fearful of discussing such issues as man and environment (*Geogrl. J.* 118 (1952), 406–25). He based his article on A. Toynbee's *Study of History* and E. Huntington's *Mainsprings of Civilization*, two books which crystallized much that they had written and thought in long and devoted careers. Spate deals at once with their limitations. Toynbee approached geography with 'a looseness of touch, amounting at times to positive misunderstanding, when dealing with the concrete facts of the environment', while conversely Huntington, writing on 'such immense movements of the human spirit as the great religions', was 'far too external in approach'. But Huntington was 'a man of his own American Century', prone to measure civilization by cars, radios, and public libraries, whereas Toynbee belonged to 'a Western Europe beginning to doubt its own liberal humanism'. Spate found no difficulty in showing the geographical fallacies and misconceptions of Toynbee, although the latter reveals far greater sympathy and understanding of people in all ages than Huntington who, having found it impossible to measure 'idealism, altruism, honesty, self-reliance, originality and artistic appreciation', uses 'literacy, education, the use of libraries and gifts for philanthropic purposes' as his yardstick. It was on such data that Huntington made his grades of civilization: where would Norway be, Spate asks, without Ibsen and hydroelectricity?

Toynbee may seek to minimize and Huntington to stress physical factors, yet both, in Spate's view, fall into the fallacy that there is or can be such a thing as environment 'taken by itself', for 'the world without man is not our world, though it is the world of geology, geodesy, seismology, geomorphology and other studies'. Having established to the point of insistence that there is no human environment without man, it is still possible to believe in strict physical control of human activities. Spate argues that 'neither environmentalism nor possibilism is adequate without large qualification'. There are psycho-physiological effects of climate on individuals, expressed through social life although, as Febvre says, it is the idea formed of environment which matters as much as the brute fact: in short, the facts are as approached. But they exist. Britain is an island with a useful 21 mile anti-tank ditch, and the sea is therefore still a barrier even if it is vastly more significant as a highway. Some element of determinism remains in Spate's view, through all the possibilism which appears to be a kind of conquest of everything unfavourable or challenging in any environment. And though Spate does not exactly say this, in the debate 'probabilism' seems an evasive term. Long after the wilder fancies of Huntington are forgotten, something may remain of his wrestling with the problems of man and environment in such a wide range of conditions.

Spate appears to have enjoyed his communion with both writers but obviously does not expect readers to accept his ideas and observations. In this he differs from many other geographers. He concludes by saying:

The approaches of the two may be loosely, but perhaps not inaccurately, summarized as aesthetic in Toynbee, scientific in Huntington ... properly considered these are not mere opposites; like our authors, they interpenetrate. There is no art which has not its scientific element, be it only in technique; nor is technique alone adequate to the scientist if he is to attain his full stature. There must be an element of intuition, itself subject to the scientific analysis of psychology and not thereby devolved, any more than the heart of a landscape is devalued by an understanding of the earth-processes which have moulded it. Perhaps the final limitation is that Toynbee broods too exclusively on Myth; Huntington on Statistic; but these are themselves but parts of Poetry and Science, which together make our vision of reality. Their apprehension as a unity is not yet, if indeed Man ever attains it.

Perhaps the study by Spate may induce in some geographers the thought that it is good at times for any scholar to consider how his work fits into a wider conspectus of the world, to ask himself—for example—what the relation of man and environment may mean or what value may lie in such phrases as 'the study of distributions' or—still more provocative of debate—what magical enlightenment may come from 'the regional concept', a revered phrase of former years. The easiest course is to avoid such considerations altogether, to cling to some concept that promises fruitful research, especially if it is new. Nevertheless the old problems of life in the world will never go away. Toynbee had a theory that much of the advance in the world's thought came from men who had gone through a time of withdrawal into contemplation followed by a return to normal living. The unity of all life with the inorganic as well as the organic world was the inspiration of the scientific advance of the nineteenth century, but this is no simple concept, rather a challenge to every thinking person and not least to the geographer whose concern, however remote and in some cases apparently forgotten, is for the whole world.

### From the mid-1950s

Academic progress from 1945 was swift, and geography became a favoured study at all educational levels. To the world at large the conquest of Everest in 1953 under the auspices of the Joint Himalayan Committee of the Royal Geographical Society and the Alpine Club was the realization of a hope held for many generations: by a happy coincidence the announcement was made on the day Queen Elizabeth II was crowned, 2 June 1953. Sir John Hunt and Sir Edmund Hillary gave a summary account of the ascent at three separate meetings in London during the autumn of 1953 (*Geogrl. J.* 119 (1953), 385–99) and various members of the party gave lectures throughout the country to large audiences. Great as the achievement was, to a historian of the Royal Geographical Society it is one more dramatic event in a long story of human determination to know the whole world and to interpret its mysteries. Few can share in such enterprises though they give untold interest to millions and reveal scientific data to a small number. Meanwhile the ordinary work of geographers went on, and here it is proposed to look first at the prospects from the early fifties in education, physical, regional, and what is broadly called human geography.

Educational progress had long been a concern of the RGS, and in reviewing H. R. Mill's *An Autobiography* (1951), O. J. R. Howarth (*Geogrl. J* 118 (1952), 89–90) said: 'To one who read his "The realm of Nature" as a first geographical textbook ... it is no matter for surprise that its author looked upon it as "the best thing I have done".' Mill associated himself with Keltie, Chisholm and Mackinder (these are the colleagues whom he specifies) in the earlier labours of procuring adequate teaching of geography.' By the mid-1950s, however, few university teachers were writing textbooks for junior pupils in schools, and a new corps of authors, mostly teachers in schools, had arisen, though a number of books by university writers were expected to appeal to school students taking Advanced Level courses. In 1955 the RGS Education Committee was able to report that some of the fears expressed five years earlier (p. 58) had proved groundless both in schools and in technical education. As before, the Education Committee was a joint enterprise of the Society and the Geographical Association.

The report of 1955 on 'Geography in Education' is a sequel (*Geogrl. J.* 121 (1955), 190–6) to that of 1950 on 'Social Studies in Schools' (p. 58). Geography had now been given a status in Civil Service examinations for the Administrative class similar to that of other major subjects, and people who had degrees and other qualifications in geography were entering a wide variety of occupations. This was shown, for example, in the 'Careers for geographers' pamphlet (p. 58). The Committee argued that geography helped to satisfy the natural desire of adults and children to know more about the world they lived in, and added that such knowledge aided international understanding. It also encouraged the habit of keen and accurate observation, partly through map work. An intellectual challenge was given by the need to understand the varied data from the natural and the human spheres, with the interpretation of their inter-relationships. Following Mackinder they regarded geography as an 'art' as well as a 'science', dealing with facts that are 'contemporary and real ... significant for, and relevant to, the life of every individual and every community'. Teachers worried by the 'convenient but artificial division of knowledge', aware that what they taught might seem to bear little relation to anything else students might learn, could find in geography a subject that was centripetal; capable of 'performing a unifying function' and so able to offset 'any narrow and premature specialisation'. (They did not, however, raise the question of such ill-advised specialization in the geography courses given in schools, a subject on which some university teachers could be eloquent.) Finally, having made the point that geography was 'explanatory' not only globally but locally, they said that 'geographers are contributing something of value to the discussion of many contemporary problems, whether they relate, for example to the planning of our countryside, the conservation of natural resources, the team survey of colonial areas, or the complexities of international affairs.'

In technical education the outlook was similarly promising (*Geogrl. J.* 124 (1958), 222–4). In 1958 the Education Committee noted that 15–20 per cent of the time in Liberal Studies courses leading to the Diploma in Technology was given to geography. That students in colleges of technology had their own special interests was recognized in the following statement: 'It is the business of geography to describe and evaluate the interrelationships and interplay of human and natural factors in the varied environments of the world. These in themselves form the setting in which modern technological developments are taking place, a setting of which every educated person is cognizant to a greater or lesser degree.' The Committee observed that four benefits could be conferred by courses in geography. Of these the first was an understanding of contemporary affairs in industry, and also of European, Commonwealth, and world problems. The second was a reasonably enlightened attitude towards change, for example in the use of new resources, irrigation, and land improvement. A third benefit lay in the ability to investigate physical and human aspects of areas, to assess regional 'personality'. Finally, technical students would be taught to make useful and relevant studies themselves: for example, they might study an industry in its regional setting, noting its locational aspects such as raw materials and power, as well as markets. As a basis such lines of study could give much to any intelligent person. The Committee commented, for example, that detailed field survey could 'lead to a lifelong interest'. This may take many forms, as tutors in adult education often discover. Apart from the appeal of canals and railways, the interest that developed in industrial archaeology in the 1960s was engendered by some of the teaching given in technical colleges. One abiding aim in teaching geography is to induce at least some people to look more closely at their own environment.

## PHYSICAL GEOGRAPHY

Educationally the outlook seemed promising, and the last two paragraphs may perhaps have given the impression of a happy band of pilgrims marching together in harmony towards the light. A less serene picture comes from the geomorphological scene in the 1950s. In 1954 W. M. Davis's *Geographical Essays* was published and reviewed by S. W. Wooldridge (*Geogrl. J.* 121 (1954), 89–90). As noted on p. 70, the work of Davis and his followers, including Cotton had come under vigorous attack, particularly from A. N. Strahler, who wrote in 1950 that it was

entirely qualitative without any penetrating analysis of erosional processes ... this treatment appealed ... to persons with little training in basic physical sciences, but who like scenery and the outdoor life; as a cultural pursuit. Davis's method was ... excellent and entirely adequate as part of the basis for understanding human geography, but as a branch of natural science ... superficial and inadequate.

Also in 1950 John Leighly had said that 'Davis's great mistake was the assumption that we know the processes involved in the development of land forms. We don't. ...' Other criticisms came from Kirk Bryan and Carl Sauer. Having said that American writers revealed 'a distinctly bitter or rancorous note which many of us, still heavily in debt to Davis, must resent and indeed, flatly repudiate', Wooldridge maintained that Davis's work was 'classic' and that Davis had confounded many of his critics by indicating that they had not studied his statements closely, noting his 'careful qualifications and glosses'. Wooldridge noted that

a mainspring of Davis's thinking during the last decade of the nineteenth century was the need to provide a disciplinary fibre for geography as a subject.... modern geography covers a wider field than Davis was inclined to allow. What seems to have been lost sight of both in Britain and America is the superbly 'clean' and intellectually attractive quality of land form study as Davis taught it. In its quasi-mathematical form and style of reasoning it has something of the quality of economic analysis.

Having said that he would wish to make it compulsory for university teachers of geography to read Davis's essays (many have done so), he also says of dons that

the 'mess of socio-economic pottage' which they so indefatigably produce fails to evoke either general interest or respect. Human or social geography has yet to find any-one of the stature of W. M. Davis to give it coherence and organization and, until this is achieved, it cannot be a matter for wonder that appreciable numbers of geographers, not content either with amassed information as such, or the pedantic quest for mere 'areal differentiation', find intellectual stimulus and sustenance in geomorphology.

That they should do so is excellent. Unfortunately Wooldridge did not answer the challenge given to the work of Davis by the Americans any more than Cotton had (p. 70) a few years earlier.

To what extent a local field study in geomorphology might be regarded as 'typical' was frequently a matter of argument in the 1950s. In 1955, for example, A. H. W. Robinson (*Geogrl. J.* 121 (1955), 33–50) wrote a paper on 'The harbour entrances of Poole, Christchurch and Pagham' based on field-work and historical map evidence. Later in the same year W. V. Lewis (*Geogrl. J.* 121 (1955), 385–6) raised the question of comparability, for while he saw some resemblances to other harbours on the south coast and on the west coast of Wales, and while he agreed with Robinson's conclusions 'for the examples he cites', he also thought that 'the instances he has chosen are not typical of conditions at larger harbour entrances'. At this time Lewis was working towards a coastal study covering at least the whole of the south coast, unfortunately never completed as he was accidentally killed in 1961. Possible certainty on coastal movement of beach materials was forecast by C. A. Kidson *et al.* in 1958, using radioactive methods of detecting the movement of shingle over the sea bed and along the shore (*Geogrl. J.* 124 (1958), 210–18). This would appear to be in accord with the wishes of the American critics of W. M. Davis (see above) who wished to have generalization, as for example in this case of coastal development, based on scientific experiment.

Another effort to deal with fundamentals of geomorphological evolution came in D. L. Linton's 'Problems of Tors' in 1955 (*Geogrl. J.* 121 (1955), 470–87). Based on field evidence from Dartmoor, the Pennines, the Cheviots, the Grampians, and the extreme north of Scotland (but not from Ireland), in all of which tors were found (of which some are splendidly illustrated by the author's own diagrams and photographs), and also on the absence of tors from other areas such as Snowdonia, the Lake District, the higher areas of south-west Scotland, and the western Highlands of Scotland, Linton reaches the conclusion that

the absence of tors from the western uplands is undoubted and it is hard to escape the conclusion that this is so because tors are part and parcel of an order of things that has been swept away from these areas by the heavy drubbing they received from the ice. When tors are found they rise from hilltops and uplands of smooth and rounded aspect with forms and features that we associate with mature and post-mature development in the cycle of normal erosion. These landscapes can only be regarded as survivals from, in the local sense, pre-glacial times, some of them possibly from a period which is pre-glacial in any sense.

He goes on to say that the western line delimiting tors is closely related to the 'areas of intense glacial structure'. Tors had been observed in many differing climatic areas, including the Tropics, and some of the problems that arose in Linton's paper, notably on the evolution of tors, were considered further by R. S. Waters in a paper of 1957, 'Differential weathering and erosion of oldlands' (*Geogrl. J.* 123 (1957), 503–10). Both papers raised controversial issues discussed in the correspondence pages of the *Journal* (*Geogrl. J.* 124, 289–93).

   Whatever view modern critics may take of the work done in physical geography as this was broadly interpreted during the 1950s, it was certainly a time when people cared for such studies. Eva Taylor, in reviewing six books and brochures dealing with the International Geophysical Year (1957–8) (*Geogrl. J.* 123 (1957), 514–17), commented that 'truly these are great days for those eager for knowledge in its own right'. She gave as an example the work of H. H. Lamb on Antarctic meteorology, in which he suggested that 'undiscovered geographical features—crests, depressions, valleys—probably held the key to many anomalies of weather, including the locally violent gales, met on its continental margins'. Fears that the melting of glaciers and ice sheets might cause a rise of sea level that could overwhelm areas of immense wealth may be remote, but within a period of fifty years the atmosphere might become so much warmer, through the increasing use of coal and oil, that this process might be accelerated. And while Sir Harold Jeffreys, a fine geophysicist, had rejected the idea of continental drift as mathematically impossible, some recent evidence suggested that there was a geomagnetic fit which made a trans-Atlantic movement not impossible. Professor E. G. R. Taylor showed that many world theories of geology, geodesy, climate, and other sciences were based on a slim network of actual stations and observations.

   That the theory of today may be outworn, or discarded (and this is not the same thing) tomorrow is noted by K. S. Sandford in a review of J. K. Charlesworth's 1700 page book (*Geogrl. J.* 123 (1957), 517–20), *The Quaternary Era* (1957). Writing of the third part of this monumental work covering thirty-five years of research and reading, he notes

Pleistocene stratigraphy, even of 'classical' sections, is temperamental, sometimes subjective, and correlation, especially over great distances, demands faith. One believes, and worships at some shrine, or one is agnostic and yearns for ultimate truth. The author gives the words of the prophets, and new prophets show the errors of their precursors as soon as the page proofs leave the author's hands.

Despite the vast literature on glaciation, there are still mysteries: for example, Sandford mentions the 'unsolved problems' of drumlins, the lack of any advance in the explanation of esker formation, and the unconvincing explanations so far available on the formation of cirques. When such familiar landforms defy explanation, it is not surprising that the work of W. M. Davis should in time be contested, but the achievement of certainty is beyond imagination. The work of Linton on tors, for example might fail to convince many readers but at least it opened lines of enquiry and investigation.

### CORRELATION AND SPECIALIZATION

Simple correlations between climate, fauna, and flora are unlikely to be valid, as Dr D. J. Crisp, Director of the Marine Biology Station of the University of Wales, Menai Bridge, Anglesey, showed in a fascinating paper in 1959 (*Geogrl. J.* 125 (1959), 1–19). He noted that during the past two hundred years the climate of northern Europe had changed, with a general rise in winter temperatures, to only a slight degree in Britain but by nearly 1 °C in Stockholm, almost 2 °C in northern Finland, and as much as 4 °C. to 5 °C in west Greenland and Spitzbergen. The greater circulation of warm air into northern latitudes had been accompanied by a rise in precipitation but there had been a slight lowering of mean annual temperatures over large land masses in lower latitudes. The effect on glaciers in northern Europe was well known and there was evidence that the Arctic ice was becoming thinner. Although some birds and animals had spread northwards (for example the lapwing and the polecat in Finland), simple correlations of such extensions of habitat were not explicable solely in climatic terms, for different species were competing with one another and insect populations fluctuate according to weather conditions. Similarly, fisheries are subject to fluctuations for which the explanation is to be sought in further research on food supplies, though the area now covered by warm-water species extends some 200 to 250 miles further north into waters previously populated almost exclusively by Arctic forms. Dr Crisp summarizes his argument by saying,

Numerous changes have been observed in the distribution of animals and plants during the period of climatic improvement in Europe.... The changes which can reasonably be accounted for by climatic amelioration are only of moderate extent and are mainly confined to the far north, though fluctuations, especially in marine life, have been recorded also in temperate latitudes. None of the apparently genuine effects have displaced boundaries northwards by more than about 200 miles.

He concludes by saying that there are changes in which the environment appears to have had a direct influence, such as the extension of Arctic plants, but that in other cases a new balance is worked out within a community of animals by reaction; climatic change may result in a displacement of population boundaries, for example by competition between animals favoured by warmer or by cooler climates.

Such papers lead the mind once again to the thrilling realization of nineteenth-century naturalists, and with them geographers, of the complex relations of all life in the world. Just as the detailed studies of recent changes in northern Europe raised world scientific problems, so the contemplation of past ice ages with the visual and scientifically measurable evidence of contemporary glaciation was a fact of present-day geography with possible consequences for future generations. This theme of time past, time present, and time future was even more apparent in the work of the International Geophysical Year which, as Dr D. C. Martin explained (*Geogr. J.* 124 (1958), 18–29), was a development from the two earlier International Polar years, of 1882–3 and 1932–3, though with a vastly extended research mandate covering the earth's atmosphere, surface, and interior, to which nuclear radiation was added.

Among the considerations of direct significance to geographers were those on meteorology, oceanography, and glaciation, and Antarctic discovery, urged and fostered by the RGS, now seemed more relevant than ever before. As always actual and, where possible, continuous observation was needed. This was shown, for example, in an account of the Falkland Islands and Dependencies Aerial Survey Expedition of 1955–7 entitled 'Airborne surveying in the Antarctic', by F. G. Mott, which appeared in 1958, (*Geogrl. J.* 124 (1958) 1–17) and in various articles on other aspects of glaciation during the late 1950s. In 1960 the fourth of the post-war Research Series monographs, *Investigations on Norwegian cirque glaciers*, was edited by W. V. Lewis. This was a discussion of work done over many years in which students and others had been involved. S. de Q. Robin in a review article (*Geogrl. J.* 126 (1960), 500–3), notes that the writers (there were eight of them) 'have tended to underestimate the importance of their own observations. They have been reluctant to throw overboard commonly accepted ideas which are inconsistent with their own observations, even though they point out the inconsistencies quite clearly.' Robin admits that verification comes only with time, but many university teachers who have supervised higher degree work would agree that researchers are prone to underplay their discoveries and to assess their implications.

Contemplation of the mysteries of the earth and of the universe has its unending appeal, but geographers retain their concern with the 'here and now' of life, of the living world which they themselves experience. In the 1950s one issue frequently discussed in academic circles was the interplay of regional and systematic geography, particularly as a wide range of new courses was designed for students in universities. Many accepted the view that regional work was the crowning achievement of geography while others regarded themselves as 'systematists'. This comes out in a review by A. A. Miller (*Geogrl. J.* 122 (1956), 117–18) of a monograph by S. W. Wooldridge and D. L. Linton, *Structure, Surface and Drainage in South-east England*, first published by the Institute of British Geographers in 1939 and reissued in 1955. That this work was a fine contribution to geomorphology is undoubted. Miller draws attention to the methodological views of its authors who 'emphatically insist' on 'the full and proper understanding of the total geography of the region' as 'the final objective of the geographer'. Miller comments that 'many, especially in the Continental and American schools, will protest that geomorphology is a science in its own right'. This point comes out even more strongly in another review, also by Miller (*Geogrl. J.* 123 (1957), 392–4), of *The Geographer as Scientist. Essays on the scope and nature of geography* (1956), in which he deals with the views of the book's author, S. W. Wooldridge.

The logical conclusion to the doctrine of 'wholeness' is the often stated but rarely attained distinction of geographical study as 'the region' ... [though] '... the end of our quest to see the regions of the world as wholes is not yet.' Without belittling the 'special geographers' he would have geographers study regions, not problems, but, though the whole world is Geography's province no one mind can encompass it all. So let him focus on a region of limited extent, unafraid of the change of parochialism. And since superficiality in any of the 'special' geographies is not to be tolerated it follows that the complete and balanced regional geographer must equip himself with a mastery of all specialisms, an omniscience limited only in respect of its areal range. If ... he *must* study problems, let him at least focus his studies in the region of his choice, regarding too much attention to 'otherwhere' as a weakening dispersion of effort.

This may be regarded as an extreme view, or at least as a provocative view, for many geographers were concerned with areas remote from their home environment, that is with 'otherwhere', as a large number of useful and welcomed regional texts show. But the real challenge comes in the idea that a geographer should be able to deal competently with the entire range of the subject in all its physical and human aspects. That this was possible had been assumed earlier but the contributions from research by various specialists had made such a position untenable. Tritely expressed, geography was entering the phase of growth known as 'learning more and more about less and less', of increasing specialization. In 1961 R. W. Steel, in an analysis of 150 papers published by the Institute of Geographers during fourteen years, shows that, though many do not fall easily into a single category, geomorphology, claimed the largest numbers (34), economic geography (33), historical geography (27) and social, including urban, geography (23). This he notes (Steel 1961 : 137) probably reflects, 'in broad terms, the main trends and interests in post-war geography in Britain'. He adds that the interest of researchers in universities and elsewhere are 'in such fields as denudation chronology, drainage evolution, the geographical analysis of historical documents, field systems, land use changes, industrial location, the distribution of population and patterns of settlement'. If technically many of these studies came into the systematic category, a large number of them were also contributions to regional geography (Steel 1961).

## REGIONAL GEOGRAPHY

Veneration for regional geography was marked before the War of 1939–45, as readers of Hartshorne's *Nature of Geography* (1939) will appreciate. A pattern of arrangement had become traditional in regional texts, which, after a brief introduction, dealt with various aspects of physical and later of human geography before proceeding to a detailed, microscopic analysis of the various regional units of whatever area the book covered. That writers of such books were expected to be competent in all aspects of geography has already been stressed. The search was for correlation, of man with environment, and this was not new but traditional, for it was a problem engaging human thought long before Mackinder, long before Humboldt and Ritter, and indeed from the beginning of civilization. To give a 'complete' study was the

aim, and not an ignoble aim. Authors had to consider the man–land relationship in all its complexity, knowing that from the first settlement the land
itself had been changed by human activity; changes might be swift and
favourable, as Cumberland showed in an article discussed below, or they
might be considerable yet inadequate to meet the challenge of population
increase and intermittent disasters, as in India and Pakistan.

Spate's *India and Pakistan: a general regional geography*, first published
in 1954, was the first regional study of the subcontinent to appear since Sir
Thomas Haldich's book of 1904. J. N. L. Baker, reviewing the new book
with a clear knowledge of its predecessor (*Geogrl. J.* 120 (1954), 493–6), says
that Spate's book shows that

today regional geography has a more logical sequence as well as a better developed
technique. Professor Spate begins with the land. Structure and relief, climate, vegetation
and soils occupy Part 1. Part 2 deals with people—population, racial, linguistic and
social features, and villages and towns. Part 3 deals with the economy—agriculture,
industry, transport and trade ... Part 4 deals with the regions of India.

This last section covers 400 pages, half the total length of the book, and deals
with the three 'macro-regions', the Mountain Rim, the Indo-Gangetic Plains,
and the Peninsular Block. There are '35 regions of the first order, 74 of the
second order, and about 225 sub-divisions of these'. This, Baker comments
may seem an 'enormous number of regions' but Spate had adopted so sensible
an approach that there was no confusion. And, as might be expected, the
backward glances of the author are 'usually less favourable than those of
Holdich' while his forward glances are apprehensive though not without hope
given 'more fertility on the land and less fertility in the homes'. The general
chapters of Parts 1–3 occupy about half the book, a fairly normal feature
of regional texts; in short, part of the treatment was systematic. Within the
fifty years that separated the two books of Holdich and Spate the changes
in India had been vast.

Change is a problem for any regional geographer, especially if his book
goes through a number of editions. Short-term change was splendidly shown
in a paper by K. B. Cumberland, 'New Zealand after twenty years: a geographer's view' (*Geogrl. J.* 125 (1959), 28–47). Cumberland in 1939 found
it 'a blatantly pioneer colonial land, incompletely transformed, inadequately
tamed, still sparsely occupied, raw, fresh and new, but youthfully exciting
and inviting'. The depression had left some wounds, including soil erosion
through rapid exploitation, but the succeeding twenty years saw 'the impressive large-scale reclamation of hitherto unused land ... intensification and
consolidation in the use of land ... conservation of resources ... refinement
and improvement of farm techniques'. With this had come an expansion in
towns, in industrial and service employment, and, despite various problems,
New Zealand had become 'a prosperous and progressive country with high
material standards of life ... paying increasing attention to non-material
aspects of national life ... slowly evolving a New World, but Pacific culture
distinctly its own'. The dynamic character of change, in landscape and life,
is a challenge to the regional geographer, not always accepted, for many
books describe conditions long since outmoded.

Both Spate's book and Cumberland's article show a concern for man and the land, for the totality of the physical and human environment. In both agricultural life was a principal consideration, though both authors were fully conscious that man made the earth, remade it, destroyed it or failed to guard against its destruction through unwise exploitation, cleared or planted forests, and in a multitude of other ways made a human landscape that gave an area personality and form. What proved to be difficult was how to deal with the range of aspects involved. Spate's book acquired a reverence given to few others for many authors gave an unconvincing and even superficial account of the physical aspects, partly because little was known of many of them, and, although some geographers regarded their subject as a 'co-ordination' dealing with relationships, there was obvious danger in drawing correlations from inadequate data. The sheer task of writing a regional geography on a basis of 'completeness' dismayed many authors.

Nevertheless the demand for regional texts remained and in fact still remains. In 1955, for example, K. H. Buchanan and J. C. Pugh's *Land and People in Nigeria* appeared; welcomed by A. T. Grove (*Geogrl. J.* 122 (1956), 383–4) as the first detailed geography of Nigeria, it followed the then classical form, having sections dealing with geology, relief, climate, vegetation and soils, followed by the 'human pattern', which included agricultural economy, forest resources, mining and industry, public and social services. Two years later, R. J. Harrison-Church's *West Africa* was described by R. Mansell Prothero as 'the first advanced work on West Africa to be written in English'; it contained sections on the physical basis and resources, and half the book was 'devoted to the major territorial divisions in turn', all seventeen of which were treated according to 'a more or less rigid and well-known formula-introduction, historical outline, climate, geology and relief etc.' (*Geogrl. J.* 123 (1957), 537).

## SPECIALIZATION: THREE EXAMPLES

### (a) Medical

Specialization appeared to offer a clearer research field than regional enquiry; for example, medical geography was revived from the late 1950s. In 1956 A. L. Banks (*Geogrl. J.* 122 (1956), 167–75), professor of human ecology at Cambridge University, drew attention to the modern recognition of health needs in towns, seen for example in the preservation of green belts and the provision of recreational facilities; rural residents have their needs also and the author gave a favourable account of a government scheme in Egypt where 'under a social worker with a basic training in agriculture' there are 'simple clinic facilities, staffed with doctor, midwife and sanitarian'. Three years later Banks (*Geogrl. J.* 125 (1959), 199–216) gave a paper on 'The study of the geography of disease'. Like the earlier paper this was of a somewhat general character and the discussion which followed brought out several interesting points. L. D. Stamp said that medical geography could be treated at four levels. At the world level some notable work had been financed by the American Geographical Society, directed by Dr Jacques May who had 'the ideal qualifications—the geographer at heart, the medical man by training'.

The second level of study was national, on which A. T. A. Learmonth had worked in India and Pakistan; in his account the aim had been 'to show that there are disease regions i.e. . . . where the association of diseases is distinctive; to delimit them positionally; and to analyse their underlying causes in the light of both geographical and medical literature'. The third level was regional, and on this Stamp drew from his experience of 'healthy' and 'unhealthy' areas of Burma, while Learmonth mentioned the work of A. Geddes in India (p. 39). The fourth level was local, shown by Dr John Snow in the 1854 cholera outbreak. He found that all his Soho patients used water from one pump in Broad Street, and when this was closed the outbreak ceased.

At the Lisbon Congress of the International Geographical Union in 1949 a commission on Medical Geography was approved: reports of its work were sent to later congresses at Washington in 1952 and Rio de Janeiro in 1956. In 1961 Dr Learmonth gave a paper on 'Medical geography in India and Pakistan' (*Geogrl. J.* 127 (1961), 10–26) introduced by slides which included 'several Mysore villages and one urban slum, with widely varying health conditions'. While every village is unique, just as there are 'repetitive patterns in land use settlement pattern and in the human geography generally' so 'repetitive patterns certainly exist in the regional distribution and regional association of disease'. Learmonth admitted that many of the statistics were inaccurate but observed that 'death-control is tending to occur markedly earlier than birth-control'. In the subcontinent as a whole:

Malaria is controlled over almost all this vast and populous area and eradication is now planned. Tuberculosis may be controlled by BCG within a few years. But the intestinal diseases, the universal infectious diseases . . . (including influenza and cool-season pneumonia) . . . and especially the still deplorably high infant mortality rates are surely subjects for an attack integrated with all-round developments, on the lines of the Community Development Projects. . . . In a sense, social medicine is community development . . . social medicine is regional planning.

All this work on medical geography was welcomed by research workers concerned with health, including anthropologists (some of whom were studying blood groups) and others beside medical specialists such as epidemiologists. In 1962 R. M. Prothero spent several months with the World Health Organization investigating malaria in relation to labour migration in Africa, following some work in 1954–5 on migrant labour in general (*Geogrl. J.* 128 (1962), 479–93). He reached the conclusion that:

The relevance of geographical study and of the geographical approach can be demonstrated in the field of malaria eradication, but there is no doubt that it is equally important in the study of other diseases and in planning for their eradication. . . . Geography can gain most in influence by concerning itself with the practical affairs of men and with the problems of the environments in which they live. I am not suggesting that these should be its only concerns, but I feel that geographical studies are more worthwhile when they are so concerned.

Meanwhile in Britain research by G. M. Howe, supported by the Medical Research Committee of the Royal Geographical Society, resulted in the publication of the *National Atlas of Disease Mortality* in the United Kingdom in 1963. Howe explained that the maps showed population density for 1956, with six classes of mortality: all causes, cancer, circulatory system, respiratory, digestive, and miscellaneous (*Geogrl. J.* 130 (1964), 15–31). The standardized mortality rate was calculated for all the 28 metropolitan boroughs, 89 county boroughs, and the urban and rural districts of the 101 administrative counties that then existed. Male and female deaths were considered separately. Considerable variations were observed: the standardized mortality rate was highest in Salford and the position was almost as serious in several other Lancashire towns, and also in Birkenhead and Stockport in Cheshire, Halifax, Dewsbury and Bradford in West Yorkshire, towns of Tyneside and Teeside, Stoke-on-Trent, Wolverhampton, and Walsall. The lowest rates occurred in the south and east of England, and in the towns of Canterbury, Reading, Oxford, Ipswich, Bath, and Norwich. Many points of interest was raised in the discussion and in a review by Jacques May (*Geogrl. J.* 130 (1964), 268–70); it was noted that statistics of this character should be studied as they appear. Obviously the difference between the more affluent south-east and the more industrial north-west was noted but the figures were for 1954–8, before the post-war rehousing programme was complete and before the use of smokeless fuel had become compulsory in towns.

There was a need for continuing research. Jacques May asks, for example, why certain forms of cancer should be so much more prevalent in some areas than in others and why infant mortality should be high in rural areas of South Wales, in Central Scotland, and in Glasgow (unfavourably prominent in all the medical statistics) as well as in the towns of Durham county and Staffordshire. Abundant material was available, provided it could be located, through the friendly co-operation of workers in geography with those in social and preventive medicine. A new and expanded edition of Howe's atlas appeared in 1970, and a reviewer, having noted that it 'provides epidemiologists, workers in public health and in the social services, and other medical and geographical research workers, with an invaluable reference source', also points out that overall mortality rates are more favourable in the south and east than in the north and west. Bronchitis mortality is still highest in the north, and lung cancer high in industrial areas, though stomach cancer, in Britain as in the rest of the world, is declining as prosperity increases (*Geogrl. J.* 136 (1970), 596–7).

### (b) Land use

Land use remained a subject of interest and, indeed, of general public concern in the 1960s. In 1961, Alice Coleman gave a paper on 'The second Land Use Survey: progress and prospect' with a map showing that by April 1961 almost all of England and Wales had been allocated to field-workers (*Geogrl. J.* 127 (1961), 168–89). Warmly welcomed and supported by L. D. Stamp, the new survey was planned on a $2\frac{1}{2}''$ to 1 mile scale with a classification of land use far more elaborate than that used for the survey made thirty years earlier. Miss Coleman opened by saying that the Stamp survey 'coincided with the nadir of the great inter-war depression. Although this enhanced its interest as a historical document, it shortened its life as a current historical source.' Stamp went further and commented that perhaps the new survey came at the zenith of British agriculture (one would like to think this was not so).

Ten years later Miss Coleman explained her scheme of preparing a 'Wild-scape Atlas', using a vegetation classification originally formulated by L. D. Stamp and slightly modified by consultation with ecologists (*Geogrl. J.* 136 (1970), 190–205). Some experimental work had been done on a few sheets of the Second Land Use Survey, and the new classification would be far more comprehensive than the pioneer work of the Smith brothers and a few others early in the twentieth century (pp. 20–1). The cost of publication, however, remains a formidable problem. By 1970, which was European Conservation Year, interest in the problems of land use and nature conservation was widespread, and the opportunities for research apparently unlimited.

Land use was at times an emotional issue, and the cool academic approach adopted by R. H. Best, author of numerous articles as well as *The Major Land Uses of Great Britain* (1959) and, with J. T. Coppock, *The Changing Use of Land in Britain* (1962) was particularly welcome. In 1965 Best gave a paper on 'Recent changes and future prospects of land use in England and Wales' (*Geogrl. J.* 131 (1965), 1–12), in which he noted that

contrary to popular belief, the post-war years have seen no sustained increase in the rate at which farmland is being transferred annually to urban uses. On the contrary, the highest rate of turnover was in the three years immediately following the end of hostilities. Subsequently, the rate of loss has fluctuated sharply, often, it would seem, in response to government action in the economic sphere.

Social need was an influence on land use; for example, from 1955 to 1960:

The main area of land transfers from agricultural to urban uses occurs in a belt of country extending from Lancashise and the West Riding of Yorkshire, diagonally through the Midlands, to London and the south west in Sussex. In this area, a transfer rate exceeding one per cent for the five years under consideration has often been characteristic.

This is clearly the 'coffin belt' of earlier notoriety (p. 40) and Best modifies it to a dumb-bell because between the West Midland and London conurbations the transfer rate was low, less than 1 per cent (exactly as J. N. L. Baker and E. W. Gilbert wished, p. 40). Best's findings appear to be confirmed by a later study on the containment of urban England, noted on p. 92. Other studies concerned with land use include D. Thomas's 'London's Green Belt: the evolution of an idea, (*Geogrl. J.* 129 (1963), 14–24), later developed into a book with the same title (1970). He showed that control had been effective in preventing the unlimited sprawl of building that would have resulted under free enterprise, though inevitably some compromises had to be made to meet the needs of London. As a reviewer commented, this book 'will prove of value to geography students for its clear demonstration of how to sample land use, how to link field observations with the history of land use controls, and how to compare the accuracy of similar survey'.

'Going to the Common' is a pleasure of childhood to many and in 1955 a Royal Commission was appointed on Common Land. Among its members were L. D. Stamp and W. G. Hoskins, the historian, who together wrote *The Common Lands of England and Wales* published in 1963. In 1964 Stamp gave a general paper on the common lands and village greens of England and Wales in which he begins by explaining that nobody appeared to know what common land was (*Geogrl. J.* 130 (1964), 457–69). In fact there were eight main types, of which the first is land subject to common rights for all or part of the year and the second land grazed in common, for example, upland grazing in the Pennines. A third category covers land in which the owner holds the land subject to the right of the 'commoner', or a group of commoners, to take the produce or to use its grazing. The fourth type is village greens of which 1400 are known, though the total is much higher. The fifth type, manorial waste, is open to public access but in many places is derelict as all the legal arrangements have become extinct. A sixth type is open wastes besides highways; there are also common lands provided at various times for a diversity of purposes, now obsolete, such as the relief of the poor, a 'goose green', even a 'common quarry' or 'stone allotment' for road repairs. Stamp showed the need for control of commons as many were derelict or in a bad state.

Major commons included the New Forest, the Forest of Dean, and Dartmoor, and more than three-quarters of all common land was rough grazing, mainly in upland areas. Some was woodland, including Burnham Beeches and Epping Forest, and some common land was used, or could be used, as nature reserves. Village greens were of varied origin and use. Some gave a defensive site against marauders: others served as common pasture for cattle, sheep, horses, and poultry. They might also serve as social centres or be used for occasional fairs and markets. Hoskins observed that in the twelfth and thirteenth centuries agricultural land was so valuable that towns were built on Commons, as at Liverpool, Petersfield in Hampshire, and Hedon in Yorkshire. Later, in the seventeenth century Harrogate was built on common land, of which the Stray is a fine relic, and at Newcastle-on-Tyne the Town Moor still exists. Despite all the legal uncertainties affecting common lands, Hoskins notes that they appear to survive, not without encroachment in some places and Stamp calculated that 4 per cent of England and Wales was common land in 5382 pieces, ranging in size from tiny village greens to extensive moorland commons.

### (c) Biographical

Many readers of the *Journal* must have welcomed studies of past geographers as good reading, especially those written with elegance and wit. There was, for example, an article by O. F. G. Sitwell on John Pinkerton (*Geogrl. J.* 138 (1972), 470–9), 'an armchair geographer of the early nineteenth century', whose *Modern Geography* of 1802 'closely resembles the national handbooks prepared by the British Admiralty ... during the Second World War with their coverage of geology and physical features, the coast, climate, soil and vegetation, the people, historical outlines, government, administration and law, public health, growth and distribution of population, drainage and reclamation, agriculture, fisheries, mining and industry, commerce and finance, the mercantile marine, the ports, roads, railways and waterways'. Sitwell suggests that the compendious geographies of the eighteenth century provided information useful to those in positions of political authority. At the same time Pinkerton's geographical descriptions showed a wish both to identify national personality and to accumulate new data not only for itself

but also as a basis for inductive, comparative study. That unthinking condemnation of eighteenth- and nineteenth-century geographers is based on
ignorance is apparent also in a paper by Alan Downes (*Geogrl. J.* 137 (1971),
379–87), engagingly titled 'The bibliographic dinosaurs of Georgian geography (1714–1830)', in which he shows that modern geography did not grow
solely out of the work of Humboldt and Ritter but that the ideas of many
earlier writers contributed to its growth.

Naturally the interest of RGS turned towards African explorers on the
centenary of Livingstone's death in 1873, though Livingstone was only one
of a number of geographers who had contributed material on Africa, among
whom was R. D. Cooley, 1795–1883, a founder of the Hakluyt Society in
1846 and a writer on East Africa who, says Dr Roy Bridges (*Geogrl. J.* 142
(1976), 27–47, 274–86), was so frequently involved in controversies that 'he
is remembered, if at all, as the man who arrogantly told explorers that they
had not seen what they thought they had seen'. He never went to Africa himself but he was widely read and knew more than many British geographers
of the work of foreign travellers and scholars, including Portuguese and Arab
travellers. He was critical of the work of Burton, Livingstone, and others,
and fought battles on paper over the snow-clad mountains of Kilimanjaro
and Kenya and the sources of the Nile. Some of his views were widely shared,
notably his fears that the RGS was unwise to make popular heroes of
explorers, but his contribution to African questions, says Dr Bridges

was not entirely negative. He did have a genuine interest in the continent and its peoples
and helped to direct the attention of the RGS there and to develop the wider preoccupation of the British with its problems. His attempts to resolve linguistic complexities and
his generally sensible approach to African social and historical questions put him in
some ways ahead of his time. Moreover, despite his irascibility, annoyance and
pedantry, he was a genuine scholar. The Hakluyt Society, at least, remains as a tangible
proof that this was so.

Apparently the RGS treated him with forbearance and charity.

Even Livingstone had been subject to reassessment: the centenary lecture
of 7 May 1973 by Professor George Shepperson shows that his purpose was
'to bring Christianity and commerce, civilization and progress—as he, in his
Victorian context understood them—to central and eastern Africa' (*Geogrl.
J.* 139 (1972), 305–19). D. J. Siddle, writing of Livingstone as 'a mid-Victorian
field scientist', notes that 'his image has suffered from the high expectations
of hero worship' and adds that 'political independence has made Africans
increasingly unwilling to accept the uncritical praise of a paternalistic white
missionary, no matter how well disposed he was to their welfare' (*Geogrl.
J.* 140 (1974), 72–9). He shows Livingstone as a fine observer of animal life,
an 'inveterate, even compulsive, measurer and recorder of almost everything
in his path, from the daily weather recordings to the physical dimensions
of the animals he shot', insect life and the tse-tse fly, including its predatory
activity and the effects of its bite. There was much that he failed to observe
and this Dr Siddle, having noted his fine account of the Kalahari desert
(1849), ascribes to the utopian vision of his later years, marked by optimism
leading in some areas to 'ambitious predictions as fanciful today as they were

a hundred years ago'. Livingstone apparently became an applied geographer such as one might find in a developing area, for to him:

Rivers became potential highways of commerce as well as interesting manifestations of a complex drainage system; plants are increasingly identified for their potential utility rather than their botanical interest; geological formations are viewed for their economic minerals; elephants are seen as the sources of ivory and soils are mentioned only for their fertility.

So much has been said on Livingstone that writers are probably wisest to assess his life and work as they see it, though in so doing they may, like many biographers, reveal more of themselves than the subject of their work.

Dorothy Middleton in her paper, 'The search for the source of the Nile', dealt with the expeditions, sponsored by RGS, of Burton and Speke (1856–9) and of Speke and Grant (1860–3) which located Lakes Tanganyika and Victoria and enabled the husband-and-wife team of the Bakers to reach Lake Albert in 1864 (*Geogrl. J.* 138 (1972), 209–24). 'The Nile quest', says Mrs Middleton, 'does not represent merely the personal ambitions and rivalries, the individual whims, of a handful of Victorian eccentrics; it was rather a manifestation of the urge to travel and explore which was such a feature of the Victorian scene.' Livingstone had the ambition of finding the source of the Nile to the end of his life but, to quote Mrs Middleton, 'with Speke lies the first word of this story of the Nile; the last word is with Stanley. In 1887–89 he led the Emin Pasha Relief Expedition across Africa from the Congo north to Bagamoyo and Zanzibar. He pitched camp three times on the western shore of Lake Albert and from here, in May 1888, his officers Jephson and Parke saw snow mountains to the south.' Stanley, she says, claimed the 'discovery' of these mountains for himself in April 1889 (*sic*), but the question of the Nile source was finally settled when the expedition marched 'along the Semliki, flowing round the western flank of the Ruwenzori from Lake Edward, all of which Stanley now put down on the map'. The fascinating characters involved come to life in this vivid article.

The study by I. A. Casada, 'James A. Grant and the Royal Geographical Society' (1974), deals with the companion of Speke who after a military career devoted a long retirement to work for African exploration and the RGS (*Geogrl. J.* 140 (1974), 245–53). All these studies are contributions to the history of African exploration, to the history of the RGS and of Africa, but the explorers were geographers filling in maps of unknown areas of a vast continent. Modern workers on 'perception' may well be intrigued by that possessed by these explorers, for it is impossible to understand their work without also understanding their personalities and the outlook of their time. To condemn them as 'imperialists' is to ignore the variety of their aspirations in a phase of history during which the world was being opened up as never before.

The penetration of the Canadian prairies is the subject of two entertaining papers by Irene M. Spry. The first of these (*Geogrl. J.* 125 (1959), 149–84) deals with Captain John Palliser, who was awarded the Patron's Medal by RGS in 1859. Born in 1817 into an Irish landed family of Yorkshire origin, he left Trinity College, Dublin, after four years 'with all the eagerness of a college student who casts aside his dull books and duller tutors for a burst after the partridges', and after a military career contracted 'prairie fever' which, says Mrs Spry, was then 'a common malady among young British sportsmen'. In 1847–8 he visited the wild prairie and mountain country of the upper Missouri river, in effect a hunting expedition, but in 1857 he organized a scientific expedition which had three main objects: 'to travel by canoe from Lake Superior to the Red River; to study the southern plains of the prairies,' and to report on possible passes through the Rocky mountains. Other travellers, including several from the Hudson's Bay Company, had been before and their journeys and observations were marked on a map by John Arrowsmith, but much of the earlier material was lost or regarded with suspicion and there was ample scope for further exploration. The first object was easily met, and from the second came the recognition of what is called the 'Palliser triangle', the area of the prairies with marginal or deficient rainfall where some disastrous experiments in agricultural settlement were made later by optimists. On passes through the Rockies, Palliser reported that several could be used with difficulty but there was still a long journey with unknown hazards to the coast at the mouth of the Fraser river. It was originally intended that the RGS should finance the expedition but in fact the British government did so, and Palliser, though honoured by the RGS, had a less enthusiastic reception from politicians and others but in time the value of his work was recognized.

In 1861 Dr John Rae, the Arctic explorer, visited the buffalo plains of western Canada (*Geogrl. J.* 140 (1974), 1–17). His account of the journey, given to RGS on 13 April 1863, was lost until the archivist found it in 1969. Rae made the usual astronomical observations expected of all good explorers, and his account is a fine study of the wild life, Indians, and Red River settlers in a critical period of change, with a discussion of the possible colonization of the Saskatchewan Valley. Mrs Spry notes that he also calculated that it was both quicker and cheaper to reach the Pacific coast by using the Panama route than by travelling overland. The 49th parallel had become the agreed boundary between Canada and the USA in 1846, but the problem of unifying Canada remained and the Canadian Pacific Railway was not completed until 1885, two years before the death of John Palliser.

Geographers whose primary contribution was academic provide less dramatic biographical material than those who were also explorers. Nevertheless they have been studied in a number of papers. of which a notable example is J. N. L. Baker's 'Mary Somerville and geography in England' published in 1948, partly to celebrate the centenary of the publication of her *Physical Geography*, which was warmly welcomed by A. von Humboldt (Baker 1948: Oughton 1978). Mrs Somerville (1790–1872), after whom the college for women in Oxford is named, was given the Patron's Medal of RGS for her work in 1869. When she decided in 1870 that she could not bring out any more editions the work was entrusted to H. W. Bates, then Assistant Secretary at RGS. Bates, born in Leicester in 1825, had acquired fame by the publication in 1860 of *The Naturalist on the Amazon*, written on the suggestion of Charles Darwin. From 1864 to his death in 1892 he gave devoted service to RGS, having been appointed on the recommendation of

Charles Darwin with the support of John Murray the publisher, who assured the Council that Bates was a competent business man (Mill 1930: 77). Among other geographers of the modern period, Mackinder has attracted numerous studies. His life and work have inspired a number of geographers, including C. A. Fisher, in his work on political geography (p. 64).

Mackinder was the subject of a fine obituary by E. W. Gilbert in 1947 (*Geogrl. J.* 110 (1947), 94–9) and of various other papers, including one by J. F. Unstead who in 1949 said that his special contribution to modern British geography depended on 'four characteristics: his clear and unified view of its content and functions; his gift of effective exposition; his energy and organizing power in obtaining recognition for Geography; his ability to recruit younger men and to inspire them to continue his work' (*Geogrl. J.* 113 (1949), 47–57). Unstead recalled his regret when Mackinder entered Parliament in 1910, though later he decided that 'perhaps he had already done the greater part of his work for geography'. It was to be abundantly discussed later. In 1969 E. W. Gilbert and W. H. Parker wrote an article on 'Mackinder's *Democratic Ideals and Reality* after fifty years' and showed that many of his ideas, for example on the heartland of the Old World, were of relevance in 1969 as in 1919 (*Geogrl. J.* 135 (1969), 228–31) for 'the United States is now the great sea power, whose aircraft carriers and submarines dominate the seas surrounding the world island and whose influence, economic, cultural or military, predominates among the islands and peninsulas of the perimeter. Contained within it, but probing for outlets, in the great heartland power, which the 1939–45 war decided was to be the USSR and not Germany.' Mackinder claimed to have invented the term 'manpower' and the authors say that the German 'economic miracle' would not have surprised him, for 'accumulated wealth, such as was destroyed in war mattered little; it was man-power, i.e. productivity, that mattered, and the more desperate the situation the greater the effort'. He also foresaw that Russia would be directed by 'ruthless organizers', for 'autocratic rule of some sort is almost inevitable if she is to depend on her own strength to cope with the Germans'.

Later, in 1976, B. W. Blouet (*Geogrl. J.* 142 (1976), 228–36) noted that arguably Mackinder's most important assignment during his parliamentary career was his appointment as British High Commissioner to South Russia in 1919. 'It ... was to lead to one of the most extraordinary examples of a theorist being confronted with the practical management of the object of his theorizing.' Blouet explains that Mackinder had in mind a 'cordon sanitaire' between Germany and (Bolshevik) Russia including not only Romania and Poland with the three Baltic States of Lithuania, Latvia, and Estonia but also a new South Russian state between the Caspian and the Black Sea, an independent Ukranian state and possibly a White Russian state. This however did not meet with the support of the British Cabinet and in 1920 he returned home and concentrated his public work on empire problems as chairman of the Imperial Shipping Committee (established in 1920) and the Imperial Economic Committee of 1925. Several other papers in these years dealt with other famous figures and were greatly appreciated by those interested in the historiography of geography, including supporters of the

International Geographical Union's Commission on the History of Geographical Thought.

## SOME MODERN PROBLEMS

Those who worked in universities, schools, and colleges in the 1960s and 1970s were constantly reminded by students that they must be concerned with the present age and that historical studies, though attractive and interesting, were an escape into romantic nostalgia. The RGS showed a balanced approach, though the numerous young authors who wished to write on their latest experiments in quantification had to look elsewhere than the *Geographical Journal*. Nevertheless, reviewers showed a continuing interest in new ideas, and when in 1965 *Frontiers in Geography Teaching*, a volume of essays edited by R. J. Chorley and P. Haggett, appeared, the reviewer quotes the editors' saying 'Better that geography should explode in an excess of reform than bask in the watery sunset of its former glories'. No longer should the best geography, like the best wine, 'be necessarily both French and long-matured' (*Geogrl. J.* 132 (1966), 572).

Obviously geomorphologists such as Wooldridge and others would hardly agree with R. J. Chorley that 'a geographer does not want to know that a stream is "mature" but what discharges have been recorded for it, not that a river terrace may possibly represent an interglacial event but information regarding its dimensions and composition, not that a slope may indicate the poly-cyclic origin of a valley but details of its geometry and soil characteristics'. One might perhaps want to know both what is discarded and what is added and one recalls the theory of J. K. Wright that American geographers were on occasion prone to 'bathwaterism' as a promising baby might also disappear. In his work, however, as in others, notably *Models in Geography* of 1967 edited by Chorley and Haggett, there was a realization that geography must be more than 'what geographers do' and that a 'model-based paradigm' was needed to replace the 'classificatory paradigm of orthodox geography'. This implied that geography was not unique in its method for it is logically a part of mainstream science and 'the traditional dualism between physical and human geography can no longer be sustained, as has been exemplified by the successful application of general systems analysis to both physical and human geographical problems'. It may be that in the search for pattern some of the advocates of the 'new' geography forgot that others had sought it before them, but at least a challenge was given, especially in the use of mathematical and statistical data, to look for interpretations that earlier workers, armed with less data and lacking such modern aids as computers, failed to discern.

Concern with immediate national problems led RGS to hold three meetings at which London's third airport (*Geogrl. J.* 137 (1971), 468–504), the Channel Tunnel (*Geogrl. J.* 139 (1973), 258–79), and the Maplin seaport to the east of Southend (*Geogrl. J.* 140 (1974), 364–72) were discussed. As it happens, none of these projects has been implemented so far though the study of all three provides fascinating reading. Some papers deal with what has happened, rather than with what might happen. Among these, P. Hall's *Containment of Urban England* shows that during the nineteenth century people and jobs were concentrated in metropolitan areas but from 1900 to 1950 there was a relative decentralization for people but a continued concentration of employment (*Geogrl. J.* 140 (1974), 368–417). From *c.* 1961, typically for the metropolitan areas, there was 'absolute decentralization of population and relative decentralization of employment'. And during the 1960s the large metropolitan areas began to experience an over-all loss of both population and jobs while a ring of peripheral metropolitan areas showed increases. Meanwhile land had become a valuable commodity, to be traded and even in some instances hoarded; indeed, there was an inflation of land and property values on a scale never previously experienced in British history. While the urban expansion had, therefore, been contained, the increasing suburbanization of the population resulted in longer journeys to work, more social segregation, and, for millions of people, a greater distance from urban facilities.

In 1971 E. C. Willatts gave a paper on 'Planning and geography in the last three decades' (*Geogrl. J.* 137 (1971), 311–38). Much of it was devoted to a review of achievements, including the national parks and the work of the Countryside Commission, and it was noted that by 1969 about two-fifths of those qualifying for membership of the Town Planning Institute had their initial degrees in geography. In the discussion that followed, P. Hall made the point that planners were also concerned with economics, psychology, and sociology, with such an enormous range of problems, in fact, that to meet them by providing courses in each would be impossible. M. Chisholm went further in suggesting that in the future interdisciplinary teams alone might be able to assess needs. In 1973 he gave the Eva Taylor lecture on 'Regional policies for the 1970s' (*Geogrl. J.* 140 (1974), 215–44) and argued:

Our regional problems are of such a kind, and exist on such a scale, that effectively they must be tackled as part of a national programme which happens to have specific regional impacts. In the second place it is necessary to think beyond the marriage of 'physical' (land use) planning and 'economic' (supply and demand management) planning, to the incorporation of social policies as well ... debate should centre around the question of how to enhance the long term growth potential of the economy instead of remaining linked to the limited notion of ameliorating structural imbalance in employment structures.

Here is a new regionalism looking beyond development areas to national economic problems, concerned with the lives of people, not only in masses but as individuals, in a world of economically advanced or advancing nations, having evolved a complex of economic and social structures and aspirations which may be uncertain in expression and brittle in a world of political change. The geographer may retreat into some fascinating glade of Academe (perhaps by writing on the history of RGS) or he may wish 'to serve the present age' (a phrase of Charles Wesley) and be led to question the whole content and trend of his subject. He may see hope in a new world based on Marxist philosophy, only in that case he is unlikely to penetrate beyond the review pages of the *Geographical Journal*, though he was considered in the review of geography from 1972 to 1976 by Cooke and Robson (see below).

By the 1970s many of the hopes of earlier geographers had been realized. In schools and colleges geography had become a major subject, with more candidates for university entrance than history or economics, or indeed any other subject than English or mathematics. University departments of geography were large, with most having 50 to 100 graduates a year; these sought, and generally found, a wide range of careers among which education and planning were prominent, though others included commerce, industry, and many forms of government service. The output of papers, monographs, and books increased and spread from avowedly geographical journals into many others. Fortunately there was some academic cross-fertilization, not least in the *Geographical Journal*, which received papers, books for review, and—most helpfully—contributions to its discussions following papers read at meetings from people who disclaimed, rather too eagerly in some cases, any pretence of being geographers. With such success complacency could easily and understandably follow, but most fortunately it did not and geographers in Britain continued to look forward, seeking new methods of research, looking again at familiar topics, never satisfied to accept what the (rarely revered) fathers had taught, conscious of a changing wor⸱ᴜ, and generally, though not always, willing to rise from an attitude of iconoclasm to one of construction. That the new academic edifice was not always equal to the expectations raised by its design was not surprising.

An opportunity for reviewing the growth of geography occurs every four years when the International Geographical Union holds its congresses. In 1964, when the congress was held in London, G. R. Crone wrote 'British Geography in the twentieth century' (*Geogrl. J.* 130 (1964), 197–220) and concluded by saying that in the previous sixty years there had been 'the development of a great organized body of geographical knowledge, the emergence of a large number of trained geographers and a recognition of the role which geography can play in education and in the wider national life'. To a considerable extent the approach had been empirical rather than theoretical and geographers had been 'prepared to tackle contemporary problems'. In Crone's view the geographer would be wise to 'insist that his field is the study of the environment in which we live today'. This may well involve fruitful co-operation with workers in other disciplines.

No survey of British geography was given at the New Delhi Congress of 1968, nor was it reported in the journal, though an active British contingent attended its various meetings. In 1972 R. W. Steel and J. W. Watson wrote a report for the 22nd Congress at Montreal (*Geogrl. J.* 138 (1972), 139–53) in which they drew attention to the increasing range of specialist studies made by geographers including research students and others whose work was financed by the Natural Environment Research Council established in 1965 and the Social Science Research Council from 1966, as well as by the universities. They noted that 'there had been a move away from regional geography in many departments' and that 'many geographers now seem more concerned with process than with place', some indeed 'more with statistical manipulation than with environmental reality'. Nevertheless the traditional concern with particular areas remained, notably at the three centres for special studies such as those for the Middle East at Durham, for South Asia at Cambridge, and for Latin America at Liverpool, all of which were directed by geographers, while geographers were associated with a number of similar centres in various universities. Traditional branches of geography, such as geomorphology, flourished, stimulated by the application of new methods partly derived from the 'quantitative revolution' while in social and economic geography new winds were blowing, welcomed by some but not by all geographers. New methods of research in urban geography were not only quantitative but also perceptive, for the changes in towns had been so great, and the problems following the rehousing of millions of people so much more severe than anticipated, that the creation of a happy urban society seemed to be as far off as ever. As working hours became shorter and leisure more plentiful, various geographers studied recreational facilities in co-operation with the Countryside Commission, the Nature Conservancy, and the Sports Council.

It would be agreeable rather than realistic to end this survey on a note of euphoric congratulation to all concerned with the successful development of geography in 150 years. Nevertheless, what geographers have done has given new visions of what they might do, and this view appears to underlie the deeply thoughtful paper of R. U. Cooke and B. T. Robson on 'Geography in the United Kingdom, 1972–76' prepared for the Moscow Congress of 1976. They see those four years as a time of consolidation and reflection 'an inevitable, if frustrating anti-climax after the heady enthusiasm and elevated hopes of the earlier period' (*Geogrl. J.* 142 (1976), 81–100). Advances have been considerable; for example, in the study of hydrographical and geomorphological systems 'theoretical models can be matched against the real world in attempts at prediction' with 'theoretical solutions to, for instance, problems of long-term land form evolution based on existing knowledge of processes'. While quantitative methods had opened new doors 'there is probably as much lack of comprehension between the first- and second-generation quantifiers as ever there was between the quantifiers and non-quantifiers of the 1960s'. With this went an increasing disillusion with 'positivism', rarely defined in geographical literature but regarded by Cooke and Robson as including the suggestion that 'it is natural science, rather than (for example) history, which must be seen as the paradigm of human knowledge'. In human geography this involves 'the search for invariate laws' and 'the belief that human actions can be characterized with regard purely to an external reality'.

There can be no such thing as a value-free judgement, a fact which underlies the search for approaches listed as normativism, phenomenology, and behaviourism. Some geographers have found satisfaction in seeking for 'relevance' and for detailed study of small areas and social groups with an emphasis on process rather than form. Deprivation has been closely studied and many have been genuinely concerned (like others long before them, only that is often forgotten) with social welfare. Old problems have not gone away, for there is 'increased public and political awareness of environmental problems', not least the prospect of diminishing natural resources with the dangers and difficulties of planning. Accepting specialization as inevitable, indeed desirable, to some a successful future for geography would appear to be in walking the corridors of power, with geomorphologists, climatologists, and biogeographers developing a new and better physical environment, economic and social geographers formulating strategy not merely in commercial activity but also in the living experience of people, all of whom will be frequently seen co-operating with governmental bodies, national and local.

Dare one say that all this has been argued before? The answer would appear to lie with individuals. Some will welcome such work, just as an earlier generation of geographers gave splendid service to education through their work for examining boards, the Geographical Association, and various other bodies. And for that matter, many years ago, A. J. Herbertson who gave such fine service to education was also a member of a commission on water supply. But there are dangers, splendidly stated by Cooke and Robson:

Applied work brings both benefits and dangers to the discipline. On the one hand it nourishes the subject with new research perspectives, improves research facilities through the funding of equipment and personnel, fosters contact with cognate disciplines, and promotes both the esteem in which individual geographers are held outside the profession and the provision of employment for its graduates. On the other hand, the pursuit of fundamental academic research of no apparent direct relevance to contemporary problems could well become starved of resources; the energies of some of the most productive researchers could become dispersed on highly empirical, routine, time-consuming exercises which have little theoretical, methodological or substantive interest and which are without innovative significance to the subject; and the important quality of independent judgement could be sacrificed at the shrine of relevance. The period 1972–76 has seen rather little new theoretical work to replenish the subject's conceptual stock. It would seem important to maintain a careful balance between the energy devoted to applied work and that devoted to the development of geographical philosophy and theory.

Much that had happened has been beyond the most sanguine hopes of academics of fifty years ago and even those of more recent times. Never before in Britain have so many people been working as geographers in education and other professions, in commerce and industry. The struggle for recognition has been won. Nevertheless, there is a ferment of questioning uncertainty about which paths to follow, a search for new methods of investigation, a wish for social commitment that in some cases is political in motive, a faith in a 'paradigm' to be accepted for a time and then replaced by another, a condemnation of much that has been done earlier and even a wish to be 'with it', 'in the swim' (or whatever phrase is current) to such an extent that the fear of being one paradigm too late becomes almost pathological. None of these attitudes is new. The world remains and the earth abides, always changing through human activity, though to a greater extent than ever before in history, environment as experienced is urban, perhaps directly so or modified into some suburban or even rururban way of life. It is hardly remarkable that there should be a greater concern than ever before with quantification, human behaviour, perception, community segregation, deprivation, and much more that is part of living experience. Social psychology, physical and mental health, educational opportunities, and other studies concerned with welfare attract geographers eager to build a better world or at least to show the need to do so. Conservation has become an emotive issue and the provisions of adequate recreational facilities for all the people a popular cause. Now as in the past the geographer is concerned with landscape and life. What light he can shed on the problem of men and the earth in the next fifty years nobody can foresee, but the ferment of ideas is itself a sign of promise though not all of them are as new as their advocates suppose and not all of them will prove as productive as they hope.

# REFERENCES

## Primary Sources

RGS Archives

a. 3 March 1891. H. J. Mackinder to J. S. Keltie.
b. 6 August 1935. H. J. Mackinder to H. R. Mill.
d. 23 October 1915. D. G. Hogarth to J. S. Keltie.
e. Preliminary meeting, RGS sub-committee on Human Geography, 9 December 1929, pencil notes apparently by A. R. Hinks.
f. Various letters from A. R. Hinks to T. G. Taylor.
g. 8 February 1920. A. G. Ogilvie to A. R. Hinks.
h. 16 October 1929. A. R. Hinks to C. B. Fawcett.
i. As e (above), 10 February 1930, first formal meeting of the sub-committee on Human Geography.

## Secondary Sources

BAKER, J. N. L. 1948. 'Mary Somerville and geography in England', *Geogrl. J.* 111/4–6: 207–22.

CANTOR, L. M. 1962. 'The Royal Geographical Society and the projected London Institute of Geography', *Geogrl. J.* 128/1: 30–5.

CRONE, G. R. 1955. *Royal Geographical Society: a record 1931–55*. RGS: John Murray.

DUNBAR, G. S. 1978. *Elisée Reclus: historian of nature*. Archon Books, Hamden Conn.

FAWCETT, C. B. 1917. 'Edale: a Pennine dale', *Scott. Geogrl. Mag.* 33: 12–25.

FLEURE, H. J. 1919. 'Human Regions', *Scott. Geogrl. Mag.* 35: 94–105.

FLEURE, H. J. 1943. 'The development of geography', *Geography*, 28/3: 69–77.

GILBERT, E. W. 1971. 'The RGS and geographical education' *Geogrl. J.* 137/2: 200–2.

GOUDIE, A. S. 1978. 'Colonel Julian Jackson and his contribution to geography', *Geogrl. J.* 144/2: 264–70.

HOSKINS, W. G., 1959. *Local History in England*, London, Longmans.

HOWARTH, O. J. R. 1951. 'The Centenary of Section E (Geography)', *Adv. Sci.* 8. 30: 151–65.

JUKES, J. B. 1962. 'On the formation of some river valleys in the south of Ireland', *Q. J. Geol. Soc.* 18: 378–403.

LEIGHLEY, J. 1977. 'Matthew Fontaine Maury 1806–1873', *Geographers: Biobibliographical Studies* (ed. T. W. Freeman *et al.*), 1: 59–63.

MILL, H. R. 1930. *The Record of the Royal Geographical Society*, RGS.

OUGHTON, M. 1978. 'Mary Somerville 1780–1872', *Geographers*, 2: 109–11.

SCARGILL, D. I. 1976. 'The RGS and the foundations of geography at Oxford', *Geogrl. J.* 142/3: 438–61.

SMITH, W. G. and MOSS, C. E. 1903. 'Geographical distribution of vegetation in Yorkshire. Part I Leeds and Halifax district', *Geogrl. J.* 21: 375–401.

SMITH, W. G. and RANKIN, W. M. 1903. 'Geographical distribution of vegetation in Yorkshire. Part II, Harrogate and Skipton district', *Geogrl. J.* 22: 149–78.

STEEL, R. W. 1961. 'A review of I.B.G. publications, 1946–60', *Transactions and Papers, Institute of British Geographers, Publication no. 29*, 129–47.

STODDART, D. R., 1975. 'The RGS and the foundations of geography at Cambridge', *Geogrl. J.* 141/2: 216–39.

WARD, R. G. 1960. 'Captain Alexander Maconochie, R.N., K.H., 1787–1860', *Geogrl. J.* 126/4: 459–68.

# 2 Geography in Education

N. J. GRAVES  *Institute of Education University of London*

## INTRODUCTION

THE study of geography within the education systems of the United Kingdom is, in historical terms, relatively recent. Although geographical textbooks may be found dating back to the classical period, as an elementary school subject geography was not widespread until the mid-nineteenth century; in secondary schools it was not a popular subject until the 1930s and as a university subject its rapid development dates from the years following World War II. In this chapter I shall examine the development of geographical education historically, attempt an assessment of its present state, indicate the nature of current research in the field, and hazard a guess as to its future prospects.

## THE HISTORICAL DEVELOPMENT OF GEOGRAPHY IN EDUCATION

If the terms geography and education are used in a wide sense to mean any information about the world and the process of initiating the younger generation into such knowledge, then clearly a number of individuals did receive a form of geographical education at the hands of private tutors in the years which followed the Dark Ages, but evidence is limited. What books were available on education, seldom referred to geography, an exception being *The Governor* by Sir Thomas Elyot (1531) which saw geography as useful in the understanding of the events of ancient history. This has to be seen in the context of the times in which England, and indeed the whole of Western Europe, were trying to catch up with the knowledge which had been available to the scholars of the ancient world. To a large extent this meant that the curriculum of the 'grammar' schools in existence was a classical one: the trivium (grammar, rhetoric, and dialetic) and the quadrivium (arithmetic, geometry, music, and astronomy).

According to Robinson (1951) it was not until the seventeenth century that some systematic attempt at teaching geography in schools could be found, and even then in a limited number of the dissenting academies. These were the schools set up to cater for the sons of the wealthier nonconformist families. Having been established recently and not being restricted by articles of association in which the curriculum had been specified by the founders in years gone by (the case of many old grammar schools), they were able to develop a relatively modern curriculum in which the natural sciences, modern languages, and some geography featured. As the ancient universities were at that time in decline, these dissenting academies were often the only educational institutions where studies were of a relatively high level. Geography, where it was taught, was of both the description and the scientific type, though in the latter case this meant understanding the concepts of latitude and longitude, their significance as well as how they were derived. The dissenting academies were not unfortunately to last beyond the early years of the nineteenth century, as in the period which followed the French Revolution of 1789 they were suspected of having revolutionary sympathies and were consequently closed. The only other institution which might be said to have dispensed some geographical knowledge was the University of Cambridge, where Newton is known to have edited Latin versions of Varenius' *Geographia Generalis* in 1672 and 1681 for the use of his students (Dickinson 1969). But not too much should be made of this fact, since Newton's interests clearly lay mainly in the fields of physics, mathematics, and astronomy.

Elementary education in England began to take off in the first half of the nineteenth century when the British and Foreign School Society and the rival National Society for Promoting the Education of the Poor in the Principles of the Established Church were given state grants to help them in their gigantic task of educating the masses (Lester Smith 1965). Although the provision of schools was inadequate and the curriculum limited, geography was one of the subjects which acquired a place in that curriculum. It is difficult to be certain of the rationale behind this development since little has been published on the curriculum of that period, but the Normal Schools (teacher training colleges), set up by the various bodies responsible for elementary education, provided geography in their curriculum; it may be therefore inferred that teachers who had acquired a knowledge of the subject were keen to teach it. Further, in 1858 a set of regulations setting standards for elementary school instruction were published in which geography was indicated as compulsory for pupil-teachers (Gordon and Lawton 1978). In 1867, under the 'Revised Code' which set out conditions under which grants were to be paid to schools, extra grants were offered if elementary schools included in their curriculum one 'specific' subject such as history, geography, and grammar, over and above the basic skills of literacy and numeracy. The Cross Commission (1886–8), which was inquiring into the workings of the Elementary Education Acts of England and Wales, agreed on the broad curriculum for elementary schools and this included geography 'specially of the British Empire'. One begins to see as the nineteenth century draws on, the impact of economic and political change. It seems as though no one officially concerned with elementary education seriously doubted the value of geography as a subject of the curriculum. Indeed, in the early years of the twentieth century, the 1905 *Handbook of Suggestions for the Consideration of Teachers and others concerned with the Work of Public Elementary Schools*, devotes a whole chapter to the teaching of geography. By 1931, the report of the Consultative Committee on the Primary School, though mentioning history and geography as desirable elements in the curriculum, did not insist that these should necessarily be taught in separate lessons. The influence of the 'progressive' education movement was tending to loosen the hitherto fairly strong structure of the elementary school curriculum. By 1959, the *Handbook of Suggestions for the Consideration of Teachers* addressed itself to Primary School teachers and though it contained a chapter on 'Geography and natural history', a note of interpretation on the curriculum chapters states: 'The chapters that follow deal with learning under the headings that have become familiar in the schools. These headings are not intended as items for a timetable, but are chosen as a convenient way of considering what might be studied in the primary school' (cited in Gordon and Lawton 1978). In the Plowden Report (1967), the argument against subject teaching is put much more strongly: 'Rigid division of the curriculum into subjects tends to interrupt children's trains of thoughts and of interest and to hinder them from realizing the common elements in problem solving. These are among the many reasons why some work, at least should cut across subject divisions at all stages in the primary school.' Teachers seemed to take this advice to heart, for when Her Majesty's Inspectors reported on the state of geography in schools in 1974 they were unable to find much geography being taught in primary schools (DES 1974), and a more recent general survey of primary schools indicated that the amount of work done in geography was limited and superficial and that 'there were substantial numbers of classes where no use was made of atlases, maps or globes' (DES 1978). It therefore seems that, starting from a position in which it was a valued part of the elementary school curriculum, geography in the post-1944 Education Act primary schools has lost its place as one of the basic curriculum subjects. This statement, of course, ignores the changing nature of geography in schools and the fact that in some primary schools excellent work is being done by teachers whose love of the subject is communicated to their pupils (Catling 1979, Jex 1979). It also ignores the fact that much useful work goes on under the guise of environmental studies, or some such other all-encompassing label. Nevertheless, it is probably true that today the influence of the 'progressive' movement in primary education has been to limit the teaching of those geographical skills and understandings which are so vital to the child growing up in a complex urbanized environment. Yet as Cracknell (1979) points out, Dewey, the educationist who is thought to have been one of the driving forces of the progressive movement, thought highly of geography as a means of developing young minds.

The growth and development of geography in secondary education is somewhat different. If secondary education is interpreted to mean education beyond that given in elementary schools, then in the nineteenth century such an education was given in the so-called 'public schools' and in the endowed grammar schools. But as the Keltie (1886) report indicated, the state of geographical education in England was not an enviable one. Either the subject was taught as a quantity of factual information to be learnt by heart or, particularly in the more prestigious schools, it was not taught at all. It was not considered a discipline for the mind, neither was it particularly useful as a means of getting students into university. The story is well known of how the Royal Geographical Society was instrumental in getting Halford Mackinder appointed to a readership at Oxford University in 1887 so that geographical education could be given some impetus at the university level (Gilbert 1972), with the object of training cadres who might in due course become the geography teachers in secondary schools. Cambridge University was also to be involved in the development of geographical education with the help of the Royal Geographical Society, though, as elsewhere, honours degrees in geography were not to be generally available until the 1920s (Stoddart 1975). With the gradual diffusion of the then 'new geography' and its man-land relationship approach, it began to be taken up in public and grammar schools, and the founding of the Geographical Association in 1893 is testimony to the developing interest in the subject among secondary school teachers. The 1902 Education Act, and its encouragement to local education authorities to create and develop secondary schools in the state sector, made possible the teaching of geography in a much larger number of secondary schools, and therefore to a greater number of pupils. By 1918 not only was geography widely taught in local authority secondary schools, but it was recognized by the Board of Education as a suitable grant-earning subject for an advanced course, that is a course beyond the School Certificate level.

The inter-war years were years of enthusiastic development of geographical education in secondary schools. The first honours geography graduates were becoming available to the schools. In university departments of education, men like James Fairgrieve of the University of London Institute of Education, were training young graduates to teach geography in an enlightened way, emphasizing an inductive approach, the use of field-work, of sample (or case) studies, of secondary sources such as maps, photographs, and the moving film (Fairgrieve 1926). The growth of geography teaching was such that by the outbreak of World War II, the supply of geography graduates was still not large enough to meet the demand from schools and many teachers of geography had been educated in other disciplines (IAAM 1939). The postwar period saw an unprecedented expansion in geographical education in secondary schools. The 1944 Education Act, with its aim of making secondary education available to all rather than to a selected few, was responsible for a large increase in the number of secondary schools, since these included grammar, technical, and modern schools. Further the school-leaving age was raised to 15 in 1947 and to 16 in 1973, so the numbers of students in schools increased, both because of the compulsion of the state and because many were staying on beyond the school-leaving age voluntarily. Since geography was by then an accepted part of the curriculum, there arose a great demand for teachers of geography which the expanding universities and colleges of education endeavoured to cater for.

The evidence for this expansion may be found in the rising number of entries for GCE and, after 1965, for CSE examinations in geography. In 1951 there were 66 448 O-level and 6 445 A-level candidates for GCE geography examinations; in 1976 (the last date for which figues are available) the figures were 188 765 and 37 064 respectively. In the CSE examination the Summer 1968 figure was 69 364 and in 1976 the number of candidates has risen to 166 332. Further evidence of growth may be found in the increase in the number of textbooks published in geography for the school market. The number of field centres used by schools has also grown, so have the various teaching aids available in the form of colour slides, film strips, overhead projector transparencies, moving films, work cards, and other multi-media teaching kits. But what kind of geography was being taught to secondary school students? In the post-war grammar schools, heirs to the pre-war secondary schools, geography was firmly established in a man–environment tradition, syllabuses being regionally based. The tradition was essentially founded on the Vidal de la Blache view of geography as the explanatory description of landscapes, be these rural or urban. The main divergence from this was some systematic teaching of geomorphology and climatology in the sixth form. In the lower school, systematic geography was taught incidentally as and when it arose in the regional description. In the new secondary modern schools, where, in the early days, the influence of examinations was to be minimal, attempts were made to break away from the subject-based curriculum in order to provide a more 'relevant' curriculum for the young people for whom secondary education was a new experience. In practice, this often meant the combining of history, geography, and civics into a combined subject labelled Social Studies. For various reasons which it is inappropriate to go into here, but in which the Royal Geographical Society played a role (Graves 1968), social studies did not succeed in establishing itself in the curriculum of secondary schools and geography flourished as in the grammar schools. In time, many O-level and later CSE courses in geography were established in the secondary modern schools.

With the publication of *Frontiers in Geographical Teaching* (Chorley and Haggett 1965) in which the so-called conceptual revolution in geography was discussed with secondary school teachers in mind, a wind of change began to blow through school geography departments. The paradigm shift which had affected university geography departments in respect to human geography in the 1950s in the USA and in the 1960s in the UK, began to have an effect on schools, largely as a result of the dissatisfaction that young graduates felt about the kind of geography then being taught in schools. Though much was written about this conceptual revolution, inevitably most teachers were ill prepared to accept, let alone understand, the nature of the changes which were proposed. Little by little, ideas began to diffuse, new textbooks were published, courses were organized by the Department of Education and Science and universities. Some examination boards, like the Oxford and Cambridge Board in 1969, modified the syllabuses for the A-level Geography examination. Subsequently other boards like the Joint Matriculation Board and the University of London Entrance and School Examinations Council followed suit and modified their A- and O-level syllabuses, which forced teachers to change the kind of geography they taught. Greater emphasis was placed on the concepts, principles, and skills of geography and less on pure information. This was a period of traumatic change for some of the older and more experienced teachers who had been trained at a time regional geography was looked upon as the heart of geographical study.

## THE CURRENT STATE OF GEOGRAPHICAL EDUCATION

As already indicated, the teaching of geography as a separate subject in the primary schools of England and Wales is the exception rather than rule. It is possible to detect a reaction to this both within the Geographical Association and within that body of Her Majesty's Inspectors responsible for geography in schools. It is difficult at present to be sure of the trend. Children in primary school are in the main taught by class teachers who are not specialists. These have a limited knowledge of geography and, unless they are in some way directed to provide learning experience for children which include geographical skills and concepts, many will not do so. Consequently how the present trend can be reversed without some energetic action is not clear. The Geographical Association has a vigorous Primary Section under the leadership of R. Barker and many of the members are geographers of some distinction. A new handbook on the teaching of geography in primary schools is about to be published. Concern about environmental education in society as a whole is likely to have some effect on primary school teachers. In general there are some grounds for hoping that the position of geography in primary schools may improve.

In secondary education the position is much brighter. Geography has maintained its place as one of the main subjects in the curriculum. Table 2.1 illustrates how it fares in the examination league compared with other subjects on the curriculum.

TABLE 2.1

*Main School Subjects*

*Examination Entries 1976 (Summer Examination only)*

(R = Rank)

|  | GCE A Level | R | GCE O Level | R | CSE | R |
|---|---|---|---|---|---|---|
| Mathematics (all) | 69 895 | 1 | 270 297 | 2 | 377 731 | 2 |
| English | 65 985 | 2 | 452 179[a] | 1 | 471 525 | 1 |
| History | 37 891 | 4 | 149 242 | 6 | 144 486 | 6 |
| French | 24 111 | 10 | 152 459 | 5 | 115 989 | 7 |
| Art | 24 829 | 9 | 112 422 | 8 | 148 820 | 5 |
| Physics | 41 803 | 3 | 137 929 | 7 | 98 276 | 9 |
| Chemistry | 34 558 | 7 | 112 221 | 9 | 63 427 | 10 |
| Biology | 32 066 | 8 | 209 559 | 3 | 153 583 | 4 |
| Economics | 35 451 | 6 | 42 162 | 10 | 109 888[b] | 8 |
| Geography | 37 004 | 5 | 188 765 | 4 | 166 332 | 3 |

[a] Language only at O-level.                [b] Economics and Social Studies.

*Source:* Statistics of Education, vol. 2, 1976.

This relatively favoured position in the curriculum has not been maintained without some struggle. First there have been the efforts made by hundreds of individual teachers to keep themselves up to date with current trends in geography. This has not been easy partly because of the rapidity of change and partly because the early emphasis on quantitative techniques in scientific geography found many teachers ill prepared to assimilate these techniques. Most had to re-orient their thinking from an idiographic to a nomothetic approach and to undergo training in elementary statistical techniques so that correlation coefficients, chi-square tests, and regression analysis became terms in a language they could understand. Secondly, with the advent of the Schools council for the Curriculum and Examinations in 1964, there had been a considerable investment in curriculum development projects in geography. In 1970 two nation-wide curriculum development projects were launched: the Geography for the Young School Leaver Project (GYSL) and the Geography 14–18 Project.

These two projects were meant to help teachers with developing the geography curriculum for two different populations of students. GYSL was launched at a time when the school-leaving age was about to be raised from 15 to 16, and, as its name implies, it was aimed to provide teachers with worthwhile curricula and curriculum materials in geography for students who would be staying involuntarily for another year at school. Geography 14–18, on the other hand, was aimed at teachers of students of average or above average ability who were looking for a more intellectually challenging geography curriculum in the age range from 14 to 18. The latter project was given time to think out its strategy and the view of geography it was to espouse, whilst the former was under pressure to produce both curricula out-lines and materials that teachers could use in the near future. Thus GYSL fairly rapidly produced the outline of a course on aspects of human geo-graphy with three main themes: Man, land, and leisure, Cities and people, and People, place, and work; these were published eventually by Nelson which provided packs of curriculum materials consisting of information sheets and work cards, maps, pictures, film strips, and audio-tapes. Judged by the number of packs sold, the take-up of the GYSL project among teachers of geography was very high. Credit for the wide dissemination of the project must be shared between the Schools Council, local education authority advisers, the publishers, and the curriculum development team (Tom Dalton, Rex Beddis, Trevor Higginbottom, and Pamela Bowen). The team estab-lished good contacts with teachers and responded to what teachers perceived to be their immediate needs.

The Geography 14–18 team (Gladys Hickman, John Reynolds, Harry Tol-ley) looked at curriculum development as a process of social change and argued that no deep curriculum change could occur without teachers chang-ing their attitude to the curriculum. Thus their first publication *A new profes-sionalism for a changing geography* (Hickman *et al*. 1973) emphasized the need for teachers to discuss among themselves the nature of curriculum reforms so that changes would come from personal conviction about the desirability and feasibility of such changes. In this process of altering the 'teaching–learn-ing system', the curriculum development team saw their role as that of 'change-agents', that is, as facilitators who would provide stimulus and help, but not as experts handing down ready made curricula from on high. The hope was to engender a system of curriculum development in geography that would be self-activating and would not run down once the curriculum de-velopment team was withdrawn. This more radical approach was not always understood by teachers in the field, or indeed by members of the Schools Council's Geography Subject Committee, with the result that the curriculum development process got off to a slow start. The team recognized that as long as public examinations remained outside the system, the chances of change occurring were going to be limited. Consequently, negotiations went on with the Cambridge University Local Examinations Syndicate which led to the establishment of an O-level examination in geography which was very different from previous examinations in that not only were course work and a special study allowed to count for 50 per cent of the total marks, but the questions in the examination paper were essentially data-response in which the appeal to memorized information was limited. With the establishment of this O-level examination, and the publication by Macmillan of a handbook for teachers (Tolley and Reynolds 1977) and of exemplar curriculum materials, the project began to have a greater influence on schools. It is estimated that over one hundred schools are currently (1979) working with the project and that many more are influenced by it.

Although the Geography 14–18 project implies that it is concerned with an age group spanning the years 14–18, in practice, owing to the short life span of the project (1970–5), and to the problems that it encountered, it was not able to consider to any depth, the geography curriculum for the 16–18 age group. Consequently, in 1976 a new project, based on the University of London Institute of Education, was set up by the Schools Council. Known as the Geography 16–19 Project and led by Michael Naish (with Ashley Kent and Eleanor Rawling assisting) it began investigating the needs of sixth-formers and what geography could bring to them of educational value. It was concerned with the whole sixth-form population and not just that section which was preparing for the A-level examination. The project originally financed for three years has been extended for another three years, though the composition of project team will change in the last three years. After a preliminary investigation of the present state of sixth-form geography, which included a survey of teacher and student opinions, the team set about their task with vigour. They had benefited from the experience of previous projects and adopted the 'Geography 14–18' view that teachers should become curriculum developers. However, they realized that much support would be needed by teachers in this enterprise and the team provide a firm framework within which curriculum development could take place. This framework, discussed with teachers, consists of a set of aims and objectives for sixth-form geography, a man–environment view of geography, four main themes around which the curriculum is to be articulated ((1) The natural environment—the challenge for man, (2) Use and mis-use of natural resources, (3) Man–environment issues of global concern, (4) Managing man-made environments and systems), and a suggestion that curriculum units should be prepared at three scales: the local, the regional, and the global.

An exemplar curriculum unit was produced by the team and subsequently about sixteen units were produced by groups of teachers in the schools associated with the project. These are being evaluated by teachers and students using forms provided by the project team. An A-level examination based on the project's view of geography in the sixth-form course has been negotiated with the University of London School Examinations Council; the first candidates are expected to sit for it in 1982. Meanwhile the project team is further investigating the examination needs of those not pursuing A-level studies and evaluating the possiblity of a Certificate of Extended Education (CEE) examination in geography based on project ideas.

Among projects catering for the lower age group was the History, Geography, and Social Science Project 8–13 directed by Alan Blyth at Liverpool University, from 1971 to 1975. As its name implies, this project was concerned with a group of subjects, which could be called the Social Studies group, for the middle years of schooling. Given the then current vogue to 'integrate' subjects in the lower secondary school, the project team chose to consider the interrelations between the three subjects rather than their integration. They argued that academic disciplines, like geography, provided resources to be used and that any topic studied in the 8–13 age range was likely to make calls on more than one discipline. Precisely how the teacher articulated the relationships between disciplines in any one curriculum unit would depend on the age and intellectual maturity of the pupils; the older they were, the more sense did a subject-based course make. It also depended on the circumstances of the school, including its staffing. A list of 'key concepts' was suggested by the project team as a basis around which the curriculum units might be structured. These were: (1) Communication, (2) Power, (3) Values and Beliefs, (4) Conflict/Consensus, (5) Similarity/Difference, (6) Continuity/Change, (7) Causes and Consequences (Blyth *et al.* 1976). Examples of units developed by the team and teachers in associated schools were published, including supporting materials such as audio-tapes and film strips.

What picture then emerges of the state of geographical education in British secondary schools? Firstly, there is no truly typical situation. In England and Wales most (80%) secondary schools in the state sector are today (1979) comprehensive schools, that is, they do not select their intake on the basis of ability. But these schools may be arranged in a middle school (8–13, 9–14), Senior High School (13–18, 13–16, 14–18, 14–16), and Sixth Form College (16–19) hierarchy. Thus geography departments in secondary schools may be small or large depending on whether the school is for a limited age range or is an all-through comprehensive from 11 to 18 years. Secondly, the curriculum is increasingly being organized in such a way that up to the age of 13 or 14 pupils are exposed to a broad range of social studies subjects under such a heading as 'Humanities' which includes geography, though the subjects associated with it may vary considerably from history (usually present) to English or even Religious Education. This has the same dangers as the undifferentiated curriculum in the primary school in that, if no teacher stresses the geographical element in the curriculum, it may go by default. Thirdly, geography for the 14 to 16 age group is often an optional subject, though it is taken by a high percentage of students. It is in that age group that students need to choose between following a CSE or O-level course, though that choice may sometimes be delayed until the end of the fourth year, when the students are about to embark on their final year of compulsory schooling. Because of the difficulties that this choice causes, a decision has been taken to merge the CSE and O-level examinations into one common system. Fourthly, for those students who stay on in the sixth form, geography may be taken by those embarking on a three-subject A-level course, or by those staying for only one year. For the latter a variety of courses exist, but teachers are not satisfied with the present options of preparing candidates for CSE, O-level, and the experimental CEE examination. It is hoped that the Geography 16–19 curriculum development project may suggest a possible solution to this problem.

None of this, however, necessarily conveys to the reader the dynamic state of geographical education in Britain today. The study of geography in British Schools is a much more intellectually demanding task both for the student and the teacher than it was twenty years ago. It is also relevant to the problems and concerns of our society. It examines some of the issues which are reflected in the mass media: the planning of road networks, the location of London's third airport, the distribution of industry in the UK, the problem regions of the EEC, the development of new towns, urban renewal, urban structure and the distribution of wealth in urban areas, the nature of development in the third world, the changing pattern of trade relationships, the nature of environmental quality and the factors affecting this. In relation to the last topic, geography in schools is still concerned with what has traditionally been called physical geography (geomorphology, hydrology, and climatology) and the search for the relationship between physical processes and human actions on the environment is one which still excites a number of students. Further, the range of teaching strategies open to geography teachers is now greater than it was twenty years ago. This is not simply because educational technology has made great strides, but also because the very concept of teaching and learning has been widened. Thus, though the picture of a teacher interacting with a class of thirty students using the textbook and the blackboard may still be valid, an observer may frequently meet five groups of six pupils working on their own using assignments provided by the teacher and resources from a central resource centre in the school or school geography department. He may find a class engaged in a game whose objective is to develop an understanding of the chance and other factors influencing a farmer's decisions to plant or not to plant certain crops. He may more and more frequently find groups of students using a computer in an interactive way, so that they may learn, for example, the implications and limitations of a particular explanatory model of industrial location such as Weber's. He may find students concerned with the processes of arriving at a decision concerning the location of a reservoir for a town's water supply. If he follows a class out on field-work he may find them measuring the angle of slope of a scarp at various points down the slope, measuring the depth and texture of the soil at these points, using a quadrant to sample the vegetation, and eventually attempting to see whether there is any relationship between the variables. If the teacher wishes to learn about the various teaching strategies that he may adopt, he can turn to the periodical *Teaching Geography* published jointly by the Geographical Association and Longman, or to *Classroom Geographer* published privately by the Brighton Polytechnic. If he wants to keep up to date with developments in geography he can consult *Geography*, the journal of the Geographical Association, and the *Geographical Magazine* which has many topical and well-illustrated articles and in higher education there is a new publication, the *Journal of Geography in Higher Education*. Thus teaching and learning geography continues to be an exciting activity.

## RESEARCH IN GEOGRAPHICAL EDUCATION

The term research is a slightly ambiguous one. It seems to connote an activity leading to the discovery of new information, new ideas, new techniques, previously unsuspected relationships, and the development of new models and practices. Yet we know that this concept of pure discovery and creation seldom corresponds to reality. Most research is a painstaking process of criticizing, analysing, rethinking, extending, and developing what has gone on before. This is particularly true in an applied field like education and seldom can a research worker cry out 'Eureka!' Yet in geographical education, there is a solid record of achievement, albeit of an unspectacular nature. As with all applied research, the nature of the activity tends to change over time with what is perceived to be problematic by the researcher. Reviews of such research have appeared from time to time in *Geography* and in the *Teaching Geography Series* (Scarfe 1949, Jay 1960, Long 1964, Naish 1972).

It should be stated at the outset, that, apart from the publicly financed curriculum development projects, all research in geographical education has been undertaken by individuals or groups, so that most of the research is small scale. It is undertaken as part of higher degree research in British universities, or as part of a programme of research and development under the aegis of the International Geographical Union's Commission on Geographical Education. Thus access to the research is not always easy, though various recent publications have helped to disseminate its findings (Boden 1976, Graves 1972, 1975, 1979, Marsden 1976, Salmon and Masterton 1974, Stoltman 1976).

Broadly speaking, research has been concerned with the following areas:

1. Objectives in geographical education, that is with the philosophical issues involved in sorting out long-term aims from short-term objectives, intrinsic from extrinsic aims, and the use of objectives in curriculum planning.
2. The nature of the curriculum process and ways of devising models of curriculum planning suitable for geographical education; course construction and the structuring of content.
3. The evaluation of geographical education, involving both curriculum evaluation and, more narrowly, the evaluation of student learning through new forms of tests and examinations.
4. The understanding of the cognitive and perceptual difficulties faced by pupils and students when learning geography. This is a wide field and includes studies of the relationship between the problem-solving activities of pupils at various ages and their stage of mental development as indicated by the Piagetian model, the understanding of spatial concepts, the perception and cognition of map information and of photographic information including aerial and satellite imagery, the environmental perception of pupils and students. This is probably the area in which most experimental research is going on. It is linked to research in language development.
5. The testing of the effectiveness of teaching strategies, such as the use of games and simulation, in putting over certain social skills and ideas. Although a rich field for experimentation, this is a very difficult one to research because of the great number of variables involved. Thus, though traditional research designs have been applied to such problems, they have seldom yielded results on which confident predictions could be made.
6. The analysis of the values and attitudes implicit in geography curricula and the effect that these have on the learner. At one time, the problem was to determine whether geography could affect children's attitudes to

other ethnic groups. Now the net is cast wider and the problems are those of the range of attitudes and values which may subtly influence children's and adolescents' own values and behaviour in the community. This research is linked to that within the field of multi-cultural education.

7. The history of geographical education and the personalities and events which have influenced the development of this aspect of subject teaching.

This is not the place to go into greater detail, but perhaps there is enough to indicate the variety of activities that are going on in this field. The meeting of the Commission on Geographical Education which is to take place in Japan in 1980 will receive reports of progress in research projects on such topics as children's perceptions of other countries, the comparative knowledge of children in different continents of basic locational information, the development of a multi-lingual glossary of terms used in geographical education, the influence of eminent geographers on geography in schools, to cite but a few.

### CONCLUSION

From being a subject of limited value taught in the elementary schools of the nineteenth century, geography is now recognized as a subject of considerable educational worth. It is taught widely in secondary schools, in further and in higher education. It is seen as a means of developing young minds to cope successfully with personal problems of orientation and spatial cognition. But it also helps pupils and students to understand the spatial aspects of societal problems and to develop a set of values about environmental issues. In so far as relevance to present-day society is a criterion of educational value, there is little doubt that geography can be made relevant to the students learning it. What, however, are the portents for the future?

It is doubtful whether the unprecedented expansion which geography in education has achieved at most levels of the educational system in the post-war period, will continue. This is not a function only of the restricted economic growth which is now with us. It is also a function of the knowledge explosion and the need to accommodate more and more subjects in the curriculum. Thus competition for room has led to a rethinking of what the curriculum is about. It is clear that the use of the discipline of geography as a resource in curriculum planning and development does not mean teaching the whole of geography, neither does it mean that geography courses can always claim the time that has in the past been accorded to them. Nevertheless, this is an opportunity to trim geography curricula so that these become more effective and make a greater impact. Teachers will need to be continually adjusting their courses, for since geography as a discipline is evolving and changing, it is necessary to reappraise the significance of these changes in terms of educational aims and values.

# REFERENCES

BLYTH, A. *et al.* 1976.   Time, place and society 8–13 in *Curriculum planning in history geography and social science*. London.

BODEN, P. 1976.   *Developments in geography teaching*. London.

CATLING, S. 1979.   Geography in primary school III Areas of experience, ILEA *Geography Bulletin*, No. 6.

CHORLEY, R. J. and HAGGETT, P. 1965. *Frontiers in geographical teaching*. London.

CRACKNELL, J. R. 1979.   Putting geography back in the primary school curriculum—Wasn't John Dewey right? *Teaching Geography*, 4.3.

Department of Education and Science 1974.   *Geography in a changing curriculum*. HMSO.

Department of Education and Science 1978.   *Primary Education in England*. HMSO.

DICKINSON, R. E. 1969. *The makers of modern geography*. London.

ELYOT, T. 1531. *The Governor*, cited in Watson, F. 1909, *The beginning of the teaching of modern subjects in England*. London.

FAIRGRIEVE, J. 1926.   *Geography in School*. London.

GILBERT, 1972.   *British pioneers in geography*. London.

GORDON, P. and LAWTON, D. 1978.   *Curriculum change in the nineteenth and twentieth centuries*. London.

GRAVES, N. J. 1968.   Geography, social science and interdisciplinary enquiry. *Geogl. J.* 134.3.

GRAVES, N. J. (ed.) 1972.   *New movements in the study and teaching of geography*. London.

GRAVES, N. J. 1975.   *Geography in Education*. London.

GRAVES, N. J. 1979.   *Curriculum planning in geography*. London.

HICKMAN, G. *et al.* 1973.   *A new professionalism for a changing geography*. Schools Council, London.

Incorporated Association of Assistant Masters in Secondary Schools 1939.   *Memorandum on the teaching of geography*. London.

JAY, L. J. 1960.   Experimental work in school geography. *Geography*, 45.

JEX, S. 1979.   Urban field studies in a primary school, *Teaching Geography*, 4.4.

KELTIE, J. S. 1886.   Report of the council of the R.G.S. Supplementary papers R.G.S., I.4.

LESTER SMITH, W. O. 1965.   *Government of education*. London.

LONG, I. L. M. 1964. The teaching of geography: a review of British research and investigations. *Geography*, 49.

MARSDEN, W. E. 1976.   *Evaluating the geography curriculum*. London.

NAISH, M. C. 1972.   Some aspects of the study and teaching of geography in Britain: a review of recent British research. *Teaching Geography*, No. 18, Geographical Association, Sheffield.

PLOWDEN, B. 1967. *Report of the central advisory committee for education (England) Children and their primary schools*. HMSO.

ROBINSON, H. 1951.   Geography in the dissenting academies. *Geography*, 36.

SALMON, R. B. and MASTERTON, T. H. 1974.   *The principles of objective testing in geography*. London.

SCARFE, N. V. 1949.   The teaching of geography in schools: a review of British research. *Geography*, 34.

STODDART, D. R. 1975.   The RGS and the foundations of geography at Cambridge. *Geogrl. J.* 141.2.

STOLTMAN, J. P. 1976. *International research in geographical education.* Western Michigan University Department of Geography.

TOLLEY, H. and REYNOLDS, R. J. 1977 *Geography 14–18: a handbook for school based curriculum development.* London.

# 3 Climate

B. W. ATKINSON   *Queen Mary College, University of London*

IN ONE form or another the importance of climate to life on earth has probably been appreciated throughout the history of mankind. From the earliest periods of hunting and gathering, through the centuries of progressively competent methods of cultivation to the present day, human activities have to a varying degree been dependent upon the interactions of population, technology, and climate. Around the turn of the present century climatic determinism provided a methodological mainstay for geography—a mainstay that survived, in school geography at least, into the second half of the century. In geography at the tertiary level, however, climatic determinism was so vigorously attacked that its credibility was lost in the inter-war period and with it perhaps, the *raison d'être* of those few geographers of that period who were interested in the atmosphere and its effects on mankind. Consequently, over the last three or four decades good scientific assessment of the relationships between climate and human activities has been remarkably absent from geographical literature, the one subject supposedly devoted, at least in part, to the analysis of such relationships. The 'overkill' of climatic (and even the slightly more acceptable environmental) determinism meant that applications of climatology, such as those by Manley (1938), Miller (1957), Gregory (1957), and Garnett (1957) were in a small minority when published. Now, over two decades later, renewed appeals for research into both pure and applied climatology abound in the literature, but this time they largely emanate from senior members of the meteorological community (Schneider 1976, White 1978) and from an apparent multitude of high-powered national and international committees under the benign eye of the World Meteorological Organization (WMO). The reasons for this resurgence of interest are admirably summarized by the US Committee for the Global Atmosphere Research Programme (1975: 2):

> As the world's population grows and as the economic development of newer nations rises, the demand for food, water and energy will steadily increase, while our ability to meet these needs will remain subject to the vagaries of climate.... As we approach full utilization of the water, land, and air, which supply our food and receive our wastes, we are becoming increasingly dependent on the stability of the present seemingly 'normal' climate. Our vulnerability to climatic change is seen to be all the more serious when we recognize that our present climate is in fact highly *abnormal*, and that we may already be producing climatic changes as a result of our own activities.

Clearly, regardless of the methodological whims within geography throughout this century, ranging as they have from determinism to possibilism to probabilism to quantification and to a concern for 'relevance', the international meteorological community (frequently supported by both general scientific bodies and governments) is convinced of the importance of climate, and more particularly climatic change, and is consequently planning 'to do something about it'. Climatic extremes (such as the Sahel drought) have concentrated the minds of decision-makers and led to a welter of policy statements over the last fifteen years. In turn the policy statements have led to massive meteorological programmes—including mankind's largest observational scientific experiment. These programmes were initiated in the mid-1960s by an extensive feasibility study (Panel on International Meteorological Co-operation 1965) which led in turn to the World Weather Watch (WWW) (World Meteorological Organization 1971), the Global Atmospheric Research Programme (GARP) (Mason 1971), the GARP Atlantic Tropical Experiment (GATE) (Mason 1976, Parker 1976), the First GARP Global Experiment (FGGE) (Gilchrist 1979a), which ran throughout 1979, and finally to the World Climate Programme (WCP) (White 1978), which will begin in earnest in the 1980s.

Clearly, massive differences exist between the aims, abilities, and achievements of the geographer with an interest in the atmosphere, frequently scientifically alone within his university and polytechnic department, and those of the co-ordinated scientific teams employed in the projects listed above. Certainly the past has seen quite marked differences in their views of climate and climatology. In common with virtually every aspect of the environment (however defined!), the atmosphere and its study have had their share of methodological treatment from geographers—a treatment more often than not concerned with the nature of climatology. Some authors (Barry 1979, Lamb 1969) felt that climatology is a discipline in its own right, separate from both geography and meteorology, a view implicitly supported by Flohn (1977, 1979), who saw it as a geophysical science, and by Mörth (1979), who saw it as an interdisciplinary study. Others were perhaps less worried about its standing as a separate science but nevertheless felt the need to comment upon its nature and methods (Barrett 1970, Barry 1970, Court 1957, Hare 1955, 1957, 1966, Thornthwaite 1953a, b) and its relationship to geography (Crowe 1965, Howe 1968, Jones 1950, Miller 1957, 1972, Oliver 1967). But geographers were not alone in their methodological reflections on climatology, as is clear from the writings of Bugaev (1973), Durst (1951), Dzerdzeevskii (1966), Godske (1959, 1966), James (1970), Landsberg (1964), Sawyer (1962), Thom (1970), and Tweedie (1967).

To a large extent an author's view of climatology depended upon both his original training and the state of atmospheric and geographical science at the time of writing. For many years climatology was considered to be a form of book-keeping (for exposés see Hare 1955, 1957, 1966), providing a numerical record of the mean atmospheric conditions at particular places. In one form or another this approach proved attractive to many professional geographers of the inter-war period and gave rise to some of the classics of climatology (Miller 1931). Essentially this approach led to descriptive, regional, applied (in the sense of being but background for human geography), and physical climatologies that were at root statistical in nature. Such studies were usually undertaken by individuals within university departments of geography rather than by teams within government research institutions. Contemporary with these 'conventional' climatological studies were the efforts of the meteorologists whose prime concern was the analysis of weather as a prerequisite for short-period (2–3 days) forecasting. Within the last two decades meteorologists have been able to extend the time-period over which they feel they have some understanding of atmospheric behaviour from one or two days to weeks and months and consequently they have felt it worth while to concern themselves with climate. They see climate, however, not merely as a collection of statistics about the atmosphere or part thereof, but as the result of the workings of the global atmospheric circulation. Thus, in contrast to the 'geographical' approach outlined above, the meteorologist tends to produce analytical, systematic, pure, theoretical and, dynamical climatologies using physical rather than statistical methods. By analogy with geomorphology, statistical climatology provides the description of 'form' whereas dynamical climatology provides the analysis of 'process'. The ultimate *raison d'être* of both types of study is of course the potential application of any newly won understanding.

In a recent attempt to provide an all-embracing context for the study of climate and climatic change, Leith (1975) introduced the notion of what Schneider and Dickinson (1974) called the 'climate system'. Leith (1975: 137) wrote as follows:

We generally think of climate as dealing with the average behaviour of the land–ocean– atmosphere system over relatively long times and as not being concerned with the detailed daily fluctuations that we call the weather. The main problem of definition is in making precise the dividing line between weather and climate. On the basis of recent results on the predictability of the weather, we might somewhat arbitrarily set the dividing time scale at two weeks. One should not, however, define the climate in terms of two week time averages because these would retain a large and unpredictable noise component from the weather fluctuations. Instead we must try to make another division into what I shall call an internal system, say the atmosphere, characterized by relatively rapid fluctuations, which is embedded in an external system, including say the ocean, that provides relatively slowly changing external influences on the internal system. We may then define the climate in terms of averages over an imagined ensemble of internal states which is in equilibrium with the slowly changing external influences. The internal ensemble consists of a myriad atmospheric states all under the influence of the same solar radiation, etc.... We may consider ensemble averages changing in response to changing external conditions on a finite time scale and be able to speak of climate [*sic*] change.

Although Leith's definition is particularly appropriate to the mathematical modelling of climate, it is sufficiently broad to encompass all the previously presented views, including Thornthwaite's (1953a) opinion that climatology should not be limited to the atmosphere alone but must include the land surface as well. (Note that Thornthwaite ignored the oceans in this particular plea for topoclimatology.)

Throughout all this discussion of the nature of climate the one common theme, even to the numerical modellers, is that at some stage temporal averages (frequently together with other statistical measures) are calculated. As the lifetime and size of atmospheric circulations are directly related, any averaging in time will, of necessity, smooth out many small systems (see Atkinson 1970), unless the averaging is done relative to the system itself. In the remainder of this article we concentrate upon circulations revealed by taking temporal averages at several stationary points and not relative to a particular moving system. Application of this criterion of temporal averaging also of course excludes much of the vast literature on meteorological processes, frequently analysed through the medium of detailed case studies (Browning, Hill, and Pardoe 1974, Atkinson and Smithson 1974, 1978).

Present-day study of the 'climate-system' follows the scientifically traditional paths of empirical (both pure and applied) and theoretical enquiry. Of the two approaches geographers have favoured the former but their overall contribution (with some notable exceptions) has been hindered by limited manpower which itself largely results from the traditionally minor role of atmospheric study within geography (Atkinson 1975). For example, of over 100 professional geographers in the University of London, only four have a primary interest in the atmosphere. Much of the English-speaking world follows a similar pattern but in countries such as Japan and the USSR the situation is slightly better. Theoretical analyses of climate, on the other hand, are almost exclusively undertaken in university/polytechnic departments of physics, geophysics, mathematics, or meteorology or in government institutions such as the British Meteorological Office. We now consider the nature of climatological studies as undertaken by, respectively, geographers and other scientists.

Even a casual perusal of geographical journals suggests a marked absence of literature about the atmosphere. An inspection of nine English-language journals (*Geographical Journal, Geography, Trans. Instit. Brit. Geog., Annals. Assoc. Amer. Geog., Geog. Rev., Canadian Geog., N.Z. Geog., Australian Geog., Australian Geog. Studies*) over the decade 1970–79 revealed that, on average, each journal contained less than one climatological article pear year. This is due to three factors: the small number of climatologists; the fact that not all of them published; and that those who did publish tended to do so either in other journals, frequently non-geographical, or in book form. Most of the latter were textbooks or comprehensive reviews rather than research monographs. Despite this relative paucity of material, in absolute terms there is currently probably more climatological literature being produced by geographers than at any previous time. In summarizing their contributions it is convenient to subdivide the subject into the following parts: regional-, physical-, synoptic-, boundary-layer-, and applied-climatology, together with climatic change.

Although regional climatology and climatic classification are no longer as popular as in the 1930s and 1940s, Landsberg's *World survey of climatology*, appearing in twelve volumes throughout the 1970s, and Chandler and Gregory's *The climate of the British Isles* (1976) are major reminders that the former lives on. Regional climatologies also still appear in the periodical literature (Chang 1962, Wittwer 1972–4, Hastings and Turner 1965, Markham 1972, Mitchell 1976). Climatic classification received a new impetus (Burgos and Vidal 1951, Howe 1953) after the appearance of the classical papers by Penman (1948) and Thornthwaite (1948). Their elucidation of potential evapotranspiration allowed a rational, physically based approach to classification through water budgets. Yet Köppen was not forgotten in the United States (James 1966, Kramer 1963), and other ways of classification have been investigated within the last decade (McBoyle 1971, Oliver 1970). Despite these efforts, regional climatological description for its own sake no longer appears to be a major research objective.

Physical climatology is primarily concerned with the heat and water budgets, usually of the globe or continental areas. The budgets are most frequently calculated for the earth's surface rather than for the atmosphere and as such they form the major part of Thornthwaite's (1953a) 'topoclimatology'. Indeed, the Laboratory of Climatology, in Centerton, NJ, USA, initiated by Thornthwaite, has been both a major stimulus for and producer of budget studies. Elsewhere Chang (1970) and Terjung (1970) have analysed the global distribution of radiation whereas Mani and Chacko (1972) (India), Maykut and Church (1973) (Point Barrow, Alaska), Ojo (1970) (West Africa), and Greenland (1973) (New Zealand) exemplify small-scale radiation studies. Water-budget studies, again mainly on a global and continental scale, are illustrated by Ceplecha (1969–71) in Australia, Hare and Hay (1971) and Packer and Sangal (1971) in North America, Yoshino (1971) in Asia, and Tuller (1968) for the world as a whole. We should note that these few specified studies are but a small fraction of the number produced world-wide over the past thirty years.

The case for synoptic climatology has been argued at length by Barry and Perry (1973). In their comprehensive volume they reviewed all the major literature on the field first explicitly identified by Jacobs (1946). They concentrated upon data, methods (essentially classification), and applications in an attempt to promulgate synoptic climatology, which they see as a link between 'understanding of the global circulation of the atmosphere ... [and] ... our knowledge of local and regional scale phenomena' (Barry and Perry 1973: xvi).

We are also fortunate that boundary-layer climates have recently been admirably summarized by Oke (1978). Although the climates of the lowest kilometre of the atmosphere may also be, of course, at one and the same time, physical, topo-, local-, meso-, and regional climates, it is their occurrence in a boundary layer that is their most fundamental characteristic. After outlining the physical basis of boundary-layer climates Oke considered, in turn, natural and man-modified environments. The former comprises climates of non-vegetated and vegetated surfaces, topographically affected climates, and the climates of animals. The latter covers the conscious and unconscious modification of boundary-layer climate with reference to, among other topics, urban climates. Over the past two decades urban climatology has attracted many geographers, no doubt inspired by Chandler's (1965) *Climate of London*. Since the appearance of that volume, three *Technical Notes* of the World Meteorological Organization (1970) (Chandler 1976, Oke 1974), together with the important results of the massive Metropolitan Meteorological Experiment (METROMEX) (Changnon, Huff, Schickedanz, and Vogel 1977) have much increased our knowledge of atmospheric processes within urban areas. The existence of journals, such as *Boundary Layer Meteorology* and *Atmospheric Environment*, solely or largely devoted to the lowest layers of the atmosphere bears witness to the magnitude of research output in this field.

A similar situation exists in the investigation of climatic change. Despite the existence of journals such as *Quaternary Research* and *Palaeo-geography, palaeo-climatology, palaeo-ecology*, the flood of research literature has given rise to yet another, namely *Climatic Change*. The subject matter therein has been a happy hunting ground for many scientists in many disciplines, geographers being no exception. Virtually all the work by geographers has been concerned with the accurate description of climatic change on many time scales, from the seasonal anomalies such as the British summer drought of 1976, through the yearly anomalies such as the Sahel drought of the early 1970s to the longer term changes such as documented by Lamb (1972, 1977), Manley (1974), and Smith (1974). The Sahel drought in particular triggered scares of a possibly long-lived desertification process which resulted in a flurry of empirical and theoretical work throughout the 1970s. Geographers played a part in this work, notably in the review of present knowledge of deserts and desertification (Hare 1977). The literature on climatic change is now enormous, covering a wide range of disciplines. It is impossible to begin to do justice to it in this brief article and the reader is referred to the volumes by Lamb (1972, 1977), Lockwood (1979), Gribbin (1978) and Pittock, Frakes, Jenssen, Peterson, and Zillman (1978).

We noted earlier that, in the backlash against determinism, geographers had hesitated to conduct applied climatological studies. They were not alone in their hesitancy, particularly in the United Kingdom. Concern over the general lack of applied climatology prompted the Natural Environment Research Council (NERC) to inquire into present climatological research and future requirements (NERC 1976). The report strongly recommended substantial support for applied climatological studies with particular emphasis on agriculture, water, and energy supply. But even prior to these exhortations geographers had not been idle: they had both reported and researched. Mather (1974), Oliver (1973), and Smith (1975) have produced comprehensive reviews of a vast range of multi-disciplinary literature covering the relationships between climate and agriculture, water resources, transport, industry (particularly pollution), and the community in general. Within a more restricted area Jackson (1977) has analysed relationships between climate, water, and land use in the tropics. The strength of applied climatology is indicated by the fact that some facets of the science are so well established as to help support separate journals such as the *Journal of Applied Meteorology*, the *International Journal of Bioclimatology*, and *Agricultural Meteorology*.

In concluding this section on climatological work by geographers it is appropriate to mention some of those works which, for one reason or another, do not readily fit into the divisions used earlier. For example, the ecologically oriented studies by Hare (1970) and Hare and Ritchie (1972) of the boreal and tundra climates of North America, whilst having both regional and applied facets, have a spirit and purpose of their own. The same is true of Ives and Barry's (1974) compendium on arctic and alpine environments. General texts, such as the reflective piece by Crowe (1971), and the encyclopaedic volumes by Stringer (1972a, b), together with the more regionally-based treatments by Lockwood (1974), Nieuwolt (1977), and Yoshino (1975), tend to cover most aspects of climatology, except perhaps the applied.

Slightly different is Barrett's (1974) book, which is based upon a particular data source and provided the first comprehensive assessment of the contribution to climatological study of meteorological satellite data. Finally the vital role of air–sea interaction in climate and climatic change has been elaborated by Perry and Walker (1977) in their review of the diverse literature on the ocean-atmosphere system.

Clearly, geographical climatologists have produced a substantial amount of both research and review, particularly over the last two decades. Yet they form but one group of scientists interested in climate, and a brief review of the work of non-geographers follows (see also Atkinson 1978).

In contrast to most geographers, research scientists in governmental meteorological institutions tend to be full-time researchers and to work in well-equipped teams. Their central problems, as recognized by both WMO and national meteorological institutions alike, is the 'explanation' of the 'climate system'. This requires a massive international co-operative effort in observation, diagnosis, and the development of theory. Each of the three aspects of this major research programme is considered below.

Most, if not all, major national meteorological agencies have a dual role—to provide forecasts and to undertake research. The provision of forecasts requires the routine collection of meteorological data, the bulk of which, after immediate consumption in the preparation of forecasts, is archived. Over the years a truly massive amount of data has been accumulated, particularly by the older services such as the British Meteorological Office, and these data have been the basis for many subsequent empirical investigations— climatological or otherwise. Within the Meteorological Office, for example, *Climatological Memoranda* have been prepared over the last fifteen years on a wide range of topics, including traditional, descriptive regional climatology (e.g. Plant 1969). In addition climatological studies have appeared in the Meteorological Office *Geophysical Memoirs*, particularly in the last decade when the mean structure of the middle and high atmosphere was established (Ebdon 1977, Hamilton, Mason, and Bridge 1973, Moffitt and Ratcliffe 1972, Wright and Stubbs 1971). *Scientific Papers* of the Meteorological Office also occasionally contain significant climatology, as exemplified by the useful report on recent climatic changes in the northern hemisphere by Painting (1977). Similar types of study covering a wide range within climatology (e.g. Brown, McKay, and Chapman 1968, Korte and Colson 1972, Mitchell 1970, Schloemer 1971, Trenberth 1973) have been produced by other national meteorological institutions. Indeed a whole multitude of descriptive and diagnostic climatological material produced by government scientists resides in the libraries of the national agencies. The potential empirical researcher is well advised to make a thorough search of such libraries before starting any climatological project.

Despite the availability of these massive data banks within national archives, it became clear in the late 1950s that improvements in the weather forecasting would be possible only with an increase in both the quality and the quantity of observations. After the feasibility studies of the early 1960s (Panel on Met. Co-op. 1965) the World Weather Watch (WWW) was instituted in 1968. In any 24-hour period WWW collects and transmits to processing centres, standard meteorological observations from more than 9200 land stations making surface observations, nearly a thousand stations making upper-air observations, nine fixed-ocean weather ships and some 7400 merchant ships making surface observations only, and reconnaissance and commercial aircraft providing more than 3000 reports daily. The main aim of WWW is to improve operational forecasts, but it led quite naturally to the associated important research activity, the Global Atmospheric Research Programme (GARP), the aim of which is to provide a more fundamental understanding of the general circulation of the atmosphere and thus of the 'climate system' as a whole. Both WWW and GARP were possible only because of the availability of high-speed computers and satellites which opened up a new dimension in weather observational capacity.

The first major research project of GARP was the GARP Atlantic Tropical Experiment (GATE) (Mason 1976) which made extensive meteorological and oceanographical observations over a three-and-a-half-month period in 1974 with the aim of providing a clearer picture of the behaviour of tropical weather systems and their ultimate effect on global weather and climate. The success of GATE (see Shaw 1978) encouraged the planners of a much larger project—the First GARP Global Experiment (FGGE) (Anon. 1978, Gilchrist 1979a).

Thousands of scientists from virtually every country in the world will be using the most sophisticated tools such as earth satellites, instrumented aircraft, ships, balloons, free-floating ocean buoys, and gigantic high-speed computers to subject the entire atmosphere of the earth and the sea surface to the most intensive surveillance and study ever made. The experiment will last for one full year [from 1 December 1978 to 30 November 1979—BWA] . . . The purpose of this highly co-ordinated international effort is to ascertain the attainable limits of weather forecasting and to investigate the mechanisms underlying climatic change. (Anon. 1978: 225.)

FGGE was the first occasion on which a truly integrated system of satellites was used to observe the earth's atmosphere. Five geostationary satellites continuously monitored the equatorial and sub-tropical belts and a series of polar-orbiting satellites were used to determine the temperature structure of the atmosphere as well as to provide information on cloudiness and the temperature of the sea.

Closely associated with FGGE are several specialized experiments concerned with significant regional phenomena. In particular the Asian Monsoon Experiment (MONEX) gathered data over the west Arabian sea, the north of the Bay of Bengal, and the South China Sea in an attempt to increase our understanding of the monsoon. West Africa, like Asia, is subject to wide inter-annual variations in rainfall (such as occurred in the Sahel drought) and the West African Monsoon Experiment (WAMEX) is also attempting to understand the three-dimensional structure and mechanism of this particular monsoon. The Polar Experiment (POLEX) is designed to provide an improved data set in the polar regions, the major heat sinks in the global 'climate system'. These data will be very important for assessing the role of snow and ice cover in climate dynamics.

The end-products of these massive observational programmes will have two main uses: to provide the bases for further diagnostic studies; and to allow the testing of climatic models by comparing their computed climatic variability to the observed variability. The construction of a model of climate (including its variability) requires a deep theoretical understanding of atmospheric behaviour and it is the search for this understanding that comprises the main activity of the international meteorological community. R. M. White, Chairman of the US National Climate Board (1978: 817) has no doubt where priorities lie: 'the mathematical model offers the only basis for a rational appraisal of climatic change and, hence, must be a central focus for any climate research effort'. The case is made at greater length in the important report of an international study conference organized by the Joint Organizing Committee (JOC) of GARP (1975) which considered the physical basis of climate and climate modelling. The report outlines the climate modelling problem as follows (JOC–GARP 1975: 16):

The processes of the climate system may be expressed in terms of a set of dynamical and thermodynamical equations for the atmosphere, oceans and ice, along with appropriate equations of state and conservation laws for selected constituents (such as water substance, $CO_2$ and ozone in the air, and salt and trace substances in the ocean). Expressed in these equations are the various physical processes which determine the changes in temperature, velocity, density and pressure. In addition, such processes as evaporation, condensation, precipitation, radiation and the transfer of heat and momentum by advection, convection and turbulence, as well as the various chemical and biological processes relevant to climate are included. The mathematical modelling of climate is guided by these same physical principles, but in addition introduces a number of physical and numerical approximations necessitated by our limited ability both to observe the system and to compute its behaviour ... the problem of modelling climate is fundamentally one of constructing a hierarchy of models, each suited to the physical processes dominant on a particular time or space scale.

An important facet of climate modelling is the ability to assess the relative contribution of an individual physical process to the maintenance and evolution of climate. This is frequently done by testing the sensitivity of the statistics generated by a climate model to perturbations in the parameters which influence that particular physical process. This could involve changes to the external system, such as surface heating from industrial activity and changing atmospheric concentration of carbon dioxide resulting from fossil fuel combustion or to the internal system, such as sea surface temperature or cloud conditions.

It is convenient to divide climate models into two types, physical and dynamical. The former type, such as developed by Sellers (1969, 1973), is based upon the energy balance equation with the annual and monthly mean surface temperatures as the dependent variables. The great advantage of this type of model is that it largely eliminates the need to consider explicitly the general circulation and its component parts by parameterization of dynamical processes. Sellers (1973) estimated that because of this, the climate over a period of 10 000 years could be simulated in less than three hours' computation time. Despite these advantages Sellers admitted that a number of 'gross simplifications' were necessary and Robinson (1971) went so far as to call it the 'educational toy' approach to climatic modelling. It is hardly surprising therefore that the major research effort has concentrated upon dynamical models.

Dynamical models of the 'climate system' have their roots in the short-term weather forecast models and the later general circulation models, which are already meteorological stock-in-trade. Routine forecasting is now done by mathematical models in at least twelve countries (Döös 1970) and general circulation simulations (Gilchrist 1979b) are frequently undertaken in at least five countries, but notably at the British Meteorological Office (Corby, Gilchrist, and Rowntree 1977), the Geophysical Fluid Dynamics Laboratory in Princeton (Smagorinsky, Manabe, and Holloway 1965 and many subsequent papers describing later developments), the National Center for Atmospheric Research in Boulder (Kasahara and Washington 1967 and later papers), and the Department of Meteorology, University of California Los Angeles (Langlois and Kwock 1968). Initially general circulation modellers were quite happy if their results showed only the broadest agreement with reality, such as the simulation of Rossby waves, at all let alone their number, location, and intensity. In recent years, however, it has proved possible to simulate quite realistically the global climate of typical Januarys and Julys (Gates and Schlesinger 1977) and, in so doing, model a real, albeit seasonal, climatic change. The required development into 'climate models' is outlined in JOC–GARP (1975: 16–17) as follows:

Certainly the global general circulation models in which the large-scale transient disturbances are explicitly resolved (and which represent the present ultimate in weather models) are an essential ingredient of studies on climatic time scales extending at least to the order of a decade. The improvement and use of those models is a major objective of GARP, and efforts should be made to make them as complete as possible for the purposes of climatic research, including their use in a systematic assessment of statistical predictability on seasonal, annual and decadal time scales.

It is equally clear, however, that such models cannot be profitably used in direct integrations over the longer climatic time scales of the order of centuries or millenia and beyond. For this purpose the dynamics of both the atmosphere and the ocean must be dealt with in a less explicit or essentially statistical manner, and it is this need which gives rise to the so-called statistical–dynamical models. This class of models must include many of the same elements of the interacting atmosphere, ocean, cryosphere, land surface and biomass as do the general circulation models themselves.

The development of such climate models depends to a large degree upon the implementation of both the proposals of the World Climate Conference held in 1979 (Gilchrist 1979c) and the World Climate Programme itself (White 1978). The Conference called upon nations to take full advantage of our present knowledge of climate, to take steps to improve that knowledge, and, lastly, to foresee and prevent man-made changes of climate that might be adverse to the well-being of humanity. In an attempt to fulfil these requirements the World Climate Programme comprises in fact three programmes: the Climate Research Programme, the World Climate Data and Services Programme, and the World Climate Impact Studies Programme. The first of these will build upon the results of GARP and FGGE in its attempts to derive a theory of climate. The second will attempt to increase world-wide effectiveness of climatic services in much the same way that the

WWW increased the effectiveness of the world's weather services. The third programme probably requires more initiatives than the other two. White (1978: 820) himself noted that, 'little work has been done on such applied research problems. It is that part of the World Climate Programme that requires new types of joint effort between the WMO and other bodies. In this applied research effort we wish to document, measure and assess the effects of climate on society.' The last few words may bring a wry smile and a sense of *déjà vu* to many geographers, particularly those who never completely lost faith in climatic determinism. Nevertheless this Climate Impact Studies programme provides a major opportunity for geographers in the 1980s. It should not be lost.

# REFERENCES

ANON. 1978. FGGE—the global weather experiment. *Meteorological Magazine*, 107, 225–32.

ATKINSON, B. W. 1970. Meso systems in the atmosphere. *Canadian Geographer*, 14: 286–308.

ATKINSON, B. W. 1975. Contribution to discussion on 'The content and relationships of physical geography' by E. H. Brown. *Geogrl. J.* 141: 43–4.

ATKINSON, B. W. 1978. The atmosphere: recent observational and conceptual advances. *Geography*, 63: 283–300.

ATKINSON, B. W. and SMITHSON, P. A. 1974. Meso-scale circulations and rainfall patterns in an occluding depression. *Quart. J. R. Met. Soc.* 100: 3–22.

ATKINSON, B. W. and SMITHSON, P. A. 1978. Mesoscale precipitation areas in a warm frontal wave. *Mon. Weath. Rev.* 106: 211–22.

BARRETT, E. C. 1970. Rethinking climatology, an introduction to the uses of weather satellite photographic data in climatological studies. *Progress in Geog.* 2: 153–205.

BARRETT, E. C. 1974. *Climatology from satellites.* London.

BARRY, R. G. 1970. A framework for climatological research with particular reference to scale concepts. *Trans. Inst. Brit. Geogr.* 49: 61–70.

BARRY, R. G. 1979. *Climatology*—a dynamic science. *Climate Monitor*, ed. P. D. Jones. Clim. Res. Unit, UEA, 2–9.

BARRY, R. G. and PERRY, A. H. 1973. *Synoptic climatology: Methods and applications.* London.

BROWN, D. M., McKAY, G. A., and CHAPMAN, L. J. 1968. The climate of southern Ontario. *Clim. Stud.* No 5, Toronto, Met. Branch.

BROWNING, K. A., HILL, F. F., and PARDOE, C. W. 1974. Structure and mechanism of precipitation and the effect of orography in a winter time warm sector. *Quart. J. R. Met. Soc.* 100: 309–30.

BUGAEV, V. A. 1973. Dynamic climatology in the light of satellite information. *Bull. Amer. Met. Soc.* 54: 394–418.

BURGOS, J. J. and VIDAL, A. L. 1951. The climates of the Argentine Republic according to the new Thornthwaite classification. *Annals Assoc. Amer. Geog.* 237–63.

CEPLECHA, V. J. 1969–71. The distribution of the main components of the water balance in Australia. *Australian Geographer*, 11: 455–62.

CHANDLER, T. J. 1965. *The climate of London.* London.

CHANDLER, T. J. 1976. Urban climatology and its relevance to urban design. *Tech. Note No. 149*, World Meteorological Organization.

CHANDLER, T. J. and GREGORY, S. (Eds.) 1976. *The climate of the British Isles.* London.

CHANG, J.-H. 1962. Comparative climatology of the tropical western margins of northern oceans. *Annals Assoc. Amer. Geog.* 52: 221–8.

CHANG, J.-H. 1970. Global distribution of net radiation according to a new formula. *Annals Assoc. Amer. Geog.* 60: 340–51.

CHANGNON, S. A., HUFF, F. A., SCHICKEDANZ, P. T., and VOGEL, J. L. 1977. Summary of METROMEX. Vol. 1: Weather anomalies and impacts. *Bulletin 62*, State of Illinois Dept. Registration and Education, Illinois State Water Survey, Urbana.

CORBY, G. A., GILCHRIST, A., and ROWNTREE, P. R. 1977. United Kingdom Meteorological Office five-level general circulation model. In General circulation models of the atmosphere, ed. J. Chang, *Methods in Computational Physics*, 17: 67–110.

COURT, A. 1957.   Climatology: complex, dynamic and synoptic. *Annals Assoc. Amer. Geog.* 47: 125–36.

CROWE, P. R. 1965.   The geographer and the atmosphere. *Trans. Inst. Brit. Geogr.* 36: 1–20.

CROWE, P. R. 1971.   *Concepts in climatology.* London.

DÖÖS, B. R. 1970.   Numerical experimentation related to GARP. *GARP Pub. Series*, No. 6, Geneva.

DURST, C. S. 1951. Climate—the synthesis of weather. *Compendium of Meteorology*, ed. T. F. Malone, American Meteorological Society, Boston, Mass.

DZERDZEEVSKII, B. L. 1966.   Some aspects of dynamic climatology. *Tellus*, 18: 751–60.

EBDON, R. A. 1977.   Average temperatures, contour heights and winds at 30 millibars over the northern hemisphere. *Geophys. Mem.* 17: 120: Met. Office, London.

FLOHN, H. 1977.   Climatology—descriptive or physical science? *Bulletin of the World Meteorological Organization*, 19: 223–9.

FLOHN, H. 1979.   Climatology as a geophysical science. *Climate Monitor*, ed. P. D. Jones, Clim. Res. Unit, UEA. Norwich, 10–18.

GARNETT, A. 1957.   Climate, relief and atmospheric pollution in the Sheffield region. *Advancement of Science*, 13: 331–41.

GATES, W. L. and SCHLESINGER, M. E. 1977.   Numerical simulation of the January and July global climate with a two-level atmospheric model. *J. Atmos. Sci.* 34: 36–76.

GILCHRIST, A. 1979a.   The first GARP Global Experiment. *Meteorological Magazine*, 108: 129–34.

GILCHRIST, A. 1979b.   Concerning general circulation models. *Meteorological Magazine*, 108: 35–51.

GILCHRIST, A. 1979c.   The World Climate Conference. *Weather*, 34: 287–9.

GODSKE, C. L. 1959   Information, climatology and statistics. *Geografiska Annaler* 41: 85–93.

GODSKE, C. L. 1966.   A statistical approach to climatology. *Arch. Met. Geoph. Bioklim.* 14B, 269–79.

GREENLAND, D. 1973.   The surface energy budget and synoptic weather in the Chilton valley, New Zealand Southern Alps. *New Zealand Geographer*, 29: 1–15.

GREGORY, S. 1957.   Rainfall studies and water supply problems in the British Isles. *Advancement of Science*, 13: 347–51.

GRIBBIN, J. (ed.) 1978.   *Climatic Change.* London.

HAMILTON, R. A., MASON, B. D., and BRIDGE, G. C. 1973.   A climatology of the stratosphere over North-West Europe. *Geophys. Mem.* 17: 119. Met Office, London.

HARE, F. K. 1955.   Dynamic and synoptic climatology. *Annals Assoc. Amer. Geog.* 45: 152–62.

HARE, F. K. 1957.   The dynamic aspects of climatology. *Geografiska Annaler*, 39: 87–104.

HARE, F. K. 1966.   The concept of climate. *Geography*, 51: 99–110.

HARE, F. K. 1970.   The tundra climate. *Trans. R. Soc. Can.* 4th Ser., 8: 393–9.

HARE, F. K. 1977.   *Climate and desertification.* Univ. Toronto Instit. of Envir. Studies.

HARE, F. K. and HAY, J. E. 1971.   Anomalies in the large-scale annual water balance over northern North America. *Canadian Geographer*, 15: 79–94.

HARE, F. K. and RITCHIE, J. C. 1972.   The Boreal bioclimates. *Geog. Rev.* 62: 333–65.

HASTINGS, J. R. and TURNER, R. M. 1965.   Seasonal precipitation regimes in Baja, California, Mexico. *Geografiska Annaler*, 47A: 204–23.

HOWE, G. M. 1953. Climates of the Rhodesias and Nyasaland according to the Thornthwaite classification. *Geog. Rev.* 43: 525–39.

HOWE, G. M. 1968. Climatology and the geographer. In *Geography at Aberystwyth.* Ed. Bowen, E. G., Carter, H., Taylor, J. A., Aberystwyth 50–67.

IVES, J. D. and BARRY, R. G. 1974. *Arctic and alpine environments.* London.

JACKSON, I. J. 1977. *Climate, water and agriculture in the Tropics.* London.

JACOBS, W. C. 1946. Synoptic climatology. *Bull. Amer. Met. Soc.* 27: 306–11.

JAMES, J. W. 1966. A modified Köppen classification of California's climates according to recent data. *Californian Geographer,* 1–12.

JAMES, R. W. 1970. Air mass climatology. *Met. Rund.* 23: 65–70.

JOINT ORGANIZING COMMITTEE GARP 1975. The physical basis of climate and climate modelling. *GARP Publ. Series No. 16,* GARP, WMO.

JONES, S. B. 1950. What does geography need from climatology? *Professional Geographer,* 2: 41–4.

KASAHARA, A. and WASHINGTON, W. M. 1967. NCAR global general circulation model of the atmosphere. *Mon. Weath. Rev.* 95: 389–402.

KORTE, A. F. and COLSON, D. 1972. Synoptic climatological studies of precipitation in the plateau states from 850-millibar lows during fall. *Tech. Mem. NWS TDL-49,* Washington Nat. Ocean. Atmos. Admin. Tech. Dev. Lab.

KRAMER, F. L. 1963. The Koeppen dry/humid boundary: preliminary test of the Patton formula. *Professional Geographer,* 15: 13–14.

LAMB, H. H. 1969. The new look of climatology. *Nature,* 223: 1209–15.

LAMB, H. H. 1972. *Climate, present, past and future.* Vol. I: *Fundamentals and climate now.* London.

LAMB, H. H. 1977. *Climate: present, past and future.* Vol. 2: *Climatic history and the future.* London.

LANDSBERG, H. E. 1964. Roots of modern climatology. *J. Washington Acad. Sci.,* 54: 130–41.

LANDSBERG, H. E. (ed.). *World survey of climatology.* Elsevier. 12 volumes, various dates.

LANGLOIS, W. W. and KWOCK, H. C. 1968. Description of the Mintz-Arakawa general circulation model. *Tech. Rept. No. 3,* Numerical simulation of weather and climate. Dept. Met., UCLA.

LEITH, C. E. 1975. The design of a statistical–dynamical climate model and statistical constraints on the predictability of climate. In The physical basis of climate and climate modelling. JOC–GARP, *GARP Pub. Ser. No. 16,* 137–141.

LOCKWOOD, J. G. 1974. *World climatology. An environmental approach.* London.

LOCKWOOD, J. G. 1979. *Causes of climate.* London.

MANI, A. and CHACKO, O. 1972. Solar radiation climate of India. *Solar Energy,* 14: 139–56.

MANLEY, G. 1938. Snowfall and its relation to transport problems with special reference to northern England. *Geogrl. J.,* 92: 522–6.

MANLEY, G. 1974. Central England temperatures: monthly means 1659 to 1973. *Quart. J. R. Met. Soc.* 100: 389–405.

MARKHAM, C. G. 1972. Baja California's climate. *Weatherwise,* 25: 64–76.

MASON, B. J. 1971. The global atmospheric research programme—a contribution to the numerical simulation and prediction of the global atmosphere. *Earth Sci. Review,* 7: 165–86.

MASON, B. J. 1976. The GARP Atlantic Tropical Experiment—an introduction. *Meteorological Magazine,* 105: 221–3.

MATHER, J. R. 1974. *Climatology: fundamentals and applications.* New York.

MAYKUT, G. A. and CHURCH, P. E. 1973. Radiation climate of Barrow, Alaska 1962–66. *J. Applied Meteorology,* 12: 620–8.

MCBOYLE, G. R. 1971. Climatic classification of Australia by computer. *Australian Geogr. Studies*, 9: 1–14.

MILLER, A. A. 1931. *Climatology.* London.

MILLER, A. A. 1957. The use and misuse of climatic resources. *Advancement of Science*, 13: 56–66.

MILLER, D. H. 1957. What climatologists need from other geographers. *Professional Geographer*, 9: 8–10.

MILLER, D. H. 1972. Is climatology an aid in unifying geography? *Australian Geogr. Studies*, 10: 203–7.

MITCHELL, J. M. 1970. The effect of atmospheric aerosol on climate with special reference to surface temperatures. *Tech. Mem. EDS 18*, Washington, National Oceanographic and Atmospheric Administration.

MITCHELL, V. L. 1976. The regionalization of climate in the western United States. *J. Applied Meteorology*, 15: 920–7.

MOFFITT, B. J. and RATCLIFFE, R. A. S. 1972. Northern hemisphere monthly mean 500-millibar and 1000-500 millibar thickness charts and some derived statistics (1951–66). *Geophys. Mem.* 16: 117. Met. Off. London.

MÖRTH, H. T. 1979. Climatological research: an interdisciplinary study. *Climate Monitor*, ed. P. D. Jones, Clim. Res. Unit, UEA Norwich, 19–25.

NATURAL ENVIRONMENTAL RESEARCH COUNCIL 1976. Research in applied and world climatology. *Pub. Series B, No. 17*, NERC, London.

NIEUWOLT, S. 1977. *Tropical climatology. An introduction to the climates of the low latitudes.* London.

OJO, S. O. 1970. The seasonal march of the spatial patterns of global and net radiation in West Africa. *J. Tropical Geography*, 30: 48–62.

OKE, T. R. 1974. Review of urban climatology 1968–1973, *Tech. Note No. 134*, World Meteorological Organization.

OKE, T. R. 1978. *Boundary layer climates.* London.

OLIVER, J. 1967. *Climatology and the environmental sciences.* Univ. Coll. of Swansea, Inaugural Lecture.

OLIVER, J. E. 1970. A genetic approach to climatic classification. *Annals Assoc. Amer. Geog.* 60: 615–37.

OLIVER, J. E. 1973. *Climate and man's environment.* New York.

PACKER, R. W. and SANGAL, B. P. 1971. The heat and water balance of southern Ontario according to the Budyko method. *Canadian Geographer*, 15: 262–86.

PAINTING, D. J. 1977. A study of some aspects of the climate of the northern hemisphere in recent years. *Sci. Pap. 35*, London Met. Off.

PANEL ON INTERNATIONAL METEOROLOGICAL CO-OPERATION 1965. The feasibility of a global observation and analysis experiment. Report to Committee on Atmospheric Sciences, Nat. Acad. Sci.–Nat. Res. Council, *NAS–NRC Pub. 1290.*

PARKER, D. E. 1976. GATE—the field project in 1974. *Meteorological Magazine*, 105: 223–33.

PENMAN, H. L. 1948. Natural evaporation from open water, bare soil and grass. *Proc. Roy. Soc.* 193A: 120–45.

PERRY, A. H. and WALKER, J. M. 1977. *The ocean–atmosphere system.* London.

PITTOCK, A. B., FRAKES, L. A., JENSSEN, D., PETERSON, J. A., and ZILLMAN, J. W. (ed.) 1978. *Climatic change and variability: a southern perspective.* London.

PLANT, J. A. 1969. The climate of West Lothian. *Clim. Mem. No. 59A*, Met. Off., typescript, London.

ROBINSON, G. D. 1971. *Man's impact on climate.* Cambridge, Mass.

SAWYER, J. S. 1962. Research in synoptic and dynamical meteorology and climatology, 1941–1962. *Meteorological Magazine*, 91: 327–35.

SCHLOEMER, R. W. 1971. Terrain and climate. *Tech. Mem. EDS No. 19*, Wash., Nat. Ocean., Atmos. Admin., Environ. Data Serv.

SCHNEIDER, S. H. 1976. *The genesis strategy*. New York.

SCHNEIDER, S. H. and DICKINSON, R. E. 1974. Climate modelling. *Reviews of Geophys. and Space Physics*, 12: 447–93.

SELLERS, W. D. 1969. A global climatic model based on the energy-balance of the earth-atmosphere system. *J. Applied Meteorology*, 8: 392–400.

SELLERS, W. D. 1973. A new global climatic model. *J. Applied Meteorology*, 12: 241–54.

SHAW, D. B. (ed.) 1978. *Meteorology over the tropical oceans*. Royal Meteorological Society, Bracknell.

SMAGORINSKY, J., MANABE, S. and HOLLOWAY, J. L. 1965. Numerical results from a nine-level general circulation model of the atmosphere. *Mon. Weath. Rev.* 93: 727–68.

SMITH, C. G. 1974. Monthly, seasonal and annual fluctuations of rainfall at Oxford since 1815. *Weather*, 29: 2–16.

SMITH, K. 1975. *Principles of applied climatology*. London.

STRINGER, E. T. 1972a. *Foundations of climatology*. San Francisco.

STRINGER, E. T. 1972b. *Techniques of climatology*. San Francisco.

TERJUNG, W. H. 1970. A global classification of solar radiation. *Solar Energy*, 13: 67–81.

THOM, H. C. S. 1970. The analytical foundations of climatology. *Arch. Met. Geoph. Bioklim.* 18B: 205–20.

THORNTHWAITE, C. W. 1948. An approach toward a rational classification of climate. *Geog. Rev.* 38: 55–94.

THORNTHWAITE, C. W. 1953a. Topoclimatology. *Proc. Toronto Met. Conference*, Royal Meteorological Society, 227–32.

THORNTHWAITE, C. W. 1953b. A charter for climatology. *Bulletin World Meteorological Organization*, 2: 40–6.

TRENBERTH, K. E. 1973. Possible sea surface temperature influences on short term climate in New Zealand. *Tech. Note 218*, Wellington, N.Z. Met. Serv., typescript.

TULLER, S. E. 1968. World distribution of mean monthly and annual precipitable water. *Mon. Weath. Rev.* 96: 785–97.

TWEEDIE, A. D. 1967. Challenge in climatology. *Australian J. Science*, 29; 273–8.

U.S. COMMITTEE FOR GARP 1975. *Understanding climatic change*. National Acad. of Sciences, Wash., D.C.

WHITE, R. M. 1978. Organizing a World Climate Program. *Bull. Amer. Met. Soc.* 59: 817–21.

WITTWER, E. L. 1972–4. Seasonal changes in wind regimes of the North Australian and adjoining intertropical area. *Australian Geographer*, 12: 340–62.

WORLD METEOROLOGICAL ORGANIZATION 1970. Urban climates. *Tech. Note No. 108*, WMO, Geneva.

WORLD METEOROLOGICAL ORGANIZATION 1971. *World Weather Watch: the plan and implementation programme 1972–74*. WMO No. 296, WMO, Geneva.

WRIGHT, P. B. and STUBBS, M. W. 1971. Circulation patterns at 850, 700, 500 and 200 millibars over the eastern hemisphere from 40 °N to 40 °S during May and June. *Geophys. Mem.* 15: 114. Met. Off. London.

YOSHINO, M. M. 1971. *Water balance of monsoon Asia—a climatological approach*. Honolulu.

YOSHINO, M. M. 1975. *Climate in a small area. An introduction to local meteorology*. New York.

# 4 Water: A geographical issue

ROY C. WARD *University of Hull*

## INTRODUCTION

WATER is everywhere in our environment; as a saline solution it covers 71 per cent of the global surface to an average depth of 4000 metres; as a solid it covers a further 17 per cent of the surface to average depths as great as 2000 metres in Antarctica; in gaseous form it envelops the globe, occurring in the atmosphere over equatorial rain forest and hot sandy desert alike; as a relatively pure liquid ($H_2O$) it occurs on and under the surface of the earth as rivers and lakes, soil moisture and ground water, and is a vital requirement of every living thing. But whereas plants and animals are adapted to its abundance or scarcity, e.g. as hydrophytes and xerophytes, fish and reptiles, man must locate, develop, and conserve water supplies or move elsewhere. Since, however, water plays a fundamental role in so many natural processes, water 'management' by man has often led to disastrous consequences affecting erosion and sedimentation, flooding and desertification, pollution and eutrophication.

Understandably then, geographers, the prime environmental scientists, have long regarded the study of water as an essential part of their discipline notwithstanding the comparatively recent appearance of labels such as 'geographical hydrology' (Ward 1978a). As academic fashions have changed, so too has the nature of our geographical interest in water. Ward (1979) drew attention to parallel shifts of emphasis in a number of countries, from a dominantly physical concern with the natural water cycle to an increasing awareness of and concern with the socio-economic implications of hydrology. Today, in a very real sense, water is a unifying focus of study for geographers drawn in increasing numbers from specialisms across the broad spectrum of the discipline. At the risk of some over-simplification it may be argued that the growing contribution of geographers to the study of water in recent years has firmly established three major areas of geographical concern: the definition of the resource base, the investigation of hydrological extremes, and consideration of the increasingly complex problems associated with water quality. These three areas will now be discussed in more detail.

## Definition of the resource base

The water resource base can be properly quantified and defined only if (i) available contemporary and historical hydrological data can be related to the continuing processes of climatic change, (ii) the current programmes of data collection provide an accurate and representative sample, and (iii) the hydrological processes reflected in those data can be properly incorporated into effective predictive models.

## (*i*) *Climate and water*

Most major water management schemes for water supply, irrigation, or flood control are designed to have an operational span of 50–100 years or even longer, and hydrological design conditions are normally estimated from the projection into the future of present and past hydrological and climatological conditions. But of course it is axiomatic that climate is always changing. As Wallis (1977) observed, 'climatic change has been a property of the earth's atmosphere for as long as the earth has had an atmosphere, and it seems probable that climatic change will continue even after mankind is no longer available to record changes.'

Strictly in terms of the availability of water supplies, climatic change is unlikely to be important where present supplies are so far in excess of demand that even major fluctuations in water availability would impose no constraints upon use and, more generally, climatic change is unlikely to impose major problems if probable future changes of climate have been adequately represented in the events of the recent past. In fact, of course, many areas are already short of water and others will suffer chronic shortages within the next few decades. Moreover, climatic change may result in 'Noah' as well as 'Joseph' effects so that, for example, drastic changes in the magnitude and frequency of flooding may occur in an area where the balance between annual water supply and demand remains relatively stable. And finally it is accepted that, for many areas of the world, the past sixty years have been quite abnormal, climatologically speaking, compared to the last millennium as a whole (Kilmartin 1976).

It is suggested that there are at least three major research areas, in which geographers have traditionally participated, which will continue to pose stimulating challenges, viz. the causes of climatic change, the patterns of climatic change, and the relationships between climatological and hydrological changes.

Tickell (1977) discussed the causes of climatic change in terms of external, internal, and anthropogenic factors, although inevitably in practice such clear and convenient distinctions are often blurred by the complexity of interactions between climatological processes (i.e. in fashionable systems terminology by positive and negative feedback). However, it would appear that at the present time discussion of the causes of climatic change is dominated by the fact that man now has the facility to modify global climate or at least to trigger prematurely changes that might have been delayed in his absence. Unfortunately much still remains to be learnt about climatological processes before the impact of increases in $CO_2$ and stratospheric water vapour or the destruction of stratospheric ozone by chlorofluorocarbons can be detailed with accuracy, although Wilcox (1975) predicted that ice-melt could cause a rapid rise of sea level of 50–70 m. within the next two centuries!

An equally fundamental ignorance is reflected in the seemingly endless arguments about the pattern of past, present, and future climatic changes. The identification of 'significant' climatic cycles which might be used as a basis for forecasting future climate is still a popular pastime despite the apparently insuperable problems outlined by Wallis (1977), viz. that tests of significance are based on the climatologically unlikely assumption of non-cyclic independence in the data (Mandelbrot and Wallis 1969), that stochastic processes with low-frequency components can yield sample functions with numerous 'significant' spurious cycles (Wallis and Matalas 1971), and that 'significant' cycles are often introduced into the analysis by moving average manipulations (Slutzky 1937). Moreover, in discussing the origins of the Hurst phenomenon (i.e. persistence), Lettenmaier and Burges (1978) demonstrated that it is difficult to distinguish from existing hydrological records the difference between a generating mechanism characterized by a non-stationary mean value with stationary residual noise and a long-term persistent model with constant mean, notwithstanding the potentially large impact of such a difference on the design of major water storage schemes.

In one sense the most important question concerns the relationships between climatological and hydrological changes because, clearly, little is gained by identifying the causes and patterns of climatic change unless such changes can be transformed into changes in water supply or in flood characteristics, etc. The US National Research Council Panel on Water and Climate (National Research Council 1977) recognized that 'models that predict the effect of climatic changes on water supplies are still primitive', thereby confirming an earlier WMO (1974) evaluation, and that such models yield even more uncertain results when applied to forecasts of future climatological conditions, because of the complexity of feedback effects, than when they are applied to current hydrological data. Interestingly, an examination by Rodda *et al.* (1978) of winter rainfall and river flow in England and Wales during a period of fifty years, suggested that 'land use changes or river regulation may have more effect on water resources than climatic changes'.

### (ii) Hydrological data collection

An increased awareness of the growing pressure on water resources in many parts of the world and of the woeful inadequacy of contemporary hydrological data led to the initiation, in 1965, of the International Hydrological Decade (IHD). In particular, recognition of the need to define the details of the hydrological systems under investigation was a major factor in the development of the IHD Representative and Experimental Basin Programme and in the associated global growth in the popularity of small instrumented catchments as a basic hydrological data-collecting technique. In most cases measurements were made of precipitation, evaporation, stream discharge, and of changes in soil moisture and ground-water storages, and these measurements were used initially to quantify the water balance over time intervals ranging from one day, or less, to one month. Results from many of these studies have been presented at a number of major symposia since 1965; see American Water Resources Association (1972), International Association of Scientific Hydrology (IAHS 1965, 1970), National Research Council of Canada (1965), and Gregory and Walling (1974).

In Britain small instrumented catchments gained rapidly in popularity and by 1965 about fifty such studies were in progress (NERC 1970), many of which, like those of the Institute of Hydrology at Plynlimon in central Wales, were intended to investigate the hydrological consequences of land-use changes, especially afforestation and deforestation and, later, urbanization.

The major disadvantages of small instrumented catchments were readily recognized (e.g. Slivitzsky and Hendler 1965; Reynolds and Leyton 1967; Ward 1971) as being lack of experimental control; problems of the representativeness of the sample area; the doubtful accuracy of the data collected; the problems of handling those data, especially when so many were in the form of autographic chart records; and, not least, the high cost of such research over a prolonged period of time. While some of these disadvantages were subsequently overcome (e.g. data processing, storage, and retrieval techniques have improved considerably with the growth in computer technology) and others were initially overstated (e.g. cost), it was clear that small instrumented catchments would not, of themselves, provide all the necessary hydrological data. Earlier suggestions (Ward 1971) that hydrological problems would be solved most readily by a rational organization of interdependent research elements involving small instrumented catchments, plot studies, model development, and basic process research have been confirmed subsequently. Certainly in Britain the present approach is to use instrumented catchments more as a means of testing hypotheses developed within a more restricted experimental framework, than as a primary research tool.

Thus, within the Institute of Hydrology's catchments at Plynlimon, process studies have become progressively more important as a complement to the 'black-box' modelling of the original experiment and now involve investigations of, for example, interception at seven selected sites, pipeflow, transpiration from a tree-covered 'natural lysimeter' plot, and sediment studies in selected area (Smith 1977). In addition, the Institute of Hydrology has been conducting a long-term investigation of the physical controls of evaporation at a site in the Thetford forest. Other approaches have included the concentration upon hill-slope units within catchments (Kirkby 1978), the laboratory investigation of water movement in soil columns (Trudgill 1977) similar to that conducted by Horton and Hawkins (1965), and examination of the micromorphology of soil structure in the hope that this may eventually yield a better index of infiltration and other soil-hydrological characteristics than that currently provided by, say, measurements of infiltration through the soil surface. Some of this latter work is being done at Sheffield University using techniques similar to those developed by Jongerius and Heintzberger (1975). Physical geographers in most British universities are already participating in these developments and have the potential to make further substantial contributions.

Such changes in the techniques of data collection are a reflection partly of a better understanding of hydrological processes and partly of improvements in data storage and analysis which have accompanied the improvements in computers in recent years. Early experimental design concentrated on a lumped-parameter, black-box approach in order to reduce the quantity of data collected to manageable proportions; the capacity of modern computers makes this as unnecessary as it is undesirable and has thus greatly facilitated complementary advances in hydrological modelling.

*(iii) Hydrological modelling*

Hydrological modelling has expanded and developed rapidly and substantially during the past decade or so. These growth characteristics are reflected in the associated terminology and literature which are at times both confused and confusing. In an early paper, now recognized as a classic, Amorocho and Hart (1964) discussed hydrological modelling techniques, and more recently Clarke (1973) presented what is probably the most lucid analysis of hydrological modelling terminology, reminding us again of the essential but often (unhappily) blurred distinction between *variables* (i.e. characteristics of the hydrological system that may change with time) and *parameters* (i.e. characteristics of the hydrological system which do not change with time) and differentiating models as *stochastic* or *deterministic* depending on whether or not they contain random variables and as *conceptual* or *empirical* depending on whether they are structured on the basis of a consideration of physical processes or simply of observed relationships. Thus a conceptual model need not be deterministic and, indeed, there would be distinct advantages in many situations in *stochastic-conceptual* models which would reflect ever-present data uncertainties.

In fact, as Freeze (1975, 1978) observed, many hydrological models are *stochastic-empirical*, e.g. rainfall–runoff correlations, or time series analysis of stream-flow sequences (i.e. the typical black boxes which it is now so fashionable to deride), and of those which are conceptual the majority are *deterministic-conceptual*, e.g. infiltration models.

At the most simple level a satisfactory model is one which succeeds, and at this level the type and complexity of modelling procedures might be determined by the uses to which the predictive model will be put and the size of the geographical unit to which it will be applied. Clearly, then, the modelling demands made by the problems of predicting annual water resources' availability for a subcontinental region will differ markedly from those made by the problem of determining spatial and vertical variations of daily soil moisture content on an individual hill-slope. Certainly Thornthwaite and Mather (1955, 1957) successfully modelled the former using only monthly data on precipitation and potential evapotranspiration from widely scattered stations. In a general sense, then, model complexity reflects the size of the hydrological/geographical unit under consideration, the time duration of the prediction period, and the degree of understanding and disaggregation of the hydrological system which is considered desirable. Increasingly, hydrologists appear to be rejecting crude black-box empirical models or grossly lumped conceptual models in favour of models which incorporate a fuller understanding of the hydrological system and the processes at work within it.

This is not to say that lumped-parameter conceptual models have not yielded good results in the past nor that they have now been totally rejected. Indeed, virtually all the major, widely used conceptual models, e.g. SSARR Corps of Engineers (Rockwood 1969), Stanford Watershed (Crawford and Linsley 1966), Road Research Laboratory (1963), Hydrocomp Simulation Program (Hydrocomp 1969), Hydrologic Engineering Center (Beard 1968),

USDAHL-70 (Holtan and Lopez 1971) and Institute of Hydrology (Nash and Sutcliffe 1970), are of the lumped-parameter type. Work on lumped-parameter models at the Institute of Hydrology began in the late 1960s and continues at the present time (Institute of Hydrology 1978). The model that has been developed represents catchment behaviour by three reservoirs or storages, with fluxes of water between them governed by plausible mathematical relations incorporating the known characteristics of such transfers. Modifications of this simple lumped model are being developed for use in upland areas with substantial nature water storages, e.g. peat hags (Institute of Hydrology 1978).

Simultaneously, work is now in progress to develop comprehensive distributed (i.e. disaggregated) models firmly based on physical principles and making full use of field measurements of soil moisture characteristics and hydraulic conductivities. Part of this work relates to the Institute's Plynlimon experimental catchments and part relates to the joint development (with two other European institutions) of a model called SHE (Système Hydrologique Européen) whose purpose is to predict the hydrological consequences of land-use changes, whether or not stream-flow records exist for a catchment area.

As has already been suggested, recent advances in the development of disaggregated (distributed) models reflect partly the improved capacity and flexibility of modern computers. However, they also reflect the advances in hydrological theory which have taken place during the 1960s and 1970s, particularly those concerning the relative contributions of surface and subsurface water movement to stream-flow and the dynamic characteristics of stream-flow generation reflected in the notions of partial or dynamic contributing areas (DCA), dynamic contributing volume (DCV), and elements of the DCA and DCV which are not connected directly to stream channels (disjunct DCA/DCV) (Jones 1979). In an outstanding contribution to an excellent collection of essays in hydrology (curiously qualified throughout the volume as 'hillslope' hydrology—as if there could be some other variety!), Freeze (1978) identified the three main components of a hydrological model as (*a*) transient, saturated–unsaturated *subsurface flow* in a two-dimensional cross-section with heterogeneous, isotropic media, and with boundary conditions that allow time- and space-dependent arrival of rainfall on the upper surface, and outflow to the stream through a transient seepage face; (*b*) one-dimensional, transient, *channel flow* with boundary conditions that allow time- and space-dependent arrival of lateral inflow; and (*c*) one-dimensional treatment of the sheet-flow representation of transient *overland flow* with boundary conditions that allow time- and space-dependent arrival of rainfall and infiltration to the subsurface system, and outflow to the stream. Although each component has been modelled separately by many workers, there are comparatively few examples of models which integrate two or more components. Freeze (1978) referred to two models which couple subsurface flow and channel flow (Freeze 1972a; 1972b) and overland flow and infiltration (Smith and Woolhiser 1971).

The complexities of modelling the dynamic nature of stream-flow generation processes over any but a very small catchment area or segment of a catchment area were stressed by Betson and Ardis (1978) and Jones (1979), although important steps in this direction have already been taken by Engman and Rogowski (1974), Lee and Delleur (1976), Beven (1977), and Jayawardena and White (1977). The prospect of further developments in this field will pose stimulating challenges to physical geographers and hydrologists during the next decade.

HYDROLOGICAL EXTREMES

The interest of geographers in hydrological extremes has long manifested itself in the literature on floods and droughts and in the special attention paid to that one-third of the earth's land surface which is classified as arid or semi-arid. More recently some of these interests have been subsumed by the development of an identifiable corpus of 'natural hazards' research much of whose coherence and structure was provided by the inspiration and pioneering leadership over forty years and more of Gilbert F. White, first in the Department of Geography at the University of Chicago and subsequently as Director of the Institute of Behavioral Science at the University of Colorado (Burton, Kates, and White 1978).

White's interests, and those of his leading disciples such as I. Burton, R. W. Kates, W. R. D. Sewell, and J. R. Sheaffer, lay in the way in which man adjusts to risk and uncertainty in natural systems and the implications of that process of adjustment for public policy (White 1973). Although these themes have been pursued in Britain and elsewhere, geographers working outside the constraints of the predominantly socio-economic definition of geography adopted in the USA have broadened their fields of interest to include a greater consideration of the physical processes as well as of their social, political, and economic implications.

Floods, which are one of the most dramatic interactions between man and his environment, emphasizing as they do both the sheer force of natural events and man's inadequate efforts to control them, have understandably attracted the attention of geographers for many years. In Britain the development of geographical interest in flood problems, which was given useful impetus by the earlier work of D. M. Harding at the University College of Swansea, E. Porter at Cambridge and Leeds, and E. C. Penning-Rowsell and his colleagues at (what is now) the Middlesex Polytechnic, has recently resulted in a number of largely complementary publications.

The *Flood Studies Report* (NERC 1975) was a major five-volume publication prepared by an Institute of Hydrology team which included a number of physical geographers. The report, although by no means without its deficiencies, not the least of which is that it is difficult to use as a manual of practice, nevertheless represented a major advance and will, it is hoped, provide the basis for future flood prediction and response practices in this country. Curiously, in the document which set out the need for the Flood Studies Report, ICE (1967) relegated to the final paragraph any mention of the collection of flood damage data. However, this omission was ably rectified by the appearance of *The Benefits of Flood Alleviation: A manual of assessment techniques* (Penning-Rowsell and Chatterton 1977), which was undoubtedly the most important document of its kind yet published in Britain.

This is primarily a manual of implementation based upon the extensive researches of a small team at Middlesex Polytechnic and although again imperfect, particularly in respect of the calculation of industrial flood damages, it represents a sound basis for calculating the cost-effectiveness of possible responses to the flood hazard. *Floods: A geographical perspective* (Ward 1978c) was a general study of major flood problems which ranges from the hydrological causes of floods, and their prediction and routing, to a survey of man's response to the flood hazard. Finally, *Human Adjustment to the Flood Hazard* (Smith and Tobin 1979) developed a more limited theme, although in greater detail, largely through the medium of case studies, with a major concentration of interest on the flood problems of Carlisle and Appleby on the Cumbrian Eden.

These closely related publications reflect a welcome sharpening of focus and consolidation of interest by geographers and related scientists in environmental hazards and stress which will undoubtedly yield long-term social and economic benefits. And yet there is still much to be done because the work to date on flooding and flood hazard has achieved little more than a clearer definition of the scope and complexity of the problems that remain, and a similar situation obtains for other contrasting environmental hazards and stresses such as drought.

The occasional stresses imposed by temporary drought in normally well-watered areas produce public concern and if severe enough an upsurge of scientific activity as well. Recent examples occurred in Britain and the USA. As Walling (1978) observed, the summer of 1976 will be long remembered for the severe drought conditions in England and Wales, the sixteen-month period to August 1976 being the driest since records began in 1727. Preliminary analyses of the drought were presented by Central Water Planning Unit (1976) and in a Royal Society symposium held in 1977 (Royal Society 1977). A more comprehensive response is embodied in a major publication by the Institute of British Geographers (Doornkamp and Gregory 1980). This drought atlas considers both physical and socio-economic aspects of the drought and should provide an important data base for planners and administrators in the future. A full consideration of the long-term impacts of the 1975–6 drought in England and Wales was severely constrained by the magnitude of the autumn rains with which it terminated, the September–October period being the wettest since records began, which resulted in a remarkably rapid recovery of soil moisture deficits, ground-water levels, and stream-flows. In 1976–7 a similarly severe drought affected many parts of the USA. The effect of this event on ground-water systems was discussed in a series of papers presented to the Spring Meeting of the American Geophysical Union in 1979 (American Geophysical Union 1979).

A potentially more serious environmental situation occurs when temporary drought conditions are globally widespread. Thus in 1972 lack of rain in many countries led to poor crops and a fall of 4 per cent in world grain production. Burton, Kates, and White (1978) recorded that by 1974 world reserves of grain had fallen to a 26-day supply, compared to a 95-day supply in 1961, and suggested that simultaneous drought catastrophes in the major grain-producing nations and in the monsoonal areas of Africa and southern

Asia could create unprecedented famine conditions which would exceed the world's capacity in surplus grains.

In areas where there is a long history of recurrent drought, human adaptation to the hazard may be more efficient and scientific study of both the physical and statistical characteristics of the drought and of response to it may yield valuable information on which to base future action. One such area in which droughts of varying intensity and duration recur is Nebraska, USA, which was studied by Lawson *et al.* (1971). Another is Tanzania whose drought characteristics, both physical and human, were well documented by a number of authors in White (1974).

In still other areas the drought hazard appears to be intensifying progressively over a long period of time. The Sahel zone of Africa has recently attracted much attention in the context of the southward extension of the Saharan desert. In such areas planning and development decisions could be considerably influenced by improved knowledge of the causes of desertification and in particular of the relative effects of climatic change and human activity (Winstanley 1973; Rapp 1974; Glantz 1976; Hare 1977).

Finally, there are the areas of permanent drought, the arid lands, in which problems of adaptation and conservation have been well documented (although not necessarily as well understood) for many years. Irrigation has long been the key to arid-land development but irrigation applied without a sound scientific understanding of its hydrological repercussions has led to numerous unmitigated disasters, e.g. salinization, many of which were chronicled by Cantor (1967). New discoveries which continue to enhance the already enormous ground-water potential of the major arid areas (Harshbarger 1968; Burdon 1977) not only suggest that irrigation will continue to play a vital role in their development but also underline the need for a better scientific understanding of irrigation problems. It is therefore encouraging to note the continuing output of valuable research findings from the CSIRO Division of Irrigation Research at Griffith, NSW, the emphasis on high-efficiency water use in countries such as Israel (Lahav 1977; Weiner 1977), the recent emergence of a new journal entitled *Irrigation Science* whose aims and scope were outlined by Stanhill (1978), and the increasing attention being paid to the *impacts* of ground-water development, both physical and socio-economic, in arid lands (Keith 1977). However, it is also clear that increasing pressure on water resources, particularly in the already intensively developed arid areas such as the USA and Israel, will probably result in the diversion of water from irrigation to other more intensive and/or economically productive uses and in the development of new technology. Some of the possible strategies, e.g. industrial development, desalination, recycling of waste water, were discussed by Beaumont (1977) and Smith (1979).

WATER QUALITY

It has been suggested that the concept of water quality is both complex and subjective (Mrowka 1974): complex because, especially in the case of surface conveyances of water in, for example, streams and lakes, quality embodies both characteristics of the fluid in transit and physical attributes of the channel or conveyance; subjective because inevitably water quality is evaluated as percieved by the investigator and his particular value system, which is subject to change with time. In any event there appears to have been a steady, even accelerating, growth of interest in water quality by geographers and hydrologists during the past two decades which is likely to be continued, at least in the immediate future. Ackermann (1969) identified this trend in the USA and, like Mikesell (1974), related it to a growing concern with environmental quality, a relationship which seems to have been confirmed by the contents of recent IAHS symposia on water quality (e.g. IAHS 1975, 1977, 1978).

Some part of this interest is *passive* in the sense that water quality characteristics are identified as constraints upon certain types of activity and water use. At one extreme the supply of water for human consumption necessitates stringent limits on faecal contamination, toxic materials, substances affecting palatability, and hardness, while at the other extreme navigation can proceed in water of virtually any quality provided that it is relatively free from large masses of floating debris which may foul propellers. Between these extremes there are a range of water quality thresholds related to the use of water for fishing, industry, irrigation, and recreation and amenity.

More important, however, are the *active* interests in water quality by those concerned with deliberately changing (i.e. improving) it or, more obviously within the geographer's scope, those active interests in water quality as an index of the rate and manner of operation of certain environmental processes, both physical and socio-economic. Thus, in terms of the operation of physical processes, it is recognized that the chemical quality of both surface and ground water may reflect the flow path of that water through a drainage basin and may therefore be used either to identify the individual contribution to stream-flow of quick-flow and base-flow (Pinder and Jones 1969; Zeman and Slaymaker 1975) or to reconstruct the major flow paths of ground water from recharge areas (Back 1966; Toth 1970). Water chemistry has also been used to analyse the down-channel movement of flood waves in terms of the shape and time-distribution of the chemograph and hydrograph (Glover and Johnson 1974). Similarly, water temperature was used by Smith (1968) to analyse the phase relationships of the various components of the stream hydrograph and by Cartwright (1970) to distinguish major areas of ground-water recharge and discharge. Furthermore, studies of water quality in the various phases of the hydrological cycle, from the precipitation input to the stream-flow output, have formed the basis of attempts by a number of research workers to describe and quantify the manner of nutrient cycling within catchment areas (Cryer 1976; Zeman 1975; Likens *et al.* 1967). In addition the rates of operation and the spatial distribution of denudational processes have been deduced with increasing precision from studies of the chemical quality of ground water, particularly karst ground water (Pitty 1966), and of stream-flow (Walling and Webb 1975; Arnett 1979).

The socio-economic applications of water-quality studies relate almost exclusively to the pervasive problems of pollution, especially that originating from urban and industrial area (IAHS 1977; UNESCO 1974) which tends mostly to affect surface waters, and that from agricultural and primary extractive operations which may also affect ground water (Bullard 1965;

Dancer 1975; IAHS 1975). Plans to intensify deliberate pollution of ground water by highly toxic, particularly radioactive, wastes will demand substantially increased understanding of ground-water flow patterns at a regional scale (Maxey and Farvolden 1965). Awareness of the existing severity of water-quality degradation is reflected in the initiation of nationwide river pollution surveys in a number of countries (DOE 1971; Ficke and Hawkinson 1975) and in strategies for drastically reducing the future level of pollution in both surface and ground water (Lvovitch 1977).

CONCLUSION

Three areas of contemporary geographical interest in water have been identified and discussed. All three areas, i.e. the identification of the water resource base, the study of the hydrological extremes of flood and drought, and the concern with water quality, reflect the growing awareness by geographers that studies of the physical environment can and should play an important role in the solution of socio-economic problems and that, in terms of the study of water, the older distinctions between 'physical' and 'human' geographers are no longer very meaningful. It has not been suggested that the three topics selected for discussion here are the only ones with which geographer-hydrologists are currently concerned; neither is it implied that in the future other topics will not assume prominence. Rather it is suggested that in a global situation of increasing pressure on resources, geographers will wish to retain an overriding interest in the socio-economic implications of water management. Clearly this may be achieved most effectively through a better understanding of the physical basis of hydrology; that, in any event, water will remain a fundamental issue in the geography of the future is surely beyond doubt.

# REFERENCES

ACKERMANN, W. C. 1969.   Scientific hydrology in the United States. *The Progress of Hydrology*. University of Illinois, 50–60.

AMERICAN GEOPHYSICAL UNION 1979.   Role of groundwater systems in drought of mid-1970s (Abstracts of papers). *Trans. Am. Geophys. Union*, 58: 1012–24.

AMERICAN WATER RESOURCES ASSOCIATION 1972.   *Watersheds in Transition*. National Sympos. on Watershed Hydrology, Colorado State Univ.

AMOROCHO, J. and HART, W. E. 1964.   A critique of current methods in hydrologic systems investigation. *Trans. Am. Geophys. Union*, 45: 307–21.

ARNETT, R. R. 1979.   The use of differing scales to identify factors controlling denudation rates, in *Geographical Approaches to Fluvial Geomorphology* (ed. A. F. Pitty). Norwich: Geo Books.

BACK, W. 1966.   Hydrochemical facies and groundwater flow patterns in northern part of the Atlantic Coastal Plain. *U.S. Geol. Surv., Prof. Paper 498-A*.

BEARD, L. R. 1968.   Hypothetical flood computation for a stream system. *The Use of Analog and Digital Computers in Hydrology*, 258–67. IAHS–UNESCO.

BEAUMONT, P. 1977.   Resource management: case study of water. *Prog. in Phys. Geog.* 1: 528–36.

BETSON, R. P. and ARDIS, C. V. 1978.   Implications for modelling surface-water hydrology, in *Hillslope Hydrology* (ed. M. J. Kirkby), 295–323. John Wiley: Chichester.

BEVEN, K. 1977.   *TOPMODEL—A physically-based variable contributing area hydrologic modelling program*. Geog. Working Paper 183. Univ. of Leeds.

BULLARD, W. E. 1965.   Acid mine drainage pollution control demonstration program: uses of experimental watersheds. *Sympos. of Budapest*, 190–200. IAHS.

BURDON, D. J. 1977.   Flow of fossil groundwater. *Quart. J. Engg. Geol.* 10: 97–124.

BURTON, I., KATES, R. W. and WHITE, G. F. 1978.   *The Environment as Hazard*. New York: OUP.

CANTOR, L. M. 1967.   *A World Geography of Irrigation*. Edinburgh: Oliver and Boyd.

CARTWRIGHT, K. 1970.   Groundwater discharge in the Illinois basin as suggested by temperature anomalies. *Water Resources Res.* 6: 912–18.

CENTRAL WATER PLANNING UNIT 1976.   The 1975–76 drought: A hydrological review. *C.W.P.U. Tech. Note* 17.

CLARKE, R. T. 1973.   A review of some mathematical models used in hydrology, with observations on their calibration and use. *J. Hydrology*, 19: 1–20.

CRAWFORD, N. H. and LINSLEY, R. K. 1966.   *Digital simulation in hydrology Stanford watershed model IV*. Stanford Univ. Dept. of Civil Engg. Tech. Report 39.

CRYER, R. 1976.   The significance and variation of atmospheric nutrient inputs in a small catchment system. *J. Hydrology* 29: 121–37.

DANCER, W. S. 1975.   Leaching losses of ammonium and nitrate in reclamation of sand spoils in Cornwall. *J. Environmental Quality*, 4: 499–504.

DOE 1971.   *Report of a River Pollution Survey of England and Wales, 1970*. Department of Environment and the Welsh Office. London: HMSO.

DOORNKAMP, J. C. and GREGORY, K. J. 1980. *Atlas of Drought in Britain: 1975–76*. London: Institute of British Geographers.

ENGMAN, E. T. and ROGOWSKI, A. S. 1974.   A partial area model for stormflow synthesis. *Water Resources Res.* 10: 464–72.

FICKE, J. F. and HAWKINSON, R. D. 1975.   The national stream quality accounting network (NASQUAN)—some questions and answers. *U.S. Geol. Surv. Circ.* 719.

FREEZE, R. A. 1972a.    Role of subsurface flow in generating surface runoff. 1. Base-flow contributions to channel flow. *Water Resources Res.* 8: 609–23.

FREEZE, R. A. 1972b.    Role of subsurface flow in generating surface runoff. 2. Upstream source areas. *Water Resources Res.* 8: 1272–83.

FREEZE, R. A. 1975.    A stochastic-conceptual analysis of one-dimensional ground-water flow in a nonuniform homogeneous media. *Water Resources Res.* 11: 725–41.

FREEZE, R. A. 1978.    Mathematical models of hillslope hydrology, in *Hillslope Hydrology* (ed. M. J. Kirkby): 177–225. Chichester: John Wiley.

GLANTZ, M. H. (Ed.) 1976.    *Politics of Natural Disaster: The case of the Sahel drought.* New York: Praeger.

GLOVER, B. J. and JOHNSON, P. 1974.    Variations in the natural chemical concentration of river water during floodflows, and the lag effect. *J. Hydrology*, 22: 303–16.

GREGORY, K. J. and WALLING, D. E. 1974.    *Fluvial Processes in Instrumented Watersheds: Studies of small watersheds in the British Isles.* IBG Special Publ. 6. London: Institute of British Geographers.

HARE, F. K. 1977.    *Climate and Desertification.* Univ. of Toronto Inst. of Environtl. Studies. Toronto.

HARSHBARGER, J. W. 1968.    Ground-water development in desert areas. *Ground Water*, 6: 2–5.

HOLTAN, and LOPEZ, N. C. 1971.    *USDAHL-70 Model of Watershed Hydrology.* Tech. Bull. 1435. USDA Agric. Res. Service. Washington, DC.

HORTON, J. H. and HAWKINS, R. H. 1965.    Flowpath of rain from the soil surface to the water table. *Soil Science*, 100: 377–83.

HYDROCOMP 1969.    *Operations Manual (2e).* Palo Alto: Hydrocomp Inc.

IAHS 1965.    *Representative and Experimental Basins.* Sympos. of Budapest. Internat. Assoc. Hydrol. Sciences Publ. 66.

IAHS 1970.    *Results of Research on Representative and Experimental Basins.* Sympos. of Wellington. Internat. Assoc. Hydrol. Sciences Publ. 96.

IAHS 1975.    *Groundwater Pollution.* Sympos. of Moscow. Internat. Assoc. Hydrol. Sciences Publ. 103.

IAHS 1977.    *Effects of Urbanization and Industrialization on the Hydrological Regime and on Water Quality.* Sympos. of Amsterdam. Internat. Assoc. Hydrol. Sciences Publ. 123.

IAHS 1978.    *Modelling of Water Quality of the Hydrological Cycle.* Sympos. of Baden. Internat. Assoc. Hydrol. Sciences Publ. 125.

ICE 1967.    *Flood Studies for the United Kingdom.* Report of the Committee on Floods. London: Institution of Civil Engineers.

INSTITUTE OF HYDROLOGY 1978.    *Research Report 1976–8.* Wallingford: NERC.

JAYAWARDENA, A. W. and WHITE, J. K. 1977.    A finite element distributed catchment model, I. Analytical basis. *J. Hydrology*, *34*: 269–86.

JONES, J. A. A. 1979.    Extending the Hewlett model of stream runoff generation. *Area*, *11: 110*–14.

JONGERIUS, A. and HEINTZBERGER, G. 1975.    *Methods in Soil Micromorphology. A technique for the preparation of large thin sections.* Netherlands Soil Surv. Inst. Paper 10. Wageningen.

KEITH, S. J. 1977.    *The Impact of Groundwater Development in Arid Lands.* Office of Arid Lands Studies, Information Paper 10. Tucson: University of Arizona.

KILMARTIN, R. F. 1976.    Hydroclimatology: A needed cross discipline. *Trans. Am. Geophys. Union*, 57: 920 (abstract only).

KIRKBY, M. J. (ed.) 1978.    *Hillslope Hydrology.* Chichester: John Wiley.

LAHAV, R. 1977.    Israel strives for efficient use of water resources. *Water and Sewage Works*, 124: 64–5.

LAWSON, M. P., REISS, A., PHILLIPS, R. and LIVINGSTON, K. 1971. *Nebraska Droughts.* Dept. of Geog. Occ. Papers 1. Lincoln: Univ. of Nebraska.

LEE, M. T. and DELLEUR, J. W. 1976. A variable source area model of the rainfall-runoff process based on the watershed stream network. *Water Resources Res.* 12: 1029–36.

LETTENMAIER, D. P. and BURGES, S. J. 1978. Climate change: Detection and its impact on hydrologic design. *Water Resources Res.* 14: 679–87.

LIKENS, G. E., BORMANN, F. H., JOHNSON, N. M., and PIERCE, R. S. 1967. The calcium, magnesium, potassium and sodium budgets for a small forested ecosystem. *Ecology*, 48: 772–85.

LVOVITCH, M. I. 1977. World water resources present and future. *Ambio*, 6: 13–21.

MANDELBROT, B. B. and WALLIS, J. R. 1969. Some long-run properties of geophysical records. *Water Resources Res.* 5: 321–40.

MAXEY, G. B. and FARVOLDEN, R. N. 1965. Hydrologic factors in problems of contamination in arid lands. *Ground Water*, 3: 29–32.

MIKESELL, M. W. 1974. Geography as the study of environment. in *Perspectives on Environment* (ed. I.R. Manners and M. W. Mikesell): 1–23. Assoc. of Amer. Geog. Pub 13. Washington, DC.

MROWKA, J. P. 1974. Man's impact on stream regimen and quality. in *Perspectives on Environment* (ed. I. R. Manners and M. W. Mikesell): 79–104. Assoc. of Amer. Geog. Publ. 13. Washington, DC.

NASH, J. E. and SUTCLIFFE, J. V. 1970. River flow forecasting through conceptual models. *J. Hydrology*, 10: 282–90.

NATIONAL RESEARCH COUNCIL 1977. *Climate, Climatic Change and Water Supply, Studies in Geophysics.* Washington, DC: National Academy of Sciences.

NATIONAL RESEARCH COUNCIL OF CANADA 1965. *Research Watersheds.* Proc. Hydrology Sympos. no. 4: Guelph.

NERC 1970. *Hydrological Research in the United Kingdom (1965–1970).* London: The Natural Environment Research Council.

NERC 1975. *Flood Studies Report.* London: The Natural Environment Research Council.

PENNING-ROWSELL, E. C. and CHATTERTON, J. B. 1977. *The Benefits of Flood Alleviation: A Manual of Assessment Techniques.* Farnborough: Saxon House.

PINDER, G. F. and JONES, J. F. 1969. Determination of the groundwater component of peak discharge from the chemistry of total runoff. *Water Resources Res.* 5: 438–45.

PITTY, A. F. 1966. *An approach to the study of Karst water.* Occasional Papers in Geog. 5. Hull: The University.

RAPP, A. 1974. *A review of desertification in Africa—water, vegetation and man.* Report no. 1, Secretariat for International Ecology. Stockholm.

REYNOLDS, E. R. C. and LEYTON, L. 1967. Research data for forest policy: the purpose, methods and progress of forest hydrology. *Proc. 9th Brit. Commonwealth Forestry Conf.* Oxford: Commonwealth Forestry Institute.

ROAD RESEARCH LABORATORY 1963. *A Guide for Engineers to the Design of Storm Sewer Systems.* Road Note 35. London: HMSO.

ROCKWOOD, D. M. 1969. Application of streamflow synthesis and reservoir regulation—'SSARR'—program to the Lower Mekong River. *The Use of Analog and Digital Computers in Hydrology*, 329–44. IAHS–UNESCO.

RODDA, J. C., SHECKLEY, A. V., and TAN, P. 1978. Water resources and climatic change. *J. Inst. Water Engrs. and Scientists*, 32: 76–83.

ROYAL SOCIETY 1977. *Scientific Aspects of the 1975–76 Drought in England and Wales.* A meeting for discussion, 28 Oct. 1977, at The Royal Society.

SLIVITZKY, M. S. and HENDLER, M. 1965.   Watershed research as a basis for water resources development. *Research Watersheds*, 289–94. Proc. Hydrology Sympos. no. 4. Guelph.

SLUTZKY, E. 1937.   The summation of random causes as the source of cyclic processes. *Econometrika*, 5: 105–46.

SMITH, K. 1968.   Some thermal characteristics of two rivers in the Pennine area of northern England. *J. Hydrology*, 6: 405–16.

SMITH, K. 1977.   Catchment area experiments at Plynlimon. *Water Services*, July: 394–401.

SMITH, K. 1979.   Trends in water resource management. *Progress in Phys. Geog.* 3: 236–54.

SMITH, K. and TOBIN, G. A. 1979.   *Human Adjustment to the Flood Hazard.* London: Longman.

SMITH, R. E. and WOOLHISER, D. A. 1971.   Overland flow on an infiltrating surface. *Water Resources Res.* 7: 899–913.

STANHILL, G. 1978.   Editorial. *Irrigation Science*, 1: 1–2.

THORNTHWAITE, C. W. and MATHER, J. R. 1955.   The Water Balance. *Publications in Climatology*, 8: 1–86.

THORNTHWAITE, C. W. and MATHER, J. R. 1957.   Instructions and tables for computing potential evapotranspiration and the water balance. *Publications in Climatology*, 10: 185–311.

TICKELL, C. 1977.   *Climatic Change and World Affairs.* Harvard Studies in International Affairs 37. Cambridge, Mass.

TOTH, J. 1970.   A conceptual model of the groundwater regime and the hydrogeologic environment. *J. Hydrology*, 10: 164–76.

TRUDGILL, S. T. 1977.   *Soil and Vegetation Systems.* Oxford: Clarendon Press.

UNESCO 1974.   *Hydrological effects of urbanization.* Studies and Reports in Hydrology, no. 18. Paris: Unesco Press.

WALLING, D. E. 1978.   Physical hydrology. *Progress in Phys. Geog.* 2: 143–9.

WALLING, D. E. and WEBB, B. W. 1975.   Spatial variation of river water quality: a survey of the River Exe. *Trans. Inst. Brit. Geog.* 65: 155–71.

WALLIS, J. R. 1977.   Climate, climatic change and water supply. *Trans. Am. Geophys. Union*, 58: 1012–24.

WALLIS, J. R. and MATALAS, N. C. 1971.   Correlogram analysis revisited. *Water Resources Res.* 7: 1448–59.

WARD, R. C. 1971.   *Small Watershed Experiments: An appraisal of concepts and research developments.* Occ. Papers in Geog. 18. Hull: The University.

WARD, R. C. 1978a.   $Q_{max}$ or MAF?—Some reflections on hydrology today. *Geography*, 63: 301–13.

WARD, R. C. 1978b.   The changing scope of geographical hydrology in Great Britain. *Progress in Phys. Geo.* 3: 392–412.

WARD, R. C. 1978c.   *Floods: A geographical perspective.* London: Macmillan.

WHITE, G. F. 1973.   Natural hazards research, in *Directions in Geography* (ed. R. J. Chorley), 193–216. London: Methuen.

WHITE, G. F. (ed.) 1974.   *Natural Hazards: Local, National, Global.* New York: OUP.

WIENER, A. 1977.   Coping with water deficiency in arid and semi-arid countries through high-efficiency water management. *Ambio*, 6: 77–82.

WILCOX, H. A. 1975.   *Hothouse Earth.* New York: Praeger.

WINSTANLEY, D. 1973.   Rainfall patterns and general atmospheric circulation. *Nature*, 245: 190–4.

WMO 1974.   *Inter-comparison of conceptual models used in operational hydrological forecasting.* Oper. Hydrol. Rept. 7. Geneva: World Meteorological Organisation.

ZEMAN, L. J. 1975. Hydrochemical balance of a British Columbia mountainous watershed. *Catena*, 2: 35–47.
ZEMAN, L. J. and SLAYMAKER, H. O. 1975. Hydrochemical analysis to discriminate variable runoff source areas in an alpine basin. *Arctic and Alpine Res.* 7: 341–51.

# 5 Biogeography

I. G. SIMMONS    *University of Bristol*

'QU'EST-CE que la biographie?' asked the biologist F. Vuilleumier in the title of a recent (1977) paper, and such an enquiry may well form a good starting-point for this essay. For although the field within geography is healthy, having representatives in most major departments (including a number of holders of Chairs) and maintaining a reasonable level of publication, it must still rank among the lowest in conceptual awareness and methodological definition. We all know it when we see it but are hard-pressed to define its diagnostic characteristics and, in the last analysis, a piece of work is often biogeography because its author claims to be a biogeographer. There is, too, the additional difficulty that the term biogeography with its component divisions of phytogeography and zoogeography is still in active use within the field of biology (and indeed was the kind of study referred to by Vuilleumier). Here there is a lively debate about the nature of this science and whether it can be said to have a body of theories which would add up to a paradigm in Kuhn's sense of the word. Biologists appear to be agreed that biogeography is a field which arises out of the division and recombination of other sciences and hence has an integrative purpose: many biological parameters (extrinsic like geology and climate, intrinsic like genetics and speciation, competition and dispersion) belong no less and no more to biogeography than to parent disciplines. There is less agreement, though, about particular hypotheses involving the distribution of various taxonomic groups over the earth's surface. This argument has been well summarized by Stoddart (1978) and revolves mainly around the relative roles of continental mobility and effectiveness of biotic dispersal in accounting for biotic distributions, particularly of a disjunct kind. Ancestral biotas which become fragmented and exhibit monopyletic taxa are called vicariant and the phenomenon is known as *vicariance*. The method relies on distributions rather than phylogenetic trees and is especially associated with the work of Leon Croziat (1958, 1964; Nelson 1973). In some ways it appears to refute the working methods of those who believe firmly in the reconstruction of phylogeny on a systematic basis (Ball 1976), and of those whose work relies heavily on the dispersal mechanisms of individual plants and animals to carry them to such places as will give disjunct distributions (MacArthur and Wilson 1967). This 'island biogeography' theory has been one of the most praised and elegant developments of recent years (Stoddart 1977), not the least because it could be predictive and because it could be applied, since many nature reserves were biotic 'islands' (Simmons 1978). Eventually we may hope for a synthesis of many of these methodological viewpoints which ought to result in detailed environmental reconstruction of the type we now have for the Pleistocene and Holocene but with the added factor of continental mobility added to lithology, climate, and sea-level change as major contextual parameters (Raven and Axelrod 1972, 1975; Smith 1974, 1975). Most geographers who follow this debate at second hand, however, will miss the coruscations of literary quotation which some American zoologists feel compelled to display in their papers.

Apart from the connecting link provided by the considerable body of published work of D. R. Stoddart and F. Rose, this sort of work is remote from the activities of British biogeographers today, certainly in their research and very largely in their teaching. It can still be argued that the state of our biogeography teaching is analogous to that of a supermarket: a number of different shapes and colours of package are selected according to the desires of the teacher and checked out as one container marked biogeography. Since teaching tends to reflect the research environment then our investigations also have the air of a set of themes whose relationships are intuitively perceived by those in the field but which lack any comprehensive framework, hypotheses, body of theory, paradigm, or other conceptual construct to hold them together. We ought perhaps to pause and ask 'does it matter that the epistemological basis of our biogeography is so small; can the structure stand on such an undeveloped foundation?'. There is clearly ample scope for disagreement on an issue of this kind. On the one hand, it seems that much of the work done by geography is not very different from that done by workers with similar interests but who are called ecologists, environmental archaeologists, and planners, for example; the places where the geographers have published their work bear out that view. That this work is acceptable, and its authors judged competent outside the discipline of geography as well as within, runs the argument, forms its own justification. By contrast, other people find that this view is acceptable up to a point but that it does not go far enough. They maintain that geographers ought to have a viewpoint of their own, a distinct flavour so to speak; so that presumably on reading a paper everyone will say, irrespective of the journal in which it is published, 'that was written by a geographer', in a tone that is at the very least devoid of contempt. My own instinct here is with the story of the Buddhist novice who on a path in a forest encountered the aged and very venerable Abbot of a famous monastery; having greeted the Abbot respectfully he asked him, 'where is true Buddha-nature to be found?' 'Walk on,' replied the Abbot, 'walk on.' But while professional practitioners may see in an inner-directed fashion a pragmatic way ahead, I suspect our students need in this field (as in geography as a whole) a clearer set of tracks which will at any rate bring them safely to the clearing where they receive their black robes. I have suggested elsewhere (1979a) that the formulation of P. Dansereau (1957), that man creates new genotypes and that man creates new ecosystems, might form a distinctively geographical view. However, this encompasses only part of our traditions of study in geography, omitting, for instance, the recent developments in spatial geometry, and so the idea is advanced here cautiously as a recipe to be tasted rather than a dogma to be swallowed. But it would allow our interests in natural ecosystems (as a datum line against which to measure human effects), in man's impact on individual taxa as well as on ecosystems, in biotic resources, their protection and conservation, to find a place in a coherent whole, which relates to the concerns of the world about us. Nevertheless, I expect that the methodological pluralism which has characterized our biogeography during my time in it will continue, and diversity of both material and outlook must in any final reckoning be a source of strength.

This diversity has not prevented the foundation (in 1975) and running of a successful study group in Britain, composed largely of geographers but with a membership extending on the one hand to biologists and to planners and professional conservation workers on the other. This Biogeography Study Group is a formally constituted study group of the Institute of British Geographers and in 1979 had 169 members, with a regular programme of field meetings and symposia. Its success reflects not only the interest in the field but the organizing ability and energy of its first Secretary and current Chairman, James A. Taylor. The group regularly publishes a Directory (Clement 1978) of its members and their interests; the latest of these has been used as an initial guide to some of the recent work which is described below. (The classification of these investigations is however my own and not that of the BSG.) Three major groups of activities, each with subdivisions, are discernible (Table 5.1) and I shall attempt to give some idea of the flavour of the work in each category with personal comments where appropriate on where the work appears to be leading.

TABLE 5.1.

*Fields of research activities of British biogeographers 1970s*

| Field[a] | Approx. no. of workers[b] |
|---|---|
| 1. Historical biogeography | |
| (a) Quaternary ecology, including limnology | 16 |
| (b) Early cultural biogeography | 2 |
| (c) Historical ecology | 5 |
| 2. Contemporary ecosystems and their management | |
| (a) Community description and dynamics | 16 |
| (b) Impact assessment | 3 |
| (c) Management, including recreation ecology | 6 |
| (d) Remote sensing | 4 |
| 3. Biotic resources | |
| (a) General approaches | 3 |
| (b) Specific resources, e.g. mineral prospecting via biota | 2 |
| (c) Land-use ecology | 14 |

[a] The present author's own divisions.
[b] From Clement (1978); some people are entered under more than one field. Members of BSG who fall outside the scope of this essay are omitted from the Table.

## HISTORICAL BIOGEOGRAPHY

The attempt to reconstruct the past in much of its multi-faceted complexity which our colleagues in historical geography have undertaken is no less fascinating to a large group of biogeographers, who have especially concentrated on the impact of human societies upon plant (and, less often, animal) communities. Interestingly, the largest group have opted for work in the prehistoric past, using the techniques of Quaternary ecology such as pollen and diatom analysis, macrofossil analysis, and stratigraphy of Holocene deposits (Crabtree 1968, 1975; Barber 1976; Pears 1977; Jones and Cundill 1978) tied together with $^{14}$C dating. Early historical times and the biogeography of historical times (e.g. of the type of work pioneered on hedge history by the Nature Conservancy Council and the Institute of Terrestrial Ecology (Pollard 1974) and on woodland history by Rackham (1975)) have not attracted many geographers, although there are perhaps signs of increasing interest, as nearly every bog and lake in Britain seems to have had its depths plumbed for cores. Work overseas includes J. Flenley's (1979) book on the history of the equatorial rain forest, and work by F. Oldfield (1977a) and Walker and Flenley (1979) on Papua New Guinea's vegetation history.

*Quaternary ecology*

In common with much work by biologists and limnologists, geographers have helped to establish the nature of the flora and fauna of the terminal Pleistocene and also that of the transition during the early Flandrian from tundra to closed deciduous woodland. This transition has now been thoroughly investigated and its main stages are clear, although there have been residual problems over the role of pine on different soils (Oldfield 1963) and over the height of the tree-line on some of the hill areas such as North Wales and the North Pennines (Taylor 1974; Squires 1978). The impact of prehistoric man has been studied in some depth, and a symposium devoted largely to this theme formed a special issue of the *Transactions* of the Institute of British Geographers (Curtis and Simmons 1976). Study of the role of Mesolithic communities has received attention from R. L. Jones (1976a), for example, who showed that a skeleton of *Bos primigenius* in North Yorkshire was embedded in deposits which had a charcoal layer at the relevant horizon. In the uplands also, Simmons was the first to apply the archaeologist G. W. Dimbleby's ideas to organic deposits and hence to demonstrate that Mesolithic cultures might be responsible for forest clearance. Conventional wisdom at the time decreed that such recessions started with Neolithic cultures but many other earlier instances have now been shown. The advent of farming communities was responsible for much more human impact upon the landscape and this has been chronicled by geographers for a number of areas, including a particularly dense network of profiles from the North York Moors (Simmons and Cundill 1974a, 1974b; Atherden 1976; Jones 1976b, 1977, 1978) and the Lake District (Oldfield 1963), as well as investigations in Wales (Taylor 1973; Taylor and Smith 1968; Crabtree 1969) and Scotland (Edwards 1974), and difficult regions with rather unpromising post-glacial deposits such as the New Forest (Barber 1975). In common with the work of biologists and archaeologists, this work has established the importance of the Bronze and Iron Ages particularly in reducing the forest cover of the landscape, especially in the uplands. Thus the essential lineament of our upland landscapes, its openness, dates largely from prehistoric times. One of the elements of the open vegetation is often blanket bog and here it has been suggested that deforestation at various periods led to the growth of blanket peat and that its inception was not therefore always caused by climatic change (Taylor 1972). One effect of deforestation was to release silt into the drainage, and mineral layers in valley peat deposits have been laid at the feet of that process by Simmons *et al.* (1975). Another study of the interaction of climate and human activity involved the movement of the Cairngorm tree-line (Pears 1968) and O'Sullivan has investigated the history of the relict native pine forests of central Scotland (1974), 1977).

Although rooted in prehistory, the continuity of organic deposits has sometimes meant that a virtually continuous record of vegetation change has been maintained through historic times, so that events like, for example, the laying-waste of farmlands after Viking raids, the harrying of the North, charcoal smelting for metal working, and the planting of exotic trees for amenity, have all been recognized. The techniques of Quaternary ecology, in their stratigraphic context especially, have been used to reconstruct in considerable detail the fluctuations of sea-level during the Holocene (Tooley 1978). The technique of site catchment analysis, used by archaeologists to give a spatial dimension to their economic studies, has been commented upon by Davidson (1972) and applied in a rather tentative way to Mesolithic man–deer relationships by Simmons (1975).

Perhaps the most creative and prolific work of recent years has been that of F. Oldfield and his collaborators on material flux, especially in lake sediments. This started with the history of Lough Neagh, including reference to recent problems of cultural eutrophication (Oldfield *et al.* 1973). They have now gone on to establish a high standard for palaeolimnological work (Oldfield 1977b) and have reached out, in collaboration with other earth and physical scientists, to establish new techniques in Quaternary research; for example, magnetic susceptibility (Olfield *et al.* 1975; Bloemendal *et al.* 1979) which provides a rapid and cheap way of correlating numerous cores from the same deposit, and dating using $^{210}$Pb which is of great accuracy and value for recent sediments (Appleby *et al.* 1978, 1979). A considerable number of papers by this group are currently in press in a variety of ecological, geophysical, and general scientific journals.

*Cultural biogeography*

Compared with, for instance, North American colleagues, there has been little work of this kind undertaken by British geographers, for whom the combination of a cultural approach to prehistory and early history and the interaction of those societies with particular taxa in such processes as domestication, has had little appeal. It is no surprise to see our work in this field dominated by a graduate of the Berkeley department, D. R. Harris. Working initially in historical ecology (see below), Harris has developed a series of papers starting with the domestication of the goat (1962) and going on to generate a series of models for the early phases of domestication of plants and animals, stressing *inter alia* the ecosystems in which early domesticates were embedded (1969, 1973) and developing in recent work (1978) a complex approach which seems altogether superior to the 'one-cause' models which have been in contention on the archaeological stage for a long time. Harris's work has been better known and quoted among archaeologists than geographers, leading to his appointment to a Chair in that field in 1979. Harris has also published work on the nature of tropical vegetation (1974) which reflect his interest in early agriculture in those regions, and a standard paper on recent plant invasions on the arid south-west of the USA, drawing attention to the human factors in these changes (1968).

## Historical ecology

This rather general term is by convention applied to studies of the recent dynamics of plant and animal communities, as in, for instance, the tracing of a woodland's history over the last few hundred years by the analysis of lichen diversity, ground flora diversity, and tree age and uniformity as well as documentary records. The radical changes wrought by sailors and settlers on island biota were an early target of work of this kind on a meso-scale: Harris (1965) conducted such a study of the Outer Leeward Islands, and Watts (1966) looked at changes in the vegetation of Barbados 1627–1800, and of Nevis (1973) in more recent years, as well as writing more generally on the themes of persistence and change in the vegetation of oceanic islands (1970). Also dealing with the last few hundred years has been the work of J. A. Matthews (1975, 1977) on the lichenometric dating of glacier fluctuations in Norway, which has also extended to palaeo-temperature estimations for the 'Little Ice Age' in that country (1976). E. Maltby has been investigating the relationships between land-use changes and soil characteristics with emphasis on recent years, especially on UK uplands, and using soil micro-organisms as an indicator (1975, 1976). He is also currently monitoring the soil and biotic changes following the great fires of the North York Moors in recent drought years. Relatively recent biotic change has also been the theme of Jarvis's studies on the introduction to Britain of species like the Cedar of Lebanon (1974) and the genus *Cistus* (1972) as well as the wider topic of the role of N. American plants in horticultural innovation in England (1973). Work on the history of woodlands in Wiltshire and in Worcestershire by P. Gough is in progress.

## Review

No grand conclusions emerge from this survey, except perhaps the dominance and continuing popularity of palaeo-ecological studies, especially those concerning prehistoric communities, often involving collaboration with archaeologists; at least one attempt at a regional synthesis of the evidence of settlement, economy, and biotic change (Spratt and Simmons 1976) has been made. However, regional vegetation histories are probably no longer very useful, and problem-oriented research is likely to be more fruitful, involving such questions as the intensification of land use in the late Mesolithic, soil-vegetation changes during blanket bog inception, and, very importantly, close-up studies of biotic change at and in settlements and fields of prehistoric age, often using soil-pollen analysis. Historical ecology of individual stands of woodland or regions like the New Forest may be expected to gain ground in an applied sense as a background to management policies, especially those involving the discovery and protection of very ancient pieces of relict woodland.

The accurate description of plant communities has long been a concern of biologists, and geographers have of late contributed to this field, as they have to the dynamics of vegetation change, coming to recognize the multi-factorial nature of the trends and the usefulness of the dynamic-trophic model of the ecosystem. It follows that there has been an interest in what contemporary human communities have done by way of alteration of these communities and the ways in which effective management of vegetation (often involving outdoor recreation, a general field to which many UK geographers have been attracted) for various purposes may be attempted. The flow of information for such work has been considerably augmented by data from remote sensing, all the way from the hand-held camera in a light aeroplane to satellite imagery.

### Vegetation description and ecological dynamics

Since many geographers have sought to inventory vegetation in the course of their work, it is not surprising that they have also taken an interest in the various methods of describing and classifying it. The most comprehensive papers are by C. M. Harrison (1971, 1975), who has collaborated with the US geographer R. E. Frankel (1974) in comparing classification systems based on numerical analysis and on phytosociology; the latter approach has also been examined by Tivy *et al.* (1971), and the wider problems of classification and description of vegetation were commented upon earlier by Pears (1968a). Harrison's (1975) account is notable for having appeared in a manual of methods in plant ecology edited by a biologist. Other workers have analysed particular biotic communities with an emphasis on their dynamics, either natural or man-made. Eyre's (1966) paper on the vegetation of an upland in the South Pennines not only emphasized the lack of naturalness in moorland communities but served as an exemplar of his view of the whole field of geography as human ecology, a view not then particularly fashionable. His major contribution in this field has been the world-scale *Vegetation and Soils* (1968), which is perhaps one of the most quoted texts on biome description in the English-speaking world, and beyond. At a local scale, biotic communities whose evolution, status, and dynamics have been studied by biogeographers include machair (Dickinson 1972), shingle and other coastline plants (Randall 1975, 1976, 1977), numerous aspects of tropical forest and savanna vegetation (Eden 1964, 1974a, 1974b), and Tomlinson (1975, 1977) has worked with biologists in determining features of the vegetation, stratigraphy, and nutrient status of the bog vegetation of the Silver Flowe, a large mire complex in south-west Scotland. This is the place to acknowledge the immense contribution made by D. R. Stoddart to the knowledge of the flora, fauna, ecology, and conservation of coral reefs and atolls, most of which appears in the *Atoll Research Bulletin*. Studies focused upon the role of animals in ecosystem dynamics are rare, but Barkham (1975a, 1975b, 1976, 1979 (with T. D. Jennings)) has been evaluating the numbers and role of slugs in mixed deciduous woodland in Norfolk, an extension of his earlier work with the late J. Norris (1970a, 1970b) on multivariate procedures and discriminant analysis as tools of analysis in comparing the vegetation and soil vegetation relationships, of Cotswold beechwoods. He has also taken a close look at one of the curious Dartmoor oak copses, Black Tor Copse (1978), and has work in press on the population of the wild daffodil. R. Pullan (work in progress) has demonstrated an interest in the role of termites in the ecological dynamics of savanna. Work in progress analogous to these examples includes that of J. Emberlin on birch regeneration in the Inverpolly National Nature Reserve in North-west Scotland, K. Farrington on woodlands in the south Midlands, S. Huber on vegetation boundaries in the Malvern Hills, and T. Dargie on vegetation dynamics in southeast Spain. A narrower but more detailed focus is apparent in the work of P. Coker on the archaeology and phytogeography of several species of boreal and montane plants in Britain (1966, 1969), as well as members of the Ericaceae such as *Phyllodoce caerulea* and *Erica vagans* (1973, 1974). A wider approach has been adopted by some investigators in looking at the role of the biotic component in the context of the whole ecosystem. F. Courtney (1977) has published a review paper on the effects of forest on the hydrological cycle, and P. Furley has a series of papers on the relationships between slope, soil and vegetation of tropical areas in Belize (1974, 1976) as well as looking at the effect of an individual species, the Cohune palm, on the development and nature of its soil profile (1975). Compared with, say, North America, the interest shown in the biogeography of the city and urban vegetation is minimal, although J. Emberlin has participated in work on air pollution and lichen distribution, a field in which the book of Hawkesworth and Rose (1976) is well known.

*Impact studies*

The impetus and framework provided in the USA by the Environmental Impact Statements required under the terms of the 1969 National Environmental Protection Act, and its imitators in other countries, has not yet come to Britain. Impact studies have therefore tended to be *ad hoc* evaluations of particular processes as they arise, with no common methodology, and are often difficult to distinguish from the next category which involves management of ecosystems. Nevertheless, we may point out C. M. Harrison's (1974) paper on heathlands and their conservation which appeared in Warren and Goldsmith's collection of essays. Focused on Britain, and with several contributions by geographers, this book is strongly influenced by the approach to the conservation of biotic communities developed in the M.Sc. Conservation course at University College, London. C. M. Harrison also has joint work on the wear of amenity areas in south-east England, published by the Recreation Ecology Research Group, and more in press. Impact frequently causes instability, and for a geographer's examination of this concept, as well as some empirical examples, we must turn to the work of Warren and Harrison (1970) and to that of A. R. Hill. Apart from reviews of the ideas involved in concepts of ecological stability (1972, 1975), Hill has published work on the environmental impacts of land drainage in Ontario (with W. C. Found and E. S. Spence 1975, 1976) and on the role of stream biota in modifying nitrogen levels raised by cultural eutrophication (1977). More general work on the impacts of oil-related development in Scotland has been directed by B. D. Clark at Aberdeen but this is not solely ecological in its focus.

## Management, including recreation ecology

Many biogeographers in the course of their work have been called in to advise and report on the management of biotic communities, but such service rarely leads to publication. Examples of systems involved are nature conservation and outdoor recreation and there is no doubt that geographers, both academic and applied, have been attracted to work in the latter; we may speculate on the precise combinations of fashion, outdoor activity, and the chance actually to change something, which may have brought this about. The big general question in this field (and a very elusive one) is that of carrying capacity: what does the term mean and how can it be measured? Barkham (1973, 1975c) has addressed himself to these questions and to their relation to the sort of research in ecology now being undertaken, and a number of others have investigated particular problems. One example is the large project directed by J. Tivy (who in 1972 also published work on recreational carrying capacity) on the effect of recreation on freshwater loch shores, for the Countryside Commission, Scotland (Tivy and Rees 1978). She has also acted as consultant on accelerated erosion in Scottish moorland habitats, and for the Sutherland development plan, and in an environmental assessment of a pump-storage scheme. Hepburn (1977) has made an ecological assessment of National Trust properties in Kent, and Barrow (1973) has commented upon the experimental restoration project at Tarn Hows (Lake District N.P.) which he formulated. Warren and Goldsmith, in an often-quoted paper (1976) studied the ecological impact of recreation on the Isles of Scilly; together with C. M. Harrison, Warren has put forward a discussion document on a nature conservation plan for Shetland (1974). A different management problem, that of the difficulties of managing successional communities, has been tackled by B. Green (1972).

## Remote Sensing

This rapidly developing set of techniques deserves separate consideration but here we will note only a few matters particularly relevant at this point. In Biogeographical studies, the introduction and vigorous promotion of remote sensing is principally associated with L. F. Curtis. Although developing it initially in applications to pedology, he has broadened his interests to include a basic text (with E. C. Barrett 1976) and an important symposium (with Peel and Barrett 1977). One of his former research students, P. Curran, has used a variety of remote sensing techniques to monitor various seasonal changes in part of the protected wetland area of the Somerset Levels. Other users of remote sensing techniques have been, for example, M. M. Cole (1968, 1975) in connection with her work on mineral prospecting (see below also) and Barkham (with Goodfellow 1974) on the spectral transmission curves for beech canopy, and V. Brack (1975) on the use of multispectral imagery in the management of upland resources. We may expect an expansion of this work in many applications in the near future.

*Review*

One theme which emerges from these studies is the interesting balance between pure and applied research. While it is clear that fundamental research is still in a reasonably healthy state, it is equally apparent that the scale of environmental problems in the fields just discussed is such that geographers (among, and often with, others) feel they have a contribution to make; it is perhaps possible to see emerging in some of these investigations an approach to both natural and social sides of the equations posed, an area in which we might hope that geography has something special to offer.

## BIOTIC RESOURCES

The boundary between the natural world and the man-shaped one is ill-defined; in one context, for instance, vegetation and land use are very hard to separate. Thus a number of studies of the ecological approach to man's use of natural resources stress the importance of the biotic components of the resource systems, either as ends in themselves or parts of a seamless and interactive web. Not surprisingly, this has led to expressions of concern on the part of some authors, not only about local or regional environmental problems but about the over-all tenor and trend of man–environment.

*Land-use ecology*

Clearly, the concepts and models used in ecology have considerable value in the study of land use which can in many ways be regarded as manipulated ecosystems. Writing with W. B. Morgan, R. Moss has also explored the interface between ecological concepts and geography (1965); the standard paper for geographers on the ecosystem as a useful model remains that of D. R. Stoddart (1965). An attempt to look at man–environment relationships of various complexities as ecosystems, using modern ideas of energy flow, was formulated in an Open University unit by Learmonth and Simmons (1977). The application of an ecological approach to the interpretation of land-use systems and their problems has been widespread and only a few examples can be singled out here: R. Pullan on farmed savanna (1974, 1974b), Moss and Morgan (1972) on savanna–forest relationships, R. Moles (1977) on a biogeographical approach to farmland, Mather's work (1972, 1974) on areal variations of land-use productivity (in biological terms) in the Scottish Highlands and on the use of those lands as habitat for red deer, are all good examples, as are M. R. Moss's studies of the productivity of Malaya (1976, 1978). Following her earlier work (as M. M. Bower, 1961, 1962) on erosion in blanket peats, M. M. Cruickshank has joined with J. G. Cruickshank in a survey of neglected agricultural land in the Sperrin Mountains of Northern Ireland (1977); J. A. Taylor has been responsible for a large corpus of work which includes studies of upland Wales, in particular the management of hill lands, the use of peat as a resource, and the mapping of vegetation (1967, 1970, 1977, 1978; with D. Job 1978). A more generalized but analogous approach is taken by J. Tivy (1973) in her edited collection *The Organic Resources of Scotland*, to which she herself was a major contributor. The relationships between soils, land use, and rural decision-making have been explored by Young (1973, 1975), and ecological indicators of reclamation on Exmoor have been chronicled by Maltby (1975).

A number of papers on particular conservation problems of a broad ecological character have appeared: at a meso-scale, Simmons and Vale (1975) pointed out the complexities of protecting the last stands of virgin California Coast Redwood. M. Eden (1975) has commented upon the demise of the Amazonian rain forest. In the course of his work with the Nature Conservancy Council H. J. Williams has investigated the relationships between nature conservation and farming, and the effect of combining conservation and chemical defence at Porton Down. S. Byrne has been working on the problems of china clay waste in Cornwall. At a much grander scale, Warren has been heavily involved in current work on the causes and cures of desertification (1977, with J. K. Maizels). Interest in animals is unusual but Pinder and Barkham (1978) have discussed the contribution of captive breeding to the conservation of rare mammals. The collection edited by J. Winslow (1977) on Melanesia is remarkable for its sustained ecological approach to problems of development.

*Specific resources*

This field is dominated in the UK by M. M. Cole's work on the use of plants as indicators of mineral concentrations, employing both ground survey geochemical methods and applications of remote sensing (1973, 1975). She has applied these methods in many parts of the world, notably in Australia and southern Africa. Apart from this work, most analogous investigations by others have been on soil resources.

*General approaches*

Many of the writers so far mentioned have commented in one way or another on the over-all state of man–environment relations from an ecological or other biotic perspective. Papers on conservation as an attitude to man's use of the earth and its resources have appeared by Simmons (1979b) and in his textbook (1974) which, although maintaining a relatively objective set of attitudes, suggested that continuation of the present trajectory was likely to lead to large-scale ecological instabilities. A much more polemical viewpoint is put forward by S. R. Eyre in his book *The Real Wealth of Nations* (1978) which argues that the primary organic productivity of a nation is its only long-term renewable resource and that the balance of the human population with this material is critical to considerations of survival. The crucial flow in considerations of the ecology of resources is perhaps energy, both solar and fuel-based, and both authors recognize the crucial debate now proceeding over future sources. Another book on environmental management from an ecological perspective is by the geomorphologist C. C. Park (1979), and 1980 will see the appearance of J. Tivy and G. O'Hare's, *Man and the Ecosystem*.

In any resource process the role of human values is very important. In geography this has been mostly taken into account in studies of hazards and in phenomenological approaches to the subject; here we may note R. Moss's papers on responsibility and ethics with regard to the environment (1972, 1975), written particularly from a religious viewpoint.

*Review*

We might argue that geographers were concerned with the environment, its quality, and its resources, long before it became so publicly fashionable to be so. But there is relatively little evidence that, as a group, geographers took the chance of the 1960s and early 1970s to bring their work, attitudes, and skills before a wider public and a wider range of grant-aiding bodies. The research and writing discussed above is, therefore, a rather smaller corpus than might be expected from a knowledge of the history of geography and its aspirations during the nineteenth and twentieth centuries.

## TEXTBOOKS

These works act, perhaps, as summations of the state of the art at a given time although constrained by what students are thought to be able to understand and by what publishers think will make profits. The classic books of Newbigin (1936) and Anderson (1951) have been succeeded by a number of biogeography texts in Britain (interestingly, only one seems to have come out of the flourishing West Coast schools in North America: M. Kellman 1975). In the traditional mould of basic phytogeography, there is B. Seddon's *Introduction to Biogeography* (1971) with its emphasis on the origins and distribution of affinity groups within the British flora. The year 1980 should see the appearance of R. L. Jones's interdisciplinary *Biogeography*, prepared for geographers, geologists, and biologists. A major group have an ecological approach to their material, starting with P. Dansereau (1957) *Biogeography. An ecological perspective*, and including J. Tivy's (1971) *Biogeography. A study of plants in the biosphere*; M. Kellman's *Plant Geography* (1975), and N. Pears' *Basic Biogeography* (1977), written especially for beginning students who have no science background. The trophic-dynamic concept of the ecosystem is the foundation of the longer and more advanced book by D. Watts, *Principles of Biogeography* (1971). The year 1979 saw the appearance of Simmons' *Biogeography: natural and cultural* of which nothing more should be said here except that it is unlike the others.

Reflecting the lack of discriminatory boundaries, books by biologists still continue to be useful in advanced geography teaching. Polunin's broad sweep in *Introduction to Plant Geography* (1960) has not been equalled in any more recent text, for example, and the concise, lively and well-produced *Biogeography. An Ecological and evolutionary approach* (1976) by Cox, Healey, and Moore continues to be widely used. In all this publishing one feature stands out above all others: a neglect of animals and the oceans by most of the geographers. We may hope that future texts will pay more attention to these important elements of the biosphere.

## THE PLACE OF BIOGEOGRAPHY

We might start an overview of this brief survey by asking, 'is there a distinctive British school of biogeography?'. From the evidence presented here, the answer must be no; largely, I suspect, because much of the work discussed, though valid in its own terms, is not distinguishable from similar work done by practitioners in other disciplines. If we read a central journal such as the *Journal of Biogeography* (whose founding editor was D. Watts), then it is scarcely possible to tell from the papers themselves which are written by geographers, let alone their nationality; and when a geographer writes in the *Journal of Ecology* or the *Journal of Archaeological Science* his paper is not noticeably different from that of the biologist or the archaeologist. But even if there is this lack of distinguishing features in our research, it is scarcely possible to conceive of our teaching without the material on natural ecosystems and biota, and man's impact upon them, and our attitudes to man–biota relationships.

I therefore suggest two particular roles for biogeography as it stands within our subject. First, geographers in general had little to do with the great years of environmental consciousness-raising that peaked in the Stockholm Conference in 1972. The profession at that time had other concerns and, in any case, few of us had any taste for the grittier sort of polemic which was perhaps a necessary feature of those years. Even though the high tide of concern has receded, there is still a tremendous amount of work to be done in achieving an ecologically stable *modus vivendi* between man and nature; the trumpet fanfare may have faded but the quieter instruments have now to take up the theme and work it out in all its variations. One of the realizations of some of the 'public ecologists' of those years was that although the biosphere might be an envelope for human affairs, adjustments within it had to be made through the media of human cultures and that need to bring together natural and social systems within a spatial framework has the air of a geographical manifesto. So biogeography with, of necessity, a foot in both natural science and social science, if we are to understand the reasons for the ways in which man uses the biosphere, seems to me a natural point of linkage with the concern for man–environment relationships which is still a problem in the real world.

Back inside the house, there is a second role for biogeography. Its evolving content shows a strong trend towards studies which involve both man and biota. It seems particularly fitted, therefore, to be in the vanguard of a resurgence of those lineaments in our subject which have emphasized man–environment relationships. These themes have never been lost but have perhaps been overshadowed in recent years, and our new analytical skills (see for example Bennett and Chorley 1978) should contribute to an increased understanding of the complex interactions between our species and its environmental systems. No biogeographer would claim that his field ought to have exclusive claim to this section of the subject but, as. D. Watts suggested in his address to the Geographical Association in 1978, biogeography seems at present well fitted to be a take-off point. Thus our studies can feel secure in being linked to a long and honourable tradition of study within and without geography as well as having the promise of an interesting and vital future of both pure and applied work.

# REFERENCES

ANDERSON, M. S. 1951. *The Geography of Living Things.* London: EUP.

APPLEBY, P. G., BATTATBEE, R. W., and OLDFIELD, F. 1978. Alternative [210]Pb dating: results from the New Guinea highlands and Lough Erne. *Nature, Lond.* 271: 339–42.

APPLEBY, P. G., OLDFIELD, F., THOMPSON, R. and HUTTUNEN, P. 1979. [210]Pb dating of annually laminated lake sediment from Finland. *Nature. Lond.* 280: 53–5.

ATHERDEN, M. A. 1976. Late Quaternary vegetational history of the North York Moors, III. Fen Bogs. *J. Biogeogr.* 3: 115–24.

BALL, I. R. 1976. Nature and formulation of biogeographical hypotheses. *Syst. Zool.* 24: 407–30.

BARBER, K. E. 1975. Vegetational history of the New Forest: a preliminary note. *Proc. Hants Field Club Archaeol. Soc.* 30: 5–8.

BARBER, K. E. 1976. History of vegetation. In S. Chapman (ed.), *Methods in Plant Ecology.* Oxford: Blackwell, 5–83.

BARKHAM, J. P. 1973. Recreational carrying capacity. *Area* 5 (3): 218–22.

BARKHAM, J. P. 1975a. Food of slugs in mixed deciduous woodland in Norfolk, England. *Oikos,* 26 (2): 211–21.

BARKHAM, J. P. 1975c. Carrying capacity and ecological research, *Science, Technology and Environmental Management* (ed. R. D. Hey and T. D. Davies). Farnborough: Saxon House, 9–16.

BARKHAM, J. P. 1978. Pedunculate oak woodland in a severe environment: Black Tor Copse, Dartmoor. *J. Ecol.* 66: 707–40.

BARKHAM, J. P. and NORRIS, J. M. 1970a. A comparison of some Cotswold beech-woods using discriminant analysis. *J. Ecol.* 58: 603–19.

BARKHAM, J. P. and NORRIS, J. M. 1970b. Multivariate procedures in an investigation of vegetation and soil relations of two beech woodlands. Cotswold Hills, England. *Ecology,* 51: 630–9.

BARKHAM, J. P. and GOODFELLOW, S. 1974. Spectral transmission curves for a beech (*Fagus sylvatica* L.) canopy. *Acta Botanica Neerlandica,* 23 (3): 225–30.

BARKHAM, J. P. and JENNINGS, T. D. 1975b. Slug population in mixed deciduous woodland. *Oecologia* (Bren.), 20: 279–86.

BARRETT, E. C. and CURTIS, L. F. 1976. *Introduction to Environmental Remote Sensing.* London: Chapman and Hall.

BARROW, G. C. 1973. Tarn Hows Experimental Restoration Project. *Recreation News Supplement,* No. 9, 13–18.

BENNETT, R. J. and CHORLEY, R. J. 1978. *Environmental Systems. Philosophy, Analysis and Control.* London: Methuen.

BLOEMENDAL, J., OLDFIELD, F., and THOMPSON, R. 1979. Magnetic measurements used to assess sediment influx at Llyn Goddionduon. *Nature, Lond.* 280: 53–5.

BOWER, M. M., *see* Cruickshank.

BRACK, E. V. 1975. An investigation into the use of multi-spectral photography for soil surveying in upland Britain. *Proc. 4th Annual Remote Sensing of Earth Resources Conference.* Univ. Tullonoma, Tennessee: Tennessee Space Institute.

CLEMENT, D. (ed.) 1978. *Directory 2.* Biogeography Study Group.

COKER, P. D. 1966. Biological flora of the British Isles. *Sibbaldia procumbens. J. Ecol.* 54: 823–32.

COKER, P. D. 1969. *Arenaria norvegica* in Morvern. *Trans. Bot. Soc. Edin.* 50: 557–64.

COKER, P. D. 1973.   Biological flora of the British Isles. *Phyllodoce caerulea. J. Ecol.* 61: 901–14.

COKER, P. D. 1974.   The status and ecology of *Erica vagans* in Fermanagh. *J. Linn. Soc. (Bot.)* 69: 153–95.

COLE, M. M. 1968.   Observations of the Earth's resources. *Proc. Roy. Soc.* A.308, 173–82.

COLE, M. M., OWEN-JONES, E. S., and CUSTANCE, N. D. E. 1973.   Remote sensing in mineral exploration, in *Environmental Remote Sensing: Applications and Achievements* (ed. E. C. Barrett and L. F. Curtis). London: Edward Arnold, 50–6. Paper presented at Bristol Symposium on Remote Sensing, Bristol, October 1972.

COLE, M. M. OWEN-JONES, E. S., CATT, P. C., COUSANS, C. M., and CHANDLER, B. J. 1975.   An evaluation of visual and semi-automated approaches to the recognition of spectral signatures in air survey and satellite imagery of mineralised natural terrain. *XV Convegno Internazionale Tecnico Scientifico Sullo Spazio,* 263–73.

COURTNEY, F. M. 1977.   The effects of forest on the hydrological cycle: a review of existing literature. *University of Leeds School of Geography Working Paper No. 204.*

COX, C. B., HEALEY, I. N., and MOORE, P. D. 1976.   *Biogeography, an ecological and evolutionary approach.* Oxford: Blackwell Scientific Publications.

CRABTREE, K. 1968.   Pollen analysis. *Sci. Progr.* 56: 83–101.

CRABTREE, K. 1969.   Post-glacial diatom zonation of limnic deposits in N. Wales. *Mitt. Internat. Verein. Limnol.* 17: 165–171.

CRABTREE, K. 1975.   A review of methodological advances in pollen analysis in *Processes in Physical and Human Geography* (ed. R. F. Peel, M. Chisholm, and P. Haggett). London: Heinemann. 266–83.

CROZIAT, L. 1958.   *Panbiogeography.* Caracas: The Author.

CROZIAT, L. 1964.   *Space, time, form: the biological synthesis.* Caracas: The Author.

CRUICKSHANK, M. M. 1961.   The distribution of erosion in blanket peat bogs in the Pennines. *Trans. IBG* 29: 17–30.

CRUICKSHANK, M. M. 1962.   The cause of erosion in blanket peat bogs. *SGM* 78: 33–34.

CRUICKSHANK, M. M. and J. G. 1977: Survey of neglected agricultural land in the Sperrin Mountains, Northern Ireland. *Irish Geography,* 10: 36–43.

CURTIS, L. F. and SIMMONS, I. G. 1976.   Man's impact on past environments. A thematic number of *Trans. IBG* NS1(3), 1–384.

DANSEREAU, P. 1957.   *Biogeography. An ecological perspective.* New York: Ronald Press.

DAVIDSON, D. A. 1972.   Terrain adjustment and prehistoric communities, in *Man, Settlement and Urbanism* (ed. P. J. Ucko *et al.*). London: Duckworth. 17–22.

DICKINSON, G. 1972.   The evolution of the Machair grassland surface of South Uist, Outer Hebrides, Scotland. *Proc. IGU Congress Montreal.* Section III.

EDEN, M. 1964.   The savanna ecosystem, southern Rupurnini, British Guiana. *McGill University, Savanna Research Series,* Vol. 1.

EDEN, M. 1974a.   The origin and status of savanna and grassland in southern Papua. *Trans IBG* 63: 97–110.

EDEN, M. 1974b.   Palaeoclimatic influences and the development of savanna in southern Venezuela. *J. Biogeogr.* 1: 95–109.

EDEN, M. 1975.   Last stand of the rain forest. *Geog. Mag.* 47: 578–82.

EDWARDS, K. J. 1974.   A half-century of pollen analytical research in Scotland. *Trans. Bot. Soc. Edin.* 42: 211–22.

EYRE, S. R. 1966.   The vegetation of a South Pennine upland, in *Geography as Human Ecology* (ed. S. R. Eyre and G. Jones). London: Edward Arnold. 147–74.

EYRE, S. R. 1968.   *Vegetation and soils. A world picture.* London: Edward Arnold, 2nd edn.

EYRE, S. R. 1978. *The Real Wealth of Nations.* London: Edward Arnold.

FLENLEY, J. 1979. *The Equatorial Rain Forest: a geological history.* London: Butterworth.

FURLEY, P. 1974. Soil-slope plant relationships in the northern Maya mountains, Belize. *J. Biogeogr.* 1 (3): 171–286; 1 (4): 263–79.

FURLEY, P. 1975. The significance of the Cohune palm, *Ortiquya cohune*, Mart. Daregren, on the nature and in the development of the soil profile. *Biotropica,* 7, 1: 32–6.

FURLEY, P. 1976. Soil-slope plant relationships in the northern Maya mountains, Belize; Pt 3. Variations in the properties of soil profiles. *J. Biogeogr.* 3: 303–19.

GREEN, B. H. 1972. The relevance of seral eutrophication and plant competition to the management of successional communities. *Biol. Cons.* 4: 378–84.

HARRIS, D. R. 1962. The distribution and ancestry of the domestic goat. *Proc. Linn. Soc. Lond.* 173: 79–91.

HARRIS, D. R. 1965. *Plants, Animals and Man in the Outer Leeward Islands, West Indies. An ecological study of Antigua, Barbuda and Anguilla.* Berkeley and Los Angeles: University of California Press.

HARRIS, D. R. 1968. Recent plant invasions in the arid and semi-arid southwest of the United States. *Ann. Assoc. Amer. Geogr.* 56: 408–22.

HARRIS, D. R. 1969. Agricultural systems, ecosystems and the origins of agriculture, in *The Domestication and Exploitation of Plants and Animals* (ed. P. J. Ucko and G. W. Dimbleby). London: Duckworth, 3–16.

HARRIS, D. R. 1973. The prehistory of tropical agriculture: an ethnoecological model, in *The Explanation of Culture Change. Models in Prehistory* (ed. C. Renfrew). London: Duckworth, 391–417.

HARRIS, D. R. 1974. Tropical vegetation: an outline and some misconceptions. *Geography,* 59: 240–50.

HARRIS, D. R. 1978. Alternative pathways towards agriculture, in *Origins of Agriculture* (ed. C. A. Reed). The Hague: Mouton, 179–243.

HARRISON, C. M. 1971. Recent approaches to the description and analysis of vegetation. *Trans. I.B.G.* 52: 113–27.

HARRISON, C. M. 1974. The ecology and conservation of British lowland heaths in *Conservation in Practice* (ed A. Warren and F. B. Goldsmith). London: Wiley, 117–29.

HARRISON, C. M. 1975. The description and analysis of vegetation, in *Methods in Plant Ecology* (ed. S. B. Chapman). Oxford: Blackwell, 85–155.

HARRISON, C. M. and FRANKEL 1974. An assessment of the usefulness of phytosociological and numerical classificatory methods for the community biogeographer. *J. Biogeogr.* 1: 27–56.

HAWKESWORTH, D. L. and ROSE F. 1976. *Lichens as Pollution Monitors.* London: Edward Arnold, Inst. of Biol. Stud. No. 66.

HEPBURN, I. R., GREEN, B. H. and LAING, A. I. 1977. *An ecological assessment of National Trust properties in Kent.* Studies in Rural Land Use 13, Countryside Planning Unit, Wye College, Kent.

HILL, A. R. 1972. Ecosystem stability and man: a research focus in biogeography, in *International Geography* (ed. W. P. Adams and F. M. Helleiner). University of Toronto Press, 255–7.

HILL, A. R. 1975. Ecosystem stability in relation to stresses caused by human activities. *Canadian Geographer,* 19: 206–20.

HILL, A. R. 1977. The influence of nitrogen fertilizer on stream nitrate concentrations near Alliston, Ontario, Canada. *Progress in Water Technology,* 8: 91–100.

HILL, A. R., FOUND, W. C. and SPENCE, E. S. 1975. *Economic and Environmental*

*Impacts of Land Drainage in Ontario.* Geographical Monograph No. 6. Atkinson College, Department of Geography, York University.

HILL, A. R., FOUND, W. C. and SPENCE, E. S. 1976.   A study of the impacts of agricultural land drainage in Ontario. *Journal of Soil and Water Conservation*, 31: 20–4.

JARVIS, P. 1972.   Early introductions of *Cistus* species and their cultivation in England. *J. Roy Hort. Soc.* 97: 244–51.

JARVIS, P. 1974.   A history of the Cedar of Lebanon in Britain. *J. Roy. Hort. Soc.* 99: 539–46.

JARVIS, P. 1973.   North American plants and horticultural innovation in England 1550–1700. *Geogr. Rev.* 63: 477–99.

JENNINGS, T. J. and BARKHAM, J. 1976.   Feeding in woodland by the slug *Arion ater* (L). *Oikos*, 27: 168–73.

JENNINGS, T. J. and BARKHAM, J. 1979.   Litter decomposition by slugs in mixed deciduous woodland. *Holarctic Ecology*, 2: 21–9.

JONES, R. L. 1976a.   The activities of mesolithic man: further palaeobotanical evidence from north-east Yorkshire, in *Geoarchaeology* (ed. D. A. Davidson and M. L. Shackley). London: Duckworth. 355–67.

JONES, R. L. 1976b.   Late Quaternary vegetational history of the North York Moors. IV. Seamer Carrs. *J. Biogeogr.*, 3: 397–406.

JONES, R. L. 1977.   Late Quaternary vegetational history of the North York Moors. V. The Cleveland Dales. *J. Biogeogr.*, 4: 353–62.

JONES, R. L. 1978.   Late Quaternary vegetational history of the North York Moors. VI. The Cleveland Moors. *J. Biogeogr.*, 5: 81–92.

JONES R. L. 1980.   *Biogeography.* Amersham: Hulton Educational Press.

JONES, R. L. and CUNDILL P. R. 1978.   *Pollen Analysis.* British Geomorphological Research Group Technical Bulletins, No. 22. Norwich: Geo Abstracts.

KELLMAN, M. C. 1975.   *Plant Geography.* London: Methuen.

LEARMONTH, A. T. A. and SIMMONS, I. G. 1977.   Man–environment relations and complex eco-systems. Open University course D204 unit 8.

MACARTHUR, R. H. and WILSON E. O. 1967.   *The Theory of Island Biogeography.* Princeton: Princeton University Press.

MALTBY, E. 1975.   Numbers of soil micro-organisms as ecological indicators of changes resulting from moorland and reclamation on Exmoor U.K. *J. Biogeogr.* 2: 117–36.

MALTBY, E. 1976.   Soil organic matter and peat accumulation on Exmoor: a contemporary and palaeo-environmental evaluation. *Trans. IBG* NS 1 (3): 259–78.

MATHER, A. 1972.   Red deer land use in the Northern Highlands. Pt. 1 The Resource, *SGM* 88: 36–47. Pt. 2 Exploitation of the resource, *SGM* 88: 86–99.

MATHER, A. 1974.   Areal variations in land-use productivities in the Northern Highlands, *SGM* 90: 153–68.

MATTHEWS, J. A. 1975.   Experiments on the reproducibility and reliability of lichenometric dates, Storbreen Gletschervorfeld, Jotunheimen, Norway. *Norsk Geografisk Tidsskrift*, 29: 97–109.

MATTHEWS, J. A. 1976.   'Little ice age' palaeotemperatures from high altitude tree growth in S. Norway. *Nature, Lond.* 264: 243–5.

MATTHEWS, J. A. 1977.   A lichenometric test on the 1750 end-moraine hypothesis: Storbreen Gletschervorfeld, southern Norway. *Norsk Geografisk Tidsskrift*, 31: 129–36.

MOLES, R. 1977.   Biogeographical field study of farmland. *Journal of Geography in Higher Education*, 1: 20–5.

MOSS, M. R. 1978.   Potential primary productivity of Peninsular Malaysia. *J. Environ. Manag.* 171–83.

MOSS, M. R. and AIKEN, S. R. 1976. Man's impact on the natural environment of Peninsular Malaysia: some problems and human consequences. *Envir. Conserv.* 3: 273–84.

MOSS, R. 1972. Responsibility in the use of nature. *Proc. IARFA* Conference, Versailles, 1–45. (Limited circulation.)

MOSS, R. 1975. Responsibility in the use of nature. I. *Christian Graduate*, 28: 69–80.

MOSS, R. and MORGAN, W. B. 1965. Geography and ecology: the concept of the community and its relationship to environment. *Ann. Assoc. Amer. Geogr.* 55: 339–50.

MOSS, R. and MORGAN, W. B. 1972. Savanna and forest in Western Nigeria, in *People and Land in Africa South of the Sahara* (ed. R. M. Prothero). London: OUP, 20–7.

NELSON, G. J. 1973. Comments on Leon Croziat's Biogeography. *Syst. Zool.* 22: 213–20.

NEWBIGIN, M. I. 1936. *Plant and Animal Geography.* London: Methuen.

OLDFIELD, F. 1963. Pollen-analysis and man's role in the ecological history of the south-east Lake District. *Geog. Annl.* 45: 23–49.

OLDFIELD, F. 1965. Problems of mid-Post-glacial pollen zonation in part of northwest England. *J. Ecol.* 53: 247–60.

OLDFIELD, F. 1977a. Lake sediments and human activities in prehistoric Papua New Guinea, in J. H. Winslow (ed.) *The Melanesian Environment.* Canberra: ANU Press, 57–60.

OLDFIELD, F. 1977b. Lakes and their drainage basins as units of sediment-based ecological study. *Progress in Physical Geography,* 1: 460–504.

OLDFIELD, F., O'SULLIVAN, P. E. and BATTERBEE, R. W. 1973. Preliminary studies of Lough Neagh sediments. I. Stratigraphy, Chronology and Pollen-analysis, in *Quaternary Plant Ecology* (ed. H. J. B. Birks and R. G. West), British Ecological Society Symposium Volume No. 14. Oxford: Blackwell, 267–78.

O'SULLIVAN, P. E. 1974. Radio-carbon-dating and prehistoric forest clearance on Speyside (East Central Highlands of Scotland). *Proc. Prehist. Soc.* 40: 206–8.

O'SULLIVAN, P. E. 1977. Vegetation history and the native pinewoods, in *The Native Pinewoods of Scotland* (ed. R. G. H. Bunce and J. N. R. Jeffers). Institute of Terrestrial Ecology, Cambridge, 60–9.

PARK, C. C. 1979. *Ecology and Environmental Management—a geographical perspective.* Folkestone: Dawson.

PEARS, N. 1968. Some recent trends in classification and description of vegetation. *Geog. Annaler.* A 50, 162–72.

PEARS, N. 1968. Post-glacial tree-lines of the Cairngorm Mountains, Scotland. *Trans. and Proc. Bot. Soc. Edin.* 40: 536–55.

PEARS, N. 1977. Pollen analysis: a review of some developments and interpretation problems. *SGM* 93, 32–44.

PEARS, N. 1977. *Basic Biogeography.* London: Longman.

PEEL, R. F., BARRETT, E. C. and CURTIS, L. F. (ed.) 1977. *Remote Sensing of the Terrestrial Environment.* London: Butterworth. Colston Papers No. 28.

PINDER, N. J. and BARKHAM, J. P. 1978. An assessment of the contribution of captive breeding to the conservation of rare mammals. *Biol. Cons.* 13: 187–245.

POLLARD, E., HOOPER, M. D. and MOORE, N. W. 1974. *Hedges.* London: Collins.

POLUNIN, N. 1960. *Introduction to Plant Geography and some Related Sciences.* London: Longman.

PULLAN, R. A. 1974a. Farmed parkland in West Africa. *Savanna,* 3: 119–51.

PULLAN, R. A. 1974b. Biogeographical studies and agricultural development in Zambia. *Geog.* 59: 309–21.

RACKHAM, O. 1975. *Hayley Wood: its history and ecology.* Cambridge: Cambridgeshire and Isle of Ely Naturalists' Trust Ltd.

RANDALL, R. 1975.   Airborne salt deposition and its effects upon coastal plant distribution: the Monach Isles National Nature Reserve, Outer Hebrides. *Trans. Bot. Soc. Edin.* 42.

RANDALL, R. 1976.   The past and present status of sea pea in the British Isles. *Watsonia*, 11: 247–51.

RANDALL, R. 1977.   Shingle foreshores, in *The Coastline* (ed. R. S. K. Barnes). Chichester: Wiley, 49–61.

RANDELL, R. E. 1978.   *Series and Techniques of Vegetation Analysis.* Oxford University Press.

RAVEN, P. H. and AXELROD, D. I. 1972.   Plate tectonics and Australasian palaeobiogeography. *Science*, 176: 1379–86.

RAVEN, P. H. and AXELROD, D. I. 1975.   History of the flora and fauna of Latin America. *American Scientist*, 63: 420–9.

SEDDON, B. 1971.   *Introduction to Biogeography.* London: Duckworth.

SIMMONS, I. G. 1974.   *The Ecology of Natural Resources.* London: Edward Arnold.

SIMMONS, I. G. 1975.   Towards on ecology of Mesolithic man in the uplands of Great Britain. *J. Arch. Sci.* 2: 1–15.

SIMMONS, I. G. 1978.   Progress report: resource management and conservation. *Progress in Human Geography.*

SIMMONS, I. G. 1979a.   *Biogeography: natural and cultural.* London: Edward Arnold.

SIMMONS, I. G. 1979b.   Conservation of plants, animals and ecosystems, in *Man and Physical Landscape Processes* (ed. K. J. Gregory and D. Walling). Folkestone: Dawson, 241–58.

SIMMONS, I. G. and CUNDILL, P. R. 1974a.   Pollen analysis and vegetational history of the North York Moors. I. Pollen analysis of blanket peats. *J. Biogeogr.* 1: 159–69.

SIMMONS, I. G. and CUNDILL, P. R. 1974b. Pollen analysis and vegetational history of the North York Moors. II. Pollen analysis of landslip bogs. *J. Biogeogr.* 1: 253–61.

SIMMONS, I. G., ATHERDEN, M. A. CUNDILL, P. R. and JONES, R. L. 1975.   Inorganic layers in soligenous mires of the North York Moors. *J. Biogeogr.* 2: 49–56.

SIMMONS, I. G. and VALE, T. R. 1975.   Problems of the conservation of the California Coast Redwood and its environment. *Environ. Cons.* 2: 1–15.

SMITH, J. M. B. 1974.   Southern biogeography on the basis of continental drift: a review. *J. Aust. Mammal Soc.* 1: 213–30.

SMITH, J. M. B. 1975.   Living fragments of the flora of Gondwanaland. *Aust. Geogr. Stud.* 13: 3–12.

SPRATT, D. A. and SIMMONS, I. G. 1976.   Prehistoric environment and activity on the North York Moors. *J. Arch. Sci.* 3: 193–210.

SQUIRES, R. H. 1971.   Flandrian history of the Teesdale rarities. *Nature, Lond.* 229: 43–4.

SQUIRES, R. H. 1978.   Conservation in Upper Teesdale: contribution from the palaeoecological record. *Trans. IBG* 3: 129–50.

STODDART, D. R. 1965.   Geography and the ecological approach. The ecosystem as a geographical principle and method. *Geogr.* 50: 241–51.

STODDART, D. R. 1977.   Progress report: biogeography. *Progress in Physical Geography*, 1: 537–43.

STODDART, D. R. 1978.   Progress report: biogeography. *Progress in Physical Geography*, 2: 514–28.

TAYLOR, J. A. 1967.   Reconnaissance vegetation surveys and maps: (including a preliminary report of the vegetation survey of Wales (1961–66), in *Geography at Aberystwyth* (ed. E. G. Bowen, H. Carter, and J. A. Taylor). Univ. of Wales Press, 87–110.

TAYLOR, J. A. 1972. Climatic peat—a misnomer? *Fourth International Peat Congress,* Helsinki, 1: 471–84.

TAYLOR, J. A. 1973. Chronometers and chronicles: a study of the palaeoenvironments of West Central Wales. *Progress in Geography,* 5: 247–334.

TAYLOR, J. A. 1974. The role of climatic factors in environmental and cultural changes in prehistoric times. *Council for British Archaeology Symposium on the Effect of Man on the Landscape: The Highland Zone,* 6–19.

TAYLOR, J. A. 1977. The distribution and interpretation of the peat deposits of the British Isles. *Fifth International Peat Congress,* Poznan, Poland, IV, 228–43.

TAYLOR, J. A. 1978. The British upland environment and its management. *Geography,* 63: 338–53.

TAYLOR, J. A. and SMITH, R. T. 1968. The Post-glacial development of vegetation and soils in northern Cardiganshire. *Trans. IBG,* 47: 85–96.

TAYLOR, J. A. and TUCKER, R. B. 1970. The peat deposits of Wales, an inventory and interpretation. *Third International Peat Congress,* Quebec, 163–73.

TAYLOR, J. A. and JOB, D. A. 1978. Productivity and management of upland grazings on Plynlimon. *J. Biogeogr.* 5: 179–91.

TIVY, J. 1971. *Biogeography. A study of plants in the biosphere.* Reprinted 1973. Edinburgh: Oliver and Boyd.

TIVY, J. 1972. The concept of carrying capacity in relation to recreational land use, in the U.S.A. *Occasional Paper* No. 3, CCS Perth VII–58. Reprinted 1975.

TIVY, J. 1973. *The Organic Resources of Scotland: their nature and evaluation.* Editor/ Author, Chaps. 1, 6, 7, and 14. Edinburgh: Oliver and Boyd.

TIVY, J., DICKENSON, G. and MITCHELL, J. 1971. Application of phyto-sociological techniques to the geographical study of vegetation. *SGM* 87(2): 83–102.

TIVY, J. and REES, J. 1978. Recreational impact on Scottish lochside wetlands. *J. Biogeogr.* 5: 93–108.

TIVY, J. and O'HARE, G. 1980. *Man and the Ecosystem.* Edinburgh: Oliver and Boyd.

TOMLINSON, R., and BOATMAN, D. J. and HULME, P. 1975. Monthly determinations of the concentrations of sodium, potassium, magnesium and calcium in the rain and in pools on the Silver Flowe National Nature Reserve. *J. Ecol.* 63: 903–12.

TOMLINSON, R. and BOATMAN, D. J. 1977. The Silver Flowe II. Features of the vegetation and stratigraphy of Brishie Bog, and their bearing on pool formation. *J. Ecol.* 65: 531–46.

TOOLEY, M. J. 1978. *Sea-level Change: north-west England during the Flandrian stage.* Oxford: Clarendon Press.

VUILLEUMIER, F. 1977. Qu'est-ce que la biogéographie? *Cr. Soc. Bio.* 475: 41–66.

WALKER, D. and FLENLEY, G. R. 1979. Late Quaternary vegetational history of the Euga Province of Upland Papua New Guinea. *Phil. Trans. Royal Soc.* London, B286: 265–344.

WARREN, A. and HARRISON, C. M. 1970. Conservation, stability and management. *Area,* 2: 26–32.

WARREN, A. and HARRISON, C. M. 1974. A proposed nature conservation plan for Shetland, University College London. *Discussion Papers in Conservation,* No. 7, 61 pp.

WARREN, A. and GOLDSMITH, F. B. (ed.) 1974. *Conservation in Practice.* London: Wiley.

WARREN, A. and GOLDSMITH, F. B. 1976. The impact of recreation on the ecology and amenity of semi-natural areas; methods of investigation used in the Isles of Scilly. *Biol. J. Linn. Soc. Lond.* 2: 287–306.

WARREN, A. and MAIZELS, J. K. 1977. *Ecological Change and Desertification.* Background document A, Conf. No. 74, No. 7, UNEP, Nairobi and Pergamon, Oxford, 124 pp.

WATTS, D. 1966.    *Man's influence on the vegetation of Barbados, 1627–1800.* University of Hull, Occasional Papers in Geography, No. 4.

WATTS, D. 1970.    Persistence and change in the vegetation of oceanic islands. *Can. Geog.* 14: 91–109.

WATTS, D. 1971.    *Principles of Biogeography: an introduction to the functional mechanisms of ecosystems.* London: McGraw Hill.

WATTS, D. 1973.    From sugar plantation to open-range grazing: changes in the land use of Nevis, West Indies, 1950–1970. *Geogr.* 58: 65–8.

WINSLOW, J. (ed.) 1977.    *The Melanesian Environment.* Canberra: ANU Press.

YOUNG, A. 1973.    Rural land evaluation, in *Evaluating the Human Environment* (ed. J. A. Dawson and J. C. Doornkamp). London: Arnold. 5–33.

YOUNG, A. 1975.    Crop/land relationships and the nature of decision-making on land use. *FAO World Soil Resources Reports*, 45: 85–7.

# 6 Geomorphology

K. M. CLAYTON *University of East Anglia*

FOR the sixty years that geography has been widely taught at university level in the UK, geomorphology has dominated physical geography. In recent years, however, it has been increasingly challenged by other aspects of physical geography such as climatology, soils, and biogeography, while physical geography itself has declined to a much smaller proportion of total geographical activity than was the case forty years ago. These considerable changes have been masked by the over-all increase in numbers of geographers in higher education, but their impact will no doubt become increasingly apparent in the years of stability, or even decline, that lie ahead.

## Number and output

The number of geomorphologists conducting, and at least occasionally writing up, research in the UK must now be around 500. This is about double what it was thirty years ago, but probably no greater than a decade ago. It is far more likely to decline over the next twenty years than to increase. About half are teaching in higher education institutions, and over half of these (29% of the total) are in university departments of geography, or are geographers in departments such as environmental sciences. An appreciable number (almost 10%) are in other university departments (mainly geology and civil engineering) and 8% are in polytechnics. A very similar proportion (7%) are in government research establishments such as the Soil Survey, Institute of Hydrology, or the Land Resources Section of the Department of Overseas Surveys.

The biggest group, perhaps surprisingly in view of all that has been said about NERC funding over the last fifteen years, is research students, almost all of them in university departments of geography: they total around 37 per cent of those British residents listed in *Current Research in Geomorphology* 1975, (Gemmell, 1975). The number is of course swollen by those completing the writing up of their Ph.D.s beyond the allotted three-year period, but nevertheless it must represent an annual output of higher degrees of around 30–35 each year. Annual recruitment to teaching posts in higher education has probably taken 10 or even 15 geomorphologists in some past years of rapid growth, but in a steady-state situation we need no more than 4 or 5 a year to maintain the stock, and the present skewed age distribution of staff in post means that recruitment for the next fifteen or twenty years will be smaller still.

Nevertheless, at no time over the last thirty years has even the majority of British output of geomorphologists with higher degrees been able to find posts in British higher education establishments. Quite a high proportion did in the heyday of university expansion in the 1960s, and in the following expansion of the polytechnics, but some have always moved into other careers. However, the main effect of the high numbers of UK geomorphology postgraduates was to give a steady flow to the new and expanding universities of the Commonwealth, particularly perhaps to Canada, but also very significantly to Australia and New Zealand. The point at which these postgraduates cease to be British geomorphologists varies with the perception of the migrant and of the analyst, but they are not overtly included here, although this remarkable exodus has undoubtedly carried British geomorphological teaching far across the globe. In the absence of any comparable outlet today, British geomorphologists are likely to find themselves moving in larger numbers into government research and administration, into Area Water Authorities, and into industry, and it would be a narrow-minded educationalist who regarded that as bad. What is disappointing is that no one has done a proper follow-up study on postgraduates in geomorphology, particularly where they have not stayed in academia. In terms of justifying public investment in research training over the next two decades, such an analysis is urgently needed.

By no means all UK research is on the UK, but remarkably little work on the UK is by foreign geomorphologists, so that in the absence of a more thorough survey, a time-series of publications on the geomorphology of the UK is of interest. This is available since about 1963 (it is too incomplete before then) in *Geomorphological Abstracts* and its successors (*Geographical Abstracts* A, *Geo Abstracts* A) and the data are reproduced in Figure 6.1. If the latest issue of *Current Research in Geomorphology* is representative,

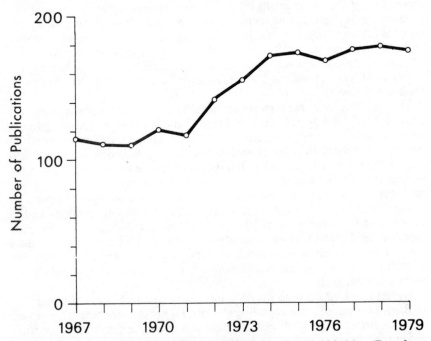

Fɪɢ. 6.1. Number of publications on the geomorphology of the British Isles. (Data from Regional Index, *Geo Abstracts* 1966–78.) Number of UK entries, 3-year running mean. [Fo. 6.3]

we might expect British papers to represent about 75 per cent of that part of UK geomorphological publications which may be assigned a regional classification—itself about 75 per cent of the total. The output of papers has increased by about 50 per cent between 1966 and 1978, with the most rapid rise in the early 1970s—the total has been stable since 1974. As an annual addition to the research literature it is quite impressive (almost four new papers each week), but as the output of 500 nominally active geomorphologists, it is appreciably less impressive—although not out of line with rates of publication in other academic subjects. Much research remains private—and uncompleted.

## Geographers as geomorphologists

In the UK, as in very many countries of the world, geomorphology is neglected by geologists. The traditional attitude of the geologists has been to emphasize old hard rocks, and even such topics as the Quaternary and sedimentology were neglected. Further, in the absence of any obvious signs of tectonic activity, there has seemed little of truly geological interest in the evolution of the British landscape, while since a satisfactory stratigraphical record appeared to end with the scattered remnants of Oligocene beds, there seemed no proper geological interest in the succeeding few million years. As with any such neglect, the result has been an intensification of activity by the geomorphologists within geography; indeed, British geomorphology has established a whole range of activities and interests of a geological nature, exploiting very efficiently the vacuum left by geological neglect. Thus some of the publications of geomorphologists are actually about sedimentology, while a high proportion of their work on the Quaternary is restricted to Quaternary stratigraphy.

There are of course no rules to restrict such activity, nor should there be. But unless this dissipation of effort is appreciated, it is difficult to comprehend the limitations and the achievements of British geomorphology today. We begin with a numerically large group holding a declining position within geography, and supported by only a few geologists, soil scientists, and civil engineers. Activity falls into three main areas: Quaternary studies, process studies, and landforms. The description and understanding of landform has declined from its pre-eminent position in the pre-war period, and Quaternary stratigraphy and studies of geomorphological processes have become fashionable and active. I shall discuss each of these in turn, beginning with those that are currently attracting most attention.

## Quaternary studies

The principal British textbooks on the Quaternary and on glacial processes and landforms are almost all written by geographers (Bowen 1978, Sugden and John 1976, Embleton and King 1975, Price 1973). These books include varying amounts of the background physics of glacier behaviour (for example see the useful review by Boulton 1977) but are in no real sense constrained by being written as geographical texts. Indeed, quite specialist work on glacial processes has been undertaken by geographers, although it would be wrong to ignore the very important role of geologists, in work on both glacial sedimentation and glacial erosion. Most work on glacial chronology has concentrated on the Devensian (last) glaciation, for the main approach to the chronology of the older glaciations has been through pollen analytical work on interglacial sediments, and relatively few geographers have applied this technique to these older sediments; it is more commonly used by them as a tool for the analysis of more recent, especially postglacial, environments. A significant school of Quaternary studies has developed from a succession of research students at Edinburgh led by J. B. Sissons, although their work has not gone unchallenged. The techniques used have included pollen analysis and $^{14}$C dating, but the morphological evidence of the ice-marginal features on aerial photographs has also been widely used. Such morphological techniques, it seems, are restricted to the last glaciation.

We may wonder whether so much effort would have gone into such work had we been able to start from a national geological survey mapping the Quaternary deposits in an adequate way. As is widely appreciated, the official survey of this country until quite recently regarded glacial drifts as unpleasant material obscuring the real rocks, and they remain unmapped at any adequate scale consistently across the whole country. Two compilations have, however, remedied to some extent the absence of a detailed national 'drift' map, that of Clayton (1963) in the Atlas of Britain, and (for the Xth INQUA Congress) the official map at 1 : 625 000 (1977). Field mapping and the publication of memoirs lags far behind; published maps still tend to show little more than lithological categories, while published contributions to chronology have a remarkable tendency to be radical if not bizarre (Poole and Whiteman 1961; Bristow and Cox 1973; Kellaway *et al.* 1975). Hence geographers have in the course of their research mapped appreciable areas (e.g. Clayton in Essex, Straw in Lincolnshire and Norfolk).

A second aspect of Quaternary research has been the reconstruction of past environments and their relationship to geomorphological processes. Glacial sedimentation, periglacial slope activity, and the behaviour of rivers at earlier stages are examples. As in all such work, emphasis has fallen on periods of rapid change and on the more extreme environments, and not everyone drawing on these accounts has made enough allowance for that. Such work is of course most useful where some quantification of work done over time elapsed can be deduced, an excellent example is the work on the Devil's Kneading-trough (Kerney *et al.* 1964). Already the fuller understanding of Quaternary environments is having some effect on the attitudes of geomorphologists to the controls shaping our contemporary landforms, and we may expect that to increase. The hint dropped by Cotton (1963) is not forgotten.

In all this activity, work on glacial landforms is surprisingly restricted. When the chronology of the dissipation of the Devensian ice sheet, or the stratigraphy of the Middle Pleistocene of lowland England are still inadequately understood, there is a tendency to adopt a selective attitude to landforms or sections, seeking to extract efficiently and without unnecessary description (or understanding) those landform or sedimentary elements that answer most directly the chronological questions being posed. Indeed, it could be argued that without some such chronological framework, little useful work could be done on the landform assemblages characterizing our glaci-ated areas.

In large part this is true, and here I am concerned more to set the record straight than to deplore the balance of current research. Yet, could we not turn to the work like that by David Linton on Scotland (e.g. Linton 1959), we would be hard put to it to describe adequately the impact of ice on the sculpture of the Scottish landscape. Arguments about the area ice covered in Zone III do not explain the pattern of Cairngorm troughs, the plateau tors, or the size and distribution of Cairngorm cirques compared, shall we say, with those on Skye. In much the same way, Linton's (1957) still too-little-known explanation of the radiating troughs of the Lake District over-turns traditional views of the evolution of the supposed Lake District dome (inferred in particular from the dip of the surrounding Carboniferous rocks) without providing or requiring any chronological framework for the Quater-nary history of the Lake District. The lack of any full understanding of the scale and intensity of Quaternary erosion in Wales demonstrates the self-imposed limitations of contemporary Quaternary studies, even by those who call themselves geomorphologists.

Outside the glaciated area (and how we continue to argue about what that might mean) research on the Quaternary has naturally concentrated on the periglacial aspects of the past. For it is deviations from contemporary pro-cesses, perhaps producing variant (and now relic) forms, that interest us most. The imbalance this might cause is obvious, for we do not know the relative rates of landscape change over colder and warmer episodes, nor do we yet have good data on the relative length of the glacial and interglacial stages. Short periods of warmth punctuating longer periods of cold is still the con-ventional view, especially since the beginning of the Anglian, and, buttressed by the assumption that change was rapid in cold periods and slow in warm, the concentration on the periglacial past seems soundly based. Yet, as R. S. Waters (1978) pointed out, the success of the periglacial phases depended on the prior preparation of regolith in interglacial periods, and the major features are not cryoplanation flats, so much as the exposed weathering front of an earlier interglacial phase.

## Geomorphological processes

In the eighteenth century Italian engineers measured the speed of flow of rivers and discovered that, as they neared the sea and the discharge rose, so speed rose slightly too. This relationship had to be rediscovered almost two hundred years later, while in the interim W. M. Davis and others had de-scribed the behaviour of rivers from source to mouth in terms of decreasing speed, and decreasing competence to carry the material that was rejected to build a flood-plain. Incidents such as this led to rejection of Davisian explana-tory description on the grounds that the explanations were insecure if not plain wrong, and to an effort to understand geomorphological processes on a new and generally much smaller scale. Outside Quaternary studies, most geomorphological work of the last twenty-five years falls into this category, and it would be regarded as the central concern of modern geomorphology by most of its contemporary practitioners.

In many ways it has been a remarkably productive period. In major areas such as studies of the dynamics of river basins, of mass-movement, and other processes on slopes, of the dynamics of limestone solution, or the complexities of weathering and soil formation, we have seen great progress in our knowledge, and British geomorphologists have played their full part. Indeed in one area, the gathering of information on the rates of operation of geomorphological (and especially of slope) processes, they have been pioneers and established many of the measurement techniques (epitomized in Young's pits). There is little point in chronicling the work here: it is available in the published record, and from time to time reviews such as those in Embleton, Brunsden, and Jones, 1978, survey the progress made.

Some comment is in order on the approach, and aims of the work, and on the contributions made by other disciplines. Prior (1978) in one of the reviews referred to above, noted that, for all our interest in mass-movement studies, geomorphologists remained less well equipped mentally, and in their laboratories, than those civil engineers who specialize in landslide research. Full understanding of the Aberfan slip and flow, or of the expensive reactivitation of slides on the M6 at Walton's Wood, comes from a combination of the theory of soil mechanics and possession of suitable test apparatus, and although both are available to many geomorphologists, they still seem to lag behind the most innovative engineering research.

Rather similar comments might be made in the area of basin or catchment studies. The level of field instrumentation and the effort expended on flow-modelling by hydrologists has frequently yielded more in the way of results than the rather less well funded research of British geomorphologists. Indeed, on the larger rivers the scale of the work makes field research more akin to engineering than conventional field experiment. Nevertheless, here, rather more than in the study of landslides, the main thrust of the engineering and hydrological work is not geomorphological, leaving problems for geomorphologists to exploit. Many an engineer will spend time deriving estimates of liquid discharge, but be unconcerned with the solid and solute load yielded by the same catchment. Indeed, at one stage the Institute of Hydrology seemed capable of being equally narrow-minded in its approach to the luxuriously instrumented pair of Plynlimon catchments, but in the end the need to maximize results from the investment and the interest and enthusiasm of geomorphologists such as Newson (1976) led a reasonable diversion of effort to process-based geomorphological research.

Gregory, himself a leading exponent of such work and co-author of a standard text (Gregory and Walling 1973), has drawn attention to the twin hydrological and hydrogeomorphological approaches to basin studies. We need to know something about the geomorphology of the catchment if we are to forecast flood flows satisfactorily, while we need to know the flow characteristics of the catchment if we are to extend the work to an understanding of the liquid discharge characteristics and slopes, let alone consider the removal of solid and solute load. Work such as that on pipeflow in Plynlimon catchments has revised naïve conceptions of overland flow and the generation of storm hydrographs, and in turn must lead to a revised view of the geomorphological work done by flowing water on and within slopes.

Work on slope form has been more firmly based on field measurement in the last twenty years, even though we still seem unable to agree on a standard technique for data collection and analysis. It has also involved wider considerations than slope form alone, and in particular a number of authors have related slope form to soil type and attempted to link soil and slope evolution. Thus Ollier and Thomasson (1957) used the asymmetrical soil pattern of the slopes of dry valleys in the Chiltern Hills to derive an interpretation based on asymmetrical evolution under a periglacial climate, while Furley has developed this wider approach in a number of papers. A danger, as Young (1969) has pointed out, is that the evidence may lead to an interpretation which, while apparently internally consistent, is at variance with wider geomorphological considerations, for it is hardly geomorphologically possible for all concave footslopes to be zones of accumulation.

In some ways geomorphologists seem to have been left on their own, and to have made most progress, in the area of limestone studies. Solution chemistry in the real world has not attracted the enthusiasm of many chemists, and where this has occurred it has often been at the behest of a geomorphological colleague. The simplicity of the final process of removal, and perhaps the fact that most material once in solution stays that way, has helped overall studies of the rates of operation of solution processes. Bare limestone surfaces whether at the surface or below ground are well suited for micro-erosion measurements (High and Hanna 1970), while world-wide comparisons in this area somehow seem less affected by other inadequately appreciated environmental variables, and have yielded more positive results than many attempts at a process-based climatic geomorphology. Recent work locating with more precision the site of solutional activity and techniques allowing the dating of speleothems promise rapid results (Atkinson *et al.* 1978).

The coast, and our stormy British coast in particular, is a zone of high environmental energy and rapid geomorphological response. As a result coastal study has been process-based for longer than the rest of geomorphology. Even so some obvious areas of measurement have proved elusive, particularly a sediment budget approach. Here difficulties in measuring onshore/offshore movement of sediment has proved an obstacle. At the level of synthesis the descriptive inventory of British coasts by Steers (1964) is matched by the standard textbook of King (1972) and these alongside an extensive research literature give British coastal geomorphology considerable prestige. We do of course live in a rather unusual coastal environment, and one important advance of recent years has been to place the British coastal environment securely in a world context, bringing a more balanced view to coastal geomorphology.

## 'Traditional' landform studies

As Bloom recently bravely reaffirmed (1978) (and he may find it easier to do so as a geologist), the core of geomorphology remains the explanatory description of landform. No study of process, however sophisticated, is complete if it is not carried through to the related forms produced. Gradually this is coming about, but one of the results of the strong reaction against denudation chronology and Davisian geomorphology generally has been a tendency to study processes in isolation from the forms they produce. Where connections have been made, they are generally on a modest scale; one facet of a slope, one element of a catchment. It is commonly assumed that in time a wider understanding of processes will lead to a better and more general understanding of landform.

Looking back it is hard to see why the broader geomorphological work on the explanatory description of landforms, pioneered between the wars, should have degenerated into the narrow concerns of what became known as denudation chronology. Had a broad approach persisted—the relationships between geology and surface form, the work done by ice, the detection of tectonic activity, the evolution of integrated drainage basins on the till sheets of lowland Britain—then explanatory description of landforms might have survived as a significant strand of British geomorphology. Unfortunately we became fascinated by what seemed to be evidence of an episodic development of landform, and turned from the reality of major upland surfaces to the seduction of fragmented flats surveyed at ever closer intervals, yet interpreted in terms of the same ideas. Since the more detailed approach seemed to work in some areas, it was applied more widely, often regardless of the possibility of major glacial disturbance. Disparaging remarks about the close relationship of heights to the contour intervals of the published maps led to more detailed field effort and more careful surveying, a *reductio ad absurdum* inspired by a deep commitment to field-work that was one of the best features of British geomorphology at that time.

It is all rather sad, for the grand outline was properly understood. While O. T. Jones (1951) championed the case for a sub-Triassic age for the upland plains of Wales and the South Pennines, geomorphologists argued that they were late-Tertiary, indeed that they were likely to be Miocene, a view remarkably vindicated in recent years (Boulter *et al.* 1971). Certainly reservations have been expressed about the sequence of events described for south-east England by Wooldridge and Linton (1955) (for example by Thornes and Jones 1969 and Hodgson and Catt 1974), but for most areas of Great Britain there has been remarkably little progress on understanding the evolution of the landscape over the last ten million years, and it has not proved much easier to write an account of the landforms of the British Isles (the Methuen series) than it would have done twenty years ago when the idea was first mooted. The account of Late Quaternary events is more secure, but the earlier phases in the evolution of the landscape are still very incompletely understood.

Interestingly, British geomorphologists have pursued very good traditional geomorphology abroad, particularly perhaps in the tropics. It seems that the wide-open spaces of these overseas landscapes impel a scale of enquiry and an interest in landform that is seen as rather *passé* in the better known, more intimate landscapes of the British Isles. Perhaps too the conviction that Tertiary and even older landscapes survive, complete with their weathered mantles, makes explanatory description of landform a more respectable aim. In time the knowledge and the attitudes from this work may well feed back into work in the UK—indeed Thomas (1978) has recently reviewed the evidence for relict Tertiary tropical elements in the UK landscape.

The lack of theory in geomorphology—and the lack of interest in developing theory—has been deplored in several recent comments. One would hardly expect a single national group of geomorphologists to develop this on their own, and it could be claimed that Chorley's work, especially on general systems theory, has had a considerable influence on modes of geomorphological thought here and abroad. The British too are often criticized for their cosy insularity, and opinions vary as to how far the isolation of distance and language has delayed the adoption of innovative ideas, especially from the continent. In such cases a few pioneers have by intention, or even by chance, found themselves introducing the ideas and the literature to their colleagues. Thus Zeuner brought the model of climatic control of river terraces, and a whole host of related attitudes about the Quaternary, to the attention of British geomorphologists, and might even be regarded as the stimulus behind the great developments in Quaternary studies. Almost twenty years ago Yates (1963) and Coleman (1958) showed the reality of neotectonic influences in the nearby continent, but despite King's (1962) attempt to apply a New Zealand view of a mobile landscape (following a visit) neotectonics has made little headway—so far. More generally Derbyshire and others (1976 have been responsible for the dissemination of continental concepts of climatic geomorphology, although the excellent analysis by Stoddart (1969) has helped to avoid unreasonable enthusiasm for thinly substantiated ideas. One particular aspect has of course received wide attention, and the dissemination of concepts about our periglacial past owes much to the continental literature and to visits to Poland and the Canadian arctic by a number of British geomorphologists. Interestingly the expensive work of the British Antarctic Survey has had little impact on us, partly because of the rather descriptive, survey, approach adopted, and also because of the involvement of few established geomorphologists from the UK. The one visit by Linton (1963, 1964) shows what opportunities have been missed.

## Geomorphology in physical geography

So far this account has focused on geomorphological research, properly justified like all research solely on its intellectual quality. Despite the need for travel, and despite the increasing use of laboratory and field apparatus, it is cheap research. Few grants are made by NERC for geomorphological research, and support is limited to about a dozen new studentships a year. There is no National Institute for geomorphological research, and it is still not seen as part of the remit of the Institute of Geological Sciences. Part of the research effort of some NERC institutes (especially IH at Wallingford and IOS at Taunton) is concerned with geomorphology, and a few geomorphologists are employed. As we have seen, most of our geomorphologists who have completed their graduate studies are full-time teachers in universities, polytechnics, and colleges of education, and 80 per cent of these are employed in geography departments. Hence perhaps 50 per cent of their effort is devoted to teaching physical geography, and, as is usual in such circumstances, a good deal of very specialized geomorphology is traditionally taught in these courses through the influence of the research interests of the staff involved.

It is in any case of the nature of a subject such as geography that fashions in research and the pattern of areas of advancing knowledge and of neglect leave a very unbalanced subject to teach at university level. For many years physical geography was far more analytical and far less descriptive than human geography, but the conceptual and quantitative revolution in human geography has redressed that balance. We are still faced with an imbalance in physical geography, although recent advances in climatology, in micrometeorology, in hydrology, and in our understanding of soils, have made the presentation of a balanced introductory course far easier. When it comes to integration across geography, to the classic man/environment relationships, the relevance, or at least the utility, of much of our geomorphological knowledge seems pretty slender. Yet it is a brave researcher who will lay his knowledge aside.

There can be few geomorphologists in recent years who have not spent time reflecting on the proper content and approach of their teaching in physical geography. They have been stimulated by the first successes of the revolution in human geography, by falling student interest in geomorphology, and by the integrative approaches to the environment that have so quickly become fashionable. If it is wrong to fragment an integrated geography—or at least to fail to illuminate the way it can be integrated—cannot the same be said of the physical environment? Discussion and introspection on these problems is far from new, but I suspect it has been given a new edge in recent years, and it is again spilling over into our journals. (e.g. Price 1978, Gray 1978).

One temptation, and it has been adopted as a solution in several cases, is to teach an applied geomorphology. This makes a lot of sense, especially for the purposes of those students who will never specialize in geomorphology. It can also be linked with an increasing level of professional involvement by geomorphologists, not unwelcome, but hardly the triumph trumpeted so loudly as, for example, by Brunsden, Doornkamp, & Jones (1978). Indeed at the research level (if it ever fully reaches that level), as at the taught level, overemphasis of applied or of applicable geomorphology will lead to even slower progress in the fundamental and still very difficult problems of the subject. Where the relevant geomorphological ideas (and these are not merely those that might be applied) can be integrated with other facets of physical geography a very strong course can be constructed. One contribution geomorphology can bring here is an awareness of varying time scales and of speeds (and styles) of environmental response. Concepts of positive and negative feedback are useful. Soil erosion may be put into better perspective through the intermediate scale of episodic arroyo cutting and alluviation, to the long-term natural rates of denudation of landscape. The almost imperceptible changes of the evolving landscape on which we act out our lives today may bring philosophical as well as practical messages to a first-year geography student.

I would not extend this particular argument to school geography. Although at university the proportion of students opting for physical geography courses continues to decline, there is little sign of a fall in interest in sixth-form geomorphology. At that level it clearly offers intellectual stimulus, while the school version of the new human geography tends to *reduce* over-all enrolments in geography, not to increase them. Indeed, some fear that unless the quantitative enthusiasm of some young school teachers is curbed a little, a disastrous fall in geography enrolments could result. In contrast, a traditional in style, yet up-to-date, geomorphology geared to the explanatory description of landforms, and including at the appropriately smaller scale the study of process, seems to continue to attract good students. Properly preserved it avoids the temptation to take geology at A-level (for the few interesting geological ideas suitable for the sixth-former are readily incorporated into a geomorphology course) and yet it provides some sort of scientific training so often needed in our regrettably specialized sixth-form curriculum. It is the British equivalent of the 'Physical Geography 101' courses of American universities, but taught at a more rigorous level. It cannot really provide the physical background for an integrated geography, but it does provide a training of intrinsic merit.

One argument taken up recently by Worsley (1979) is the place of geomorphological research. He argues that it is outgrowing its geographical home, and that the adoption of a more geological approach would be beneficial, not least in the acquisition of appropriate resources for modern research. It has of course been true of all too many UK geology departments over the last thirty years that if they taught geomorphology at all, they taught an outdated approach unrelated to modern research developments. In turn geologists have deplored the amateurism of geographical summaries of stratigraphy, or plate tectonics. It seems difficult to achieve a balanced development of geomorphology across the two subjects it depends on, to the detriment of the subject (Clayton 1971). In the United States it is almost missing from geography departments; here it is neglected by our geologists. No one is going to change these patterns in a short time, for it has long been found administratively and intellectually convenient to adhere to conventional subject boundaries. The most rapid change is likely to spring from fully integrated departments of environmental science (where the terms geography and geology are commonly avoided) or from expansion of geomorphological activity in NERC Institutes and similar research bodies (e.g. the Soil Survey).

How far could it be a solution for geomorphology to shake off its awkward subject parents and establish itself as an independent discipline? Is this not what geophysics has done, so that it is now a listed degree course separate from both geology and physics? The analogy is quite a close one, for few geology departments have wished to teach much geophysics, while, with the lack of student demand in recent years (coupled with the new concepts of plate tectonics), physics departments have been eager to embrace geophysics. Here of course is the seduction of applied geomorphology, for only the strength of exploration geophysicists allows the base for the modest independence geophysics enjoys. Geomorphology has at present no such base, and independence might lead to the erosion of its entrenched strength in geography departments—as the United States has shown, you don't need a research geomorphologist to teach an introductory course in physical geography. We should recognize as one aim the establishment of high-quality geomorphological teaching and research in departments of geology as in geography, and the acceptance of geomorphology by the Institute of Geological Sciences; independence would do little to help that. Indeed, more would be done if geomorphologists would encourage geologists to take over some of the more geological topics they have developed in recent years. Growth of interest in both the Quaternary period and in sedimentology in geology departments shows that this could well come about, and in that case we could well develop the more balanced and broadly based geomorphology we geographers surely desire.

# REFERENCES

ATKINSON, T. C., HARMON, R. C., SMART, P. L., and WALTHAM, A. C. 1978. Palaeoclimatic and geomorphic implications of $^{230}$Th/$^{234}$U dates on speleothems from Britain. *Nature*, 272: 24–8.

BLOOM, A. L., 1978. *Geomorphology*. New York.

BOULTER, M. C., FORD, T. D., IJTABA, H., and WALSH, P. T. 1971. The Brassington Formation; a newly recognized Tertiary formation in the southern Pennines. *Nature: Physical Sci.* 231: 134–6.

BOULTON, G. S. 1977. Review of Embleton and King 1975 and Sugden and John 1976. *Earth Surface Processes*, 2: 441–2.

BOWEN, D. Q., 1978. *Quaternary geology*. Oxford.

BRISTOW, C. R. and COX, F. C. 1973. The Gipping till: a reappraisal of East Anglian glacial stratigraphy. *J. Geol. Soc. London*, 129: 1–37.

BRUNSDEN, D., DOORNKAMP, J., and JONES, D. K. C. 1978. 'Applied Geomorphology'. A British View' in Embleton, *et al.* 1978.

CLAYTON, K. M. 1963. A map of the drift geology of Great Britain and Northern Ireland. *Geogrl. J.* 129: 75–81.

CLAYTON, K. M. 1971. Geomorphology—a study which spans the geology/geography interface. *J. geol Soc. London*, 127, 471–6.

COLEMAN, A. M. 1958. The terraces and antecedence of a part of the River Salzach. *Trans. Inst. British Geogr.* 25: 119–34.

COTTON, C. A. 1963. Development of fine-textured relief in temperate pluvial climates. *New Zealand J. Geol. Geophys.* 6: 528–33.

DERBYSHIRE, E. (ed.) 1976. *Geomorphology and climate*, London.

EMBLETON, C., BRUNSDEN, D. and JONES, D. K. C. 1978. *Geomorphology: present problems and future prospects*. London.

EMBLETON, C. and KING, CUCHLAINE, A. M. 1975. *Glacial geomorphology*. London.

GEMMELL, A. (ed.) 1975. *Current research in geomorphology*. Norwich.

GRAY, M. 1978. The future of physical geography: some alternatives. *Scottish Geogr. Mag.* 94: 183–4.

GREGORY, K. J. and WALLING, D. E. 1973. *Drainage basin form and process: a geomorphological approach*. London.

HIGH, C. and HANNA, F. K. 1970. A method for the direct measurement of erosion on rock surfaces. *BCRG, Technical Bulletin*, 5. Norwich.

HODGSON, J. M. and CATT, J. A. 1974. The geomorphological significance of clay-with-flints on the South Downs. *Trans. Inst. British Geogrs.* 61: 119–29.

JONES, O. T. 1951. The drainage systems of Wales and adjacent regions. *Quart. Jour. Geol. Soc. London*, 107: 201–25.

KELLAWAY, G. L., REDDING, J. H., SHEPHARD-THORN, E. R., and DESTOMBES, J.-P. 1975. The Quaternary history of the English Channel. *Phil. Trans. R. Soc.* A 279: 189–218.

KERNEY, M. P., BROWN, E. H., and CHANDLER, T. J. 1964. The late-glacial and post-glacial history of the chalk escarpment near Brook, Kent. *Phil. Trans. R. Soc. London, B.* 248: 135–204.

KING, C. A. M. 1962. Geomorphological contrasts in New Zealand and Britain, in *Land and livelihood, geographical essays in honour of George Jobberns*, Christchurch, 48–73.

KING, C. A. M. 1972. *Beaches and coasts*, 2nd edition. London.

LINTON, D. L. 1957.   Radiating valleys in glaciated lands. *Tijdschr. K. ned. aardrij-kund. Genoot.* 74: 297–312.

LINTON, D. L. 1959.   Morphological contrasts of eastern and western Scotland, in *Geographical essays in memory of A. G. Ogilvie* (ed. R. Miller and J. W. Watson). Edinburgh, 16–45.

LINTON, D. L. 1963.   Some contrasts in landscape in British Antarctic Territory. *Geogrl. J.* 129: 274–82.

LINTON, D. L. 1964.   Landscape evolution. *Antarctic Research* (ed. R. Priestley, R. J. Adie, and G. de Q. Robin), 85–99.

NEWSON, M. D. 1976.   The physiography, deposits and vegetation of the Plynlimon catchments. *UK Institute of Hydrology, NERC Report,* 30. 44 pp.

OLLIER, C. D. and THOMASSON, A. J. 1957.   Asymmetrical valleys of the Chiltern Hills. *Geogrl. J.* 123: 71–80.

POOLE, E. G. and WHITEMAN, A. J. 1961.   The glacial drifts of the southern part of the Shropshire–Cheshire plain. *Q. J. Geol. Soc. London,* 117: 91–130.

PRICE, R. J. 1973.   *Glacial and fluvioglacial landforms.* Edinburgh.

PRICE, R. J. 1978.   The future of physical geography: disintegration or integration? *Scottish Geogr. Mag.* 94: 24–30.

PRIOR, D. B. 1978.   Some recent progress and problems in the study of mass movement in Britain, in Embleton *et al.* 1978.

STEERS, J. A. 1964.   *The coastline of England and Wales,* 2nd edition. Cambridge.

STODDART, D. R. 1969.   Climatic geomorphology, in *Water, earth and man* (ed. R. J. Chorley). London, 473–85.

SUGDEN, D. E. and JOHN, B. S. 1976.   *Glaciers and landscapes.* London.

THOMAS, M. F. 1978.   Denudation in the tropics and the interpretation of the tropical legacy in higher latitudes—a view of the British experience, in Embleton *et al.* 1978.

THORNES, J. B. and JONES, D. K. C. 1969.   Regional and local components in the physiography of the Sussex Weald. *Area,* 2: 13–21.

WATERS, R. S. 1978.   Periglacial geomorphology in Britain, in Embleton *et al.* 1978.

WOOLDRIDGE, S. W. and LINTON, D. L. 1955.   *Structure, surface and drainage in south-east England.* London.

WORSLEY, P. 1979.   Whither geomorphology? *Area,* 11: 97–101.

YATES, E. M. 1962.   The development of the Rhine. *Trans. Inst. British Geogr.* 32: 65–81.

YOUNG, A. 1969.   The accumulation zone on slopes *Zeit. für Geomorph.* 13: 231–3.

# 7 Recent Developments in Survey and Mapping

D. W. RHIND and T. A. ADAMS *University of Durham*

This paper concentrates on topographic mapping and considers first the institutional structure of the mapping industry and how this relates to geographers. Summaries of recent developments in survey, photogrammetry, and cartography, especially in Britain, are followed by an analysis of factors common to them notably the central positions of digital processing and storage of data, and the consequential increase in integration of survey and mapping. The considerable differences between British and much other mapping are outlined, and their implications for the introduction of some techniques in common use elsewhere is discussed. Finally, some speculations are made on mapping in the future.

## CONTEMPORARY GEOGRAPHY AND SURVEY AND MAPPING

It is arguable that we already have or are collecting enough information concerning the surface of the earth and human activities on it. Certainly there is a thread in recent geographical writing which suggests that empirical work is over-emphasized and theoretical concepts are under-represented in most geographical research. In terms of human resources, well in excess of £2000 million is spent each year on survey and mapping (United Nations 1976). Examples of prolific data capture abound: today, for instance, virtually the whole world is being monitored on an 18-day cycle by a Landsat satellite. More parochially, such work as the altimetric resurvey of the Tyne valley by R. F. Peel and his associates, begun after the World War II and necessary for the study of the local geomorphology because of the then parlous state of published maps, is now rendered superfluous by the ready availability of high-quality contour data. This paper, however, is predicated upon a different viewpoint: while some theory is essential if research is not to be *ad hoc* and should always be explicit rather than implicit, geographical data are needed to evaluate most theories (e.g. see Johnston 1978). Further, much of the available data concerning the natural land surface of our planet and the manifestations of human processes on it can only answer yesterday's research problems. Such data were collected for specific purposes and are ill suited for others—as Rhind and Hudson (1980) have shown in relation to land-use surveys. Finally—and quite apart from research work—there is an apparently never-ending practical requirement for more detailed, more up-to-date, or differently classified environmental data. Without them, roads are not built and changes in policy have to be based on hunches rather than facts.

Such an opening statement is necessary because of changing academic circumstances and changing technology. Though it is hard to demonstrate conclusively, it seems clear to us that academic geographers have less direct contact today with topographic data collectors in particular, than, say, at the time of publication of Debenham's *Map making* (1936). Contemporary geographical literature and advice on careers, e.g. *A matter of degree* (*Geo. Abs*, 1979), suggests that contemporary geographers are subject to greater exposure to instruments used in monitoring geomorphological processes and methods of carrying out questionnaire surveys than those employed to determine position on the earth's surface. Most British geographers now concentrate their research on Britain and countries for which data are readily available. Yet, in the absence of grid lines painted over fields, it is self-evident that the definitions of absolute and relative positions of observable features in the landscape are all based upon topographic mapping and, at the most general level, all geographical statements concerning Britain ultimately depend upon this definition of topography. British geographers have begun to take the availability of topographic data for granted and, in all the recent shifts in the geographical paradigm (Taylor 1976), have concentrated increasingly on theoretical considerations and the analysis of what are often secondary data.

In reviewing recent developments in a field which is of basic importance to geographers but one in which they have a diminishing involvement, our considerations will perforce focus chiefly upon the work of surveyors, photogrammetrists, cartographers, electronic engineers, astrophysicists, and computer scientists. Because of the great breadth of the subject, we will be highly selective and concentrate largely upon topographic data collection and display, rather than upon related cadastral mapping and the manipulation of topographic data for other purposes. For the same reason, we will in the main restrict our viewpoint to Britain: the reader who is concerned to obtain a more thorough overview should consult such publications as the twice-yearly *Proceedings of the American Congress on Survey and Mapping*, together with papers given at the regular meetings of the International Association of Geodesists, the International Society of Photogrammetrists, and the International Cartographic Association, and in the standard surveying, photogrammetric, and cartography journals. We shall also ignore all recent works in historical cartography or carto-bibliography, an area of considerable academic interest (Wallis 1976).

Topographic maps are not only of direct use but also form the base for the collection and display of much other data, such as soils. Almost all developed countries now provide topographic mapping through one or more government-owned and controlled organizations (Bomford 1979). This institutional factor has often had implications for the adoption of innovations and the availability of data. For these reasons, we now examine both the professional groups and the institutions involved in topographic survey and mapping.

## THE INSTITUTIONAL FRAMEWORK

It is conventional wisdom to consider the collection and dissemination of data defining the characteristics of the earth's surface, both man-made and natural, to be a job for identifiable groups of skilled practitioners. Thus geodesists establish primary triangulations, surveyors provide more local 'control' or carry out small, local surveys, and photogrammetrists compile the topographic detail using suitably transformed aerial photographs. Each of these groups contains exponents of the practical and more theoretical approaches; the latter are most usually found in universities. Some degree of 'professional' status is also attached to each of the groups. This is in marked contrast to the situation of those who normally present the results of survey in graphic form for wider dissemination, i.e. cartographers, even though they are often a large majority of those employed in the whole survey and mapping operation. In Britain at least, the role and status of the cartographer has long been that of a technician, as the pamphlet by the British Cartographic Society on jobs for cartographers makes clear; a declining number of academics have—at least until very recently—been involved in the theory and practice of map-making (Board and Taylor 1977). The existence of these different groups of practitioners has been formalized by the creation of chartered institutes or societies and cemented by *post hoc* definitions of the disciplines in such a fashion as to ensure end-to-end joins rather than overlap (see e.g. Keates 1973). But, as we shall see, this apparently stable and mutually agreed division of responsibilities is fast becoming undermined internationally by changes in the technology of data collection and display, by government attitudes towards national mapping organizations (Bomford 1979), and by the growing self-awareness of cartographers. Robinson, Morrison, and Muehrcke (1977) have described the growing numbers and influence of cartographers in recent years; one of their indicators is the increase in the number of students enrolled in cartography courses in the USA from less than 4000 in 45 courses in 1948–9 to over 11 500 in 450 courses in 1972–3. In Britain the bulk of training in relation to mapping courses is in survey departments or in a few geography departments, but some moves towards a more professional status for carto-graphers have been made.

In the last half decade, increasing numbers of survey and cartographic organ-izations have been critically reviewed by their peers. At the national level, the United States and the Canadian Federal Task Forces examined mapping across all the major Federal Departments in their respective countries and the Serpell Committee has reviewed the future tasks for Ordnance Survey (HMSO 1979a). Reassessments at a lower level, such as that currently being conducted in South Australia, are numerous for a variety of reasons. In almost all instances, a substantial factor has been the increasing costs and the uncertain benefits (at least in the eyes of civil servants) of topographic mapping. Making a case for mapping is not as trivial as it might appear; assessing costs and comparing them with other organizations, for instance, is not invariably straightforward since accounting procedures (such as amort-ization rates and the sectors of operation charged as a national service in comparison with those regarded as a repayment service) vary widely. Assess-ing benefits is, however, complicated beyond description and tends to rely upon cited examples or opinions. In consequence, most government survey and mapping organizations are justified in meeting yesterday's re-quirements—which have the benefit of being well known, provable, and cost-able. Until very recently, such a procedure has slowed down the introduction of technological innovations in major government-financed survey organiza-tions in Britain.

CONTEMPORARY DEVELOPMENTS

We will now consider the most significant recent developments under the traditional disciplinary banners.

*Survey and geodesy*

The basic principles of survey have remained unchanged for many centuries (Lyons 1931): in essence, the whole subject is based upon measuring or determining the best method of measuring lengths or angles (sometimes considered as ratios of distances). Most of the conventional survey instrumentation still found in geography departments, such as chains and theodolites, were devised in the seventeenth and early eighteenth centuries and have remained basic survey tools ever since. Extremely high accuracies have been obtained by such means, particularly in the measurement of angles. The Ramsden theodolite, for instance, was capable of producing accuracies of one second of an arc in the early part of the nineteenth century and, since then, the major developments in optical equipment have been increasing robustness, ease of use, and reduction in weight (Anon, 1977).

The first commercially available Tellurometer electro-magnetic distance measuring device (EDM) was introduced in 1956. Though cumbersome, extremely heavy and rather complex to operate, these devices revolutionized the concept of observing triangulation networks. Hitherto, such networks were based upon observation of many angles and azimuths and a few base line measurements obtained by catenary taping; the Hotine retriangulation of Britain in 1936 included only two distances to provide scaling. Computation and adjustment of the results from these and tellurometer surveys was exceptionally onerous until the widespread availability of computers in the late 1950s and early-to-mid-1960s. Ashkenazi *et al.* (1971), for example, have suggested that the Hotine retriangulation contains scale errors of up to 20 metres in Northern Scotland, largely through the use of computational procedures which were inevitable before the development of computers.

Until the late 1960s, control points were fixed by traverse using a theodolite and an EDM or steel tape, whilst points of detail were fixed by optical tacheometry. In 1968, Zeiss introduced an electronic tacheometer with electronic circle reading and an electro-optical range-finder (Holden 1979); facilities were also provided to record measurements automatically in the field and for transmission of data to computers. Further developments in integrated circuitry have led to smaller and lighter second-generation instruments based on microprocessors, such as the Wild TC1 and the HP Total Station; field data are stored, together with station identification, meteorological data, and map projection parameters. Plug-in computer programs are available to carry out adjustments and other computations in the field. Such equipment is currently expensive, costing about £20 000 in the UK, and this severely restricts the situations in which they are an economic proposition. None the less, their ease of use and incorporation into totally integrated survey systems (see below) suggests that they will become progressively more common: the end of the traditional field-book may then be in sight except in amateur surveying.

If recent developments in field survey have been substantial, those in geodesy are little short of spectacular—particularly if events in the near future are anticipated. Almost all of these are based upon the use of artificial satellites. Robbins (1977) has argued that their uses are currently: determination of absolute and relative positions on the earth's surface, thereby leading to improved geodetic networks, better descriptions of the shape of the earth and of tectonic movement; determination of the earth's gravity field; determination of the rotational parameters of the earth.

In simple terms, observation of the satellite orbit can tell us about the shape and gravitational field of the earth, but the satellite itself is also a platform from which point-fixing on the earth can be carried out. Three different observation procedures have been used with satellites. The first is the measurement of direction, achieved by photographing the night sky to determine satellite position from a series of ground stations (astrotriangulation). The second is the measurement of distance to the satellite by laser ranging (or trispheration) from the ground to retro-reflectors on the satellite. Both methods are currently dependent upon fine weather. In contrast, the development of methods based on the Doppler frequency shift of signals from the satellite has led to an all-weather operational position-fixing system. Originally inspired by the need for military navigation, the TRANSIT system is based upon six satellites in polar orbits. First operational in 1964 and released for civilian purposes in 1967, it can provide absolute individual point-fixing (unlike previous relative methods) to an accuracy of 1 to 1.5 metres in each coordinate, albeit after considerable data processing. Today, commercial users of TRANSIT far outnumber government or military users (Stansell 1978, quoted in Holden 1979). A replacement, the NAVSTAR Global Positioning System, is to be available in the early 1980s. Based eventually upon twenty-four satellites, this is expected to give high positioning accuracies in very short observation times: relative positions 2000 km. apart should be determinable to accuracies of around 10 cm. Within this basic positioning system, it seems likely that interpolated positions will be given virtually instantaneously by an inertial positioning system: this 'black box' effectively determines its X, Y, and Z location in relation to a NAVSTAR-provided start point, using three accelerometers and an atomic clock to provide distance changes.

All these developments relate primarily to the fixing of X and Y position on the earth's surface. Others, however, such as SEASAT, are beginning to provide much improved altimetry. The first-generation SEASAT has given maps of the geoid in ocean areas to accuracies of ± 1.5 to 2 metres and, in the early 1980s, the laser ranging in SEASAT B should give heights to the nearest 10 cm. With developments like this, real-time monitoring of changes in sea-level begins to become feasible; one consequence of such a capability would be the ability to predict the onset and effects of North Sea surges with great precision. Rapid increases in accuracy in measurement of the earth's gravity field—knowledge of which is now critical because of its detailed influence on satellite orbits—have already occurred.

The over-all consequence of all these changes is that enormous gains in speed, accuracy, and ease of operation are with us and improvements will continue—to the extent that the present-day role of a surveyor may conceivably be reduced to that of a technician (Dale 1977). Indeed, the Director of the Australian national mapping organization (Bomford 1979) has stated categorically that there is already a gross over-production in higher education of geodesists and an under-supply of those with training in digital cartography. All the developments described thus far are critically dependent on computer technology; most pieces of equipment are still much more expensive, if far more productive and accurate, than the traditional counterparts (where they exist).

*Photogrammetry and remote sensing*

In this paper photogrammetry and remote sensing are taken together though some have argued that, by definition, the latter subsumes the former; we intend to include not only measurements but also the interpretation and detection of change in any remote-sensed imagery, whether this is produced from satellites or aerial photographs. Such a viewpoint echoes that manifested in the addition of 'and Remote Sensing' to the title of the journal *Photogrammetric Engineering* in 1975. For the purposes of this paper, we consider only three important developments—the analytical plotter, orthophotomapping, and Landsat-type mapping.

Before the World War II and even after it (Thompson 1958), photogrammetry in Britain was regarded by most surveryors with suspicion; the conviction that maps of sufficient accuracy could not be made was widespread. This contrasts remarkably with the situation in Austria, France, Germany, Italy, and Switzerland. Thompson's (1958) explanation for this difference is the long-standing earlier use of photo-theodolites in these countries, thus providing an easy transition to photogrammetry, together with the existence of mountains which were extremely difficult to survey without direct use of aerial survey. Whatever the reasons, photogrammetry in Britain developed initially through the private air survey companies and through the efforts of individual university and army staff. E. H. Thompson was a seminal influence on the field (see Atkinson 1977), publishing important contributions, particularly on the mathematical basis of the discipline, from 1935 onwards. He was appointed as the second holder of the Chair of Photogrammetry and Surveying in University College, London, in 1951.

In the period up to 1976, the standard photogrammetric mapping instrument was the analogue stereoplotter; it existed in a variety of versions based on direct optical projection (such as the familiar Multiplex projector) or with mechanical or optical-mechanical projection from the stereoscopic air photographs (Wolf 1974). Thompson and Mikhail (1976) have shown that one of the developments in this field in the period 1972–6 was the attachment of simple digitizers, usually as shaft encoders, to these machines, thereby encoding a digital version of the usual map plotted on a directly linked coordinatograph. At the 1976 Congress of the International Society for Photogrammetry in Helsinki, however, seven manufacturers described new analytical plotters and, according to Dubuisson (1977), a new era had begun in photogrammetry. Analytical plotters consist essentially of a precision stereocomparator and a digital store or a coordinatograph, linked together by a minicomputer. Since they have no optical or mechanical limitations, plotters are capable of handling a wide variety of photography, including vertical, tilted, oblique, convergent, high oblique, and panoramic photographs. In addition, they can accommodate any focal-length lens and correct for any combination of systematic errors caused by lens distortion, film shrinkage, atmospheric refraction, and earth curvature. As Dubuisson (1977: 1368) put it:

In general, the analytical stereoplotter permits the performing of three dimensional plotting of projective spaces in infinite number.

Through the play of software alone i.e. without acquiring and installing new equipment, it is possible to obtain views, plans, elevations, sections and perspectives, orientated in space according to any position whatsoever, from digitizations of single points in the 'photogrammetric model'.

The first commercially available analytical stereoplotters were announced in 1964; all early plotters were expensive and required considerable amounts of manual intervention. The most recent models are greatly improved in both respects (see Hobbie 1977; Vigneron 1977; Seymour 1977; Kelly, McConnell, and Mildenberger 1977; Friedman 1977; Inghilleri 1977; Dell Foster and Folchi 1977). Konecny (1977) has compared the available plotters and Helava (1977) has predicted future developments. Though there are good reasons why such plotters are not introduced immediately into many organizations—chiefly, perhaps, the need to continue amortizing existing equipment, especially in government organizations—it seems inevitable that they will find widespread use, particularly as the costs of the computer hardware components drop below that of the relatively small amount of optical and mechanical equipment they contain. The most important consideration for our present purposes, however, is not that they are becoming progressively cheaper and easier to use or the advantages of their unquestioned flexibility: it is the key place in which on-line digital processing and storage holds in the system, rather than being merely an afterthought, as in analogue plotters with digital storage. In short, we are moving towards truly automatic mapping systems. Because the systems are so organized, the costs of providing data for non-graphic uses should be trivial and it should be the norm that such data are available in a convenient format.

The second major development in photogrammetry has had a longer gestation period and, in one sense, may now be seen as just one of the functions of an analytical plotter. Scheipflug recognized the need for an orthographically projected photograph as long ago as 1903, i.e. one which, like a conventional map, is planimetrically correct. Equipment to carry out differential rectification of photographic images to remove the effects of photographic tilt and of relief variations was produced in France in the early 1930s. The first practical piece of equipment, however, was Bean's *Orthophotoscope* in 1953; since then, a series of improved machines have become available. Blachut and van Wijk (1978) have described the results of an international orthophoto experiment carried out between 1972 and 1976, comparing ten orthophoto instruments on the basis of the quality of the photograph and of the height information each produced, together with speed and convenience of use.

Petrie (1977) has described the history, current availability, and uses of orthophotomaps. In doing this, he has pointed out that rapid developments in photo mapping from orthophotographs had only begun in the early 1970s, even though controlled mosaics of air photographs had been available for some time. The clear inference is that the serious positional, linear, and areal errors inherent in aerial photographs render them useless for most mapping purposes. Many different methods of producing orthophotographs exist and some involve attaching extra optical-mechanical equipment to existing stereoplotters. But the most flexible way by far is electronically; images from a stereoscopic pair of air photographs are continuously scanned and the two images correlated to produce a value for the parallax and hence the altitude. Under favourable circumstances, digital terrain models may be obtained from such equipment as well as the orthophoto. Perhaps the two primary advantages of the orthophoto are its speed of production and low cost in those areas where existing maps are not adequate or do not exist and—like all photomaps—its ability to depict certain types of terrain (such as marsh-lands) much better than any line map. Lambert (1971), quoted by Petrie (1977), demonstrated an increase in speed of production by a factor of three when orthophotomaps were produced instead of line maps in an Australian topographic series. Mapping agencies such as the Lands Department of South Australia have been increasingly pushed into using orthophotomaps because of the prohibitive costs of conventional line-mapping from scratch. Yet in the UK—unlike many other countries (Petrie 1977)—little use has been made of orthophotomaps, largely, it seems, because of the existence of large-scale line map coverage and the OS continuous revision policy. It is, however, entirely feasible that, should government cutbacks in finance (and consequently in manpower) affect the ability of OS to maintain their existing maps to acceptable standards, the only solution would be to move rapidly towards orthophoto-based replacements. Parenthetically, a rare but important British contribution to the orthophotomap literature is Hill's (1973) assessment of user performance on photomaps reproduced by methods varying widely in cost.

To geographers, if not to many photogrammetrists (e.g. Thompson 1971, Petrie 1970 and 1971) the most exciting developments in mapping in the last twenty years have come from the development of manned and unmanned satellites. The most well-known of these by now is probably the Landsat (for-merly ERTS) series, and Calvocoresses (1979) has summarized the possi-bilities of mapping—mostly small-scale and thematic—in character from this imagery and suggested future improvements. Numerous papers in many journals have given individual examples of mapping geological structures, soils, water depth, flood extents, and much else. It is instructive that the stat-istics of use for the period 1973–6 show that Landsat was the most popular data produced from the EROS Data Center and also that private industry was the biggest customer for this imagery. The situation is probably rather different in the UK, even before the founding of the National Contact Point for Remote Sensing at the Royal Aircraft Establishment, Farnborough.

Welch and Lo (1977) have confirmed empirically the theory-based predic-tion made by others that the possibility of economically deriving height information from Landsat images or the (manned) Skylab photography is very limited. On the other hand, Fischer, Hemphill, and Kover (1976) have reviewed the progress in mapping from remote-sensing imagery more generally and conclude that the production of many maps—both thematic and topographic (such as seven maps of Antarctica published by the British Antarctic Survey in 1974 at 1 : 250 000 scale)—would not have been feasible by other means.

Most of the use of Landsat data thus far has been of those provided in graphic form. Increasing attention is, however, being paid to the equivalent (but higher spectral-resolution) data in digital form (see e.g. Jackson *et al.*, forthcoming). In this, the data from the Landsat multi-spectral scanner are stored as individual picture elements, or pixels, in raster form—as in one form of cartographic digitizing (below). Such data are readily printed out on a variety of devices, and automatic classification procedures can be applied to them to organize the data into groups, even if—thus far—the success rate in classification is not adequate for many practical purposes (Rhind and Hudson 1980). Given the improving spatial resolution of new satellite systems (Doyle 1978a), we may expect that these success rates will improve and, as a result, lead to much wider use of the digital data.

*Cartography*

Developments in both academic and practical cartography in recent years have been rapid, although the relationship between the two sets of practitioners has often exhibited antipathy and lack of mutual understanding (compare, for instance, the numerous reviews of Robinson and Petchennik (1976) by academics and practising cartographers).

On the more applied side of British cartography, two developments stand out as being of critical importance in the 1970s. The first of these is the use of revised photo-mechanical procedures to revolutionize the creation and distribution of maps. In Britain, the Ordnance Survey is a largely centralized source of topographic maps; these are used for a wide variety of purposes (Drewitt 1973), including that of underlays to geological, soils, and other thematic maps. Unlike federations such as Germany, Nigeria, or Australia in which the component states have some (occasionally total) responsibility for basic scale mapping, OS provides maps (at 1 : 1250, 1 : 2500, and 1 : 10 000 scales) to cover the entire country (Harley 1975). A consequence of this centralization is the standardization of specifications and a close supervision of map quality (although OS publish little in the way of formal map accuracy standards). A consequence of the detailed coverage is the large number of map sheets at the basic scales—of the order of 220 000 (Thompson and Woodsford 1979). Unlike the series of 204 sheets at the 1 : 50 000 scale, few copies of each of those sheets are ever needed; all, apart from the contours of the 1 : 10 000 scale maps, are printed in monochrome. Consequently, the vast numbers of sheets and slow turnover of stocks have led to severe handling and storage problems, both at OS and in the main agents. The OS solution (Marles 1971) was to transfer the 1 : 1250 and 1 : 2500 scale maps on to 35 mm. microfilm and store each sheet as an insert on a punched card; total coverage of Great Britain can now be held in this fashion in one large filing cabinet and copies printed immediately, on demand, by every main agent. Supply of up-dates is simple since a microfilm insert card is merely posted to each main agent and the out-of-date equivalent destroyed. More recently, OS have begun to provide a similar service based on field documents from the regional offices; as a consequence, and since OS works on a continuous revision system for the basic scale maps, it is now (in theory) possible to obtain maps of all parts of the country which are out of date by no more than fifty units of change (each normally equivalent to one new house) and often to a level much better than that. The rapid conversion of the one-inch maps into first series 1 : 50 000 scale maps, carried out between 1972 and 1976 (Hardy 1974), also made substantial use of photo-mechanical techniques. Many of these new sheets were derived largely from photographically enlarged one-inch maps but the procedures developed at OS ensured that no corresponding increase in line width occurred.

Inevitably, such procedures have disadvantages: the image quality of the printed-on-demand large-scale map is not as high as that of the lithographically printed sheet. In addition, the paper used is hygroscopic and thus less suitable for field use. The first series 1 : 50 000 scale sheets are clearly interim in nature, if only because the contours cannot be converted from Imperial to metric units by such photo-mechanical means. None the less, the creation and dissemination of the standard maps of Great Britain have, within a decade, been transformed by use of these methods. Furthermore, their very success has been a contributory factor in slowing the whole-hearted adoption of the second major development in applied British cartography— that of automation or, more practically, computer-assistance.

This topic has been discussed in general terms by Rhind (1977), who itemized and evaluated the many benefits claimed for the use of computers in map-making. Few of these have yet been realized to the degree anticipated by some of the innovators, except where specific circumstances pertain, e.g. where an atlas is being made from pre-existing census data in machine-readable form (HMSO 1979b). The principles on which 'automated cartography' are based are simple and were described in workable form as early as 1964 by Bickmore and Boyle. Normally—at least in Britain—lines and area boundaries are encoded as a connected series of pairs of co-ordinates using a digitizer—a device which senses the position, relative to defined origin, of a cursor being traced over a map; more recent work has investigated automatic line-following devices (Davey, Harris, and Preston 1978; Howman and Woodsford 1978). The same output of strings of coordinates is readily obtained from stereo-plotters by attaching shaft encoders in place of the mechanical plotter (see previous section). Errors are then detected and corrected by any one of several techniques (Rhind 1974). Alternatively, existing maps or photographs may also be scanned in raster form to provide arrays of square cells; for maps, each of these cells is coded as being empty or filled with some portion of a feature whereas, in the scanning of photographs, some number on a standardized scale—indicating the 'greyness' in each cell—is recorded. Both methods of encoding have their advantages and disadvantages (see Rhind 1977) and routine conversion facilities from one 'notation' to another are highly desirable (Rhind, forthcoming, 1). Following the encoding and editing stage—subject to reasonable classification of the data—it is simple to extract individual cartographic features within defined areas, to change map scale within quite wide limits, and to transform the data on to different map projections. Maps may be produced in which line-work quality at least equals that of the best cartographer. Alternatively, it is possible from an existing data base to produce much cruder maps—which Coppock (1975) has claimed are entirely suitable for many research purposes—exceptionally quickly and cheaply.

In many respects, Britain has been a world leader in the development of computer-assisted cartography. Yet, with one exception (see below), those active in the field in Britain have generally adopted a low-budget, man/ machine interactive, and computationally unsophisticated approach to the task, at least by comparison with many government agencies in the United States. Apart from the construction of equipment, we may distinguish three different developments in Britain which are worthy of note by a world audience in this field. These are the work of the Ordnance Survey, that of the Experimental Cartography Unit, and various developments in universities. Such a list is not exhaustive: it excludes, for example, the important early work on and use of the LINMAP mapping system (Gaits 1969) in the Ministry of Housing and Local Government and (later) the Department of the Environment.

The history of the Experimental Cartography Unit has not yet been written, even though its founder and director for the first twelve years has now retired. It remains the only organization of its kind in the world known to these authors and was set up in 1967 as the result of a Royal Society initiative, being sponsored from the outset largely by the Natural Environment Research Council. The initial objectives of the Unit were to examine ways of making maps by means of computer and to improve the graphic quality of maps; it was located in the Royal College of Art to tap the graphic expertise therein and had a special liaison with the Imperial College of Science and Technology to inject the other of C. P. Snow's 'two cultures'. While most of the projects carried out by the Unit have ranged widely over the earth sciences, others have been in a variety of fields, including planning (see, for example, Rhind and Trewman 1975, Margerison 1977, ECU 1978). Comparatively few published works have emanated from the substantial work of the Unit and those which have appeared rarely revealed much of general value (e.g. Lang 1971). Yet the ECU has nevertheless had a major impact upon the British and the world cartographic fraternity, almost all of which is due to the first director, D. P. Bickmore.

It is now difficult to appreciate that, until a decade ago when the ECU demonstrated the feasibility of high quality map-making via a computer, few thought this was likely to be routine for decades. The demonstration was conclusive: it began with small, sample maps (Bickmore 1968) and concluded with the publication of a standard series IGS one-inch geology map in both drift and solid editions (Rhind 1971). A definition of the way in which systems could be set up to perform such mapping functions was also given (Evans 1970). A second important contribution of the ECU was to act as a catalyst or, more properly, as an irritant. Until a collaborative project based upon

ECU digitizing of basic scale Ordnance Survey maps in 1968/9—initiated largely at Bickmore's insistence—the national mapping agency had taken little interest in the use of the computers in mapping beyond that of reduction of survey observations and survey adjustments. The results of this collaborative work were described by Cobb (1971) and Bickmore (1971). Thereafter, Ordnance Survey and ECU were not involved in collaboration to any significant degree until the late 1970s and, indeed, often viewed each other as competitors in innovation; OS designed and implemented their own computer-based system soon after the end of this first project (Gardiner-Hill 1972). The third contribution of the ECU came after these early experiments; they—earlier than many others—realized that cartographic data would also have to serve other than graphic purposes. To do so, different methods of organization and different levels of encoding accuracy are of major importance (Rhind 1976; forthcoming, 1).

Changes in the organizational base of the ECU and, in particular, a change in 1974 towards production as well as R and D coincided with a decline in the pre-eminence of the Unit's position. Soon afterwards, it became possible to purchase from other sources 'off the shelf' computer software which was less dependent on particular machines, more modular, and faster running than that of the ECU. It is hard to avoid the conclusion that, by the mid-1970s, the initial role of the Unit had been fulfilled although certain interesting new projects such as Ecobase (Molineux 1978) were carried out thereafter.

Ordnance Survey, on the other hand, became more and more involved in automated cartography throughout the 1970s. From a slow start—sparked largely by Bickmore—they created their own system which was oriented towards the encoding of existing large-scale maps of urban areas and the regular geometric features which predominate in these areas (Gardiner-Hill 1972). By 1973/4, a pilot production line had been set up and, by late 1978, 7000 map sheets had been encoded, edited, and data-banked. All of this was done on the barest minimum of research resources and with little concern other than for the use of automation to replicate existing cartographic tasks—a point made by the Royal Society submission to the Serpell Committee (Roy. Soc. 1978). OS maps became available on tapes on a routine basis in 1975 and, in 1979, cost the same as the standard paper product. Little use, however, was made of them outside OS in the 1970s, largely because of a lack of contiguous areal coverage of digital maps and the cartographically oriented organization of the data (below). Simultaneously, it became clear that to produce maps using the OS system was between 1.3 and 1.6 times as expensive as doing it by hand and took longer (Thompson 1978). Such figures are computed on the basis of a host of assumptions—such as the data not being used for derived mapping (see Bell 1978) and their being a short-run cost, i.e. it does not take into account future savings from a digital basis to map revision. None the less, these financial statistics were an important element considered by the Ordnance Survey Review Committee set up by the Secretary of State for the Environment in May 1978 and chaired by Sir David Serpell.

The committee's terms of reference were to review the role of OS up to the end of the twentieth century, taking into account changing user needs, financial constraints, and changing technology (Smith 1978). Such a review was necessitated by the near-completion of the programme established for OS by the Davidson Committee in the 1930s; this resulted in such changes as the conversion of all large-scale plans to National Grid format, the adoption of a kilometre-based grid, and the move towards metrication in scales and contour levels. Another review of OS activities was carried out wholly internally by the civil service in 1972 (the Janes Report), predicated upon the then government's desire to cut public expenditure. The resulting proposed cuts in OS services were bitterly opposed by the Royal Society, academics, and various users, although the credit for preventing their implementation has been debated (Harley 1974, Wallis *et al* 1975). The report of the Serpell committee, however, was published on 30 October 1979: its recommendations (HMSO 1979a) include that additional expenditure be allocated to OS to permit the transfer of maps into computerized data bases, from which up-to-date and task-specific maps and other data—such as land parcels—might be created.

The key to successful use of OS data external to that organization—and hence to the spreading of costs, as well as the avoidance of multiple digitization of topographic data—lies in making the data usable for non-graphical as well as graphical purposes. Such organizations as the Post Office could make use of OS digital data as a backcloth to a digital record of their own services: in principle, maps could be updated and then re-plotted automatically, thus avoiding the costly and error-prone redrawing of the service networks on paper OS maps every time a map becomes too cluttered with changes. In addition it should be possible to overlay one data set on another and establish, for instance, the location on the ground of breaks in underground telephone lines; this is only possible with data disaggregated into much smaller units, e.g. house frontages rather than street frontages (Rhind, forthcoming, 1). OS commissioned a software house to write computer programs for this restructuring of their data; Thompson (1978, 1979) and Thompson and Woodsford (1979) have reported on these developments and on the OS plans to create data bases by digitizing the basic scale maps and, on a shorter time basis, the 1 : 50 000 scale maps. The implications of these plans, not least from the data volumes created (Adams and Rhind, forthcoming), are substantial.

This brief paper can not delve into the complexities of creating and reorganizing OS digital topographic data or into the way in which these could then be used in conjunction with other data—geology, soils, socio-economic and medical data, or much else—for administrative functions or for research purposes. Such developments, if they proceed, will necessitate considerable effort in building what have been termed geographical information systems (Rhind, forthcoming, 1). At the time of writing, further comment would be premature since no government response to the Serpell Committee proposals has been given. Despite all these *caveats*, it is clear from this description and from current work throughout the world that the future of much cartography will be predicated upon digital data bases: non-cartographic uses are now being found for cartographic data. Digital terrain models, for instance, are now becoming widely used in the USA, both for map-making and for such purposes as assessing the visual impacts of new buildings (PERS 1978).

Set alongside these quite revolutionary developments, the academic contributions to automation have been largely theoretical—although the GIMMS mapping system created by T. C. Waugh of Edinburgh University (Waugh and Taylor 1976) is an outstanding example of university work being adopted in government departments (such as Statistics Canada). Other automated mapping work to have had significant impacts has been that of the Census Research Unit at the University of Durham (Rhind, Visvalingam, Perry, and Evans 1976; Rhind, Evans, and Visvalingam, 1980; HMSO 1979b); a wall chart containing population census maps produced by this Unit was given to all secondary schools in Britain and a national population atlas has been produced. Other important work on the mapping of recreation-related data from a data bank has been carried out by the Tourist and Recreation Research Unit at the University of Edinburgh (Duffield and Coppock 1975).

Cartography as a research field and as a topic taught in most university courses diminished sharply in the 1960s, with the exception of that in the University of Glasgow; arguably, only three important books on mapping have been published in recent years by British authors (Keates 1973, Maling 1974, Harley 1975). Cooper (1979) has outlined the contemporary situation in the UK so far as cartographic education is concerned, although in one or two instances it is noticeable that teaching of some aspects of cartography is present within other courses—such as the Durham M.Sc. course (Rhind, forthcoming, 2). Most academic research by British cartographers on cartography has, however, been oriented towards questions of map design and human perception of maps (Board 1976, Brandes 1976, Board 1977, Board and Taylor 1977). Certainly the use of maps as data sources in geographical research, such as that by Clarke (1966), has greatly diminished. Since Board and Taylor have recently reviewed research into map perception, little more will be added here except to stress the extreme complexity even of psychophysical studies of maps: in general, the levels of statistical significance of results from such studies have been low or, if high, have been on problems which are arguably trivial in nature. High levels of variation in user response to the same cartographic stimuli appear to be the norm and the measures of success used in map perception—speed and accuracy—are not necessarily the only criteria of importance (although they are usually the easiest to measure). Such research is therefore difficult by any standards.

## SURVEYING AND MAPPING IN TOMORROW'S WORLD

One single trend is obvious from all the above discussion of developments. This is that a digital basis is increasingly common to all disciplines, and, in certain of them, automation is resulting in the need for new types of personnel, usually at technician level. It is manifest that the advantages of thus 'going digital' are *gestalt*—the total benefits are greater than the sum of the parts, since each data file created is, in theory, another data file which may be mapped or used in conjunction with other data. There are, of course, certain dangers in this potential profligacy and in the tailoring of research to that based solely on the data available (Rhind, forthcoming 1). None the less, the potential advantages are enormous—ranging from increased speed of production, more up-to-date maps, and much greater flexibility in what is shown and how it is depicted, to the effectively free provision of contour and other data for much geographical research. It is extremely unlikely that these developments will not continue, although the gestation period in the UK will depend upon the funds available and management policies, such as the free dissemination of computer software (as in the USA where much is available 'in the public domain'). Already, senior representatives of both OS (Smith 1978) and the USGS (Southard 1978) have proclaimed that their organizations are not mapping agencies but have a primary responsibility for supplying topographic data, much of which will be in map form.

A second generalization deducible from the previous discussion is that survey and mapping in Britain is unlike that in many other parts of the world. This is manifested in a concentration on different scale maps, good existing map coverage, a concern with map revision rather than map creation, and little use of orthophotomaps; Doyle (1978b) has declared the intention of the United States Geological Survey to construct a cartographic data base derived from their 1 : 24 000 scale maps—which contrasts starkly with the OS concentration on 1 : 1250 and 1 : 2500 scale basic surveys. To some extent, certain other European countries share some of these British characteristics, and other countries will certainly become more concerned with map revision in coming years. Irrespective of these comparisons, the main British approach to automation of topographic mapping must be through the encoding of existing maps, rather than through the collection of digital data directly from photogrammetric and field survey instruments. This is therefore a once-and-for-all cost, additional to present mapping expenditure.

If developments in the last decade have diminished the need for traditional survey skills, brought the concept of a totally automated survey and mapping system somewhat nearer (if still distant), and threatened to deluge researchers with topographic and other geographical data in digital form, what are the likely developments in the immediate future? One safe prediction is that suppliers of maps and topographic data will have to purvey these in the two forms for a considerable time to come—the clients for each format will probably be mutually exclusive. A second and equally safe prediction is that we are still very far from having solved all the problems of introducing computers to the mapping process, and this will continue to be the case in the foreseeable future: the linking together of data collected at different spatial resolutions and acceptable and economic generalization procedures are two 'unsolved' problems, especially since there is little theory behind most research into cartographic communication (Board 1977). More speculatively, the introduction of Prestel and the development of computer networks promise to make mapping of topography or selected thematic topics available to all with a telephone and a TV set; a prototype TV technology-based colour mapping system has already been built for use of the US President (Zimmermann 1978). Beyond this, the rate of change of technology is truly so remarkable in all fields of survey and mapping—from inertial positioning systems to laser etching of printing plates and ink jet printers—that prediction is certain to be invalidated and probably upstaged. We can be sure, however, that such developments will continue to force changes in attitudes and organization of the traditional practitioners engaged in survey and mapping; the human response to these enforced changes is much less certain.

*Acknowledgements*

The authors thank Mr G. J. F. Holden for kindly providing them with a copy of his report on technology change in the survey industry. Thanks are also due to former colleagues in the Department of Land Surveying at the North East London Polytechnic for their help and advice on survey, mapping and related matters. This paper was compiled while one of the authors (DWR) was in the comparative tranquillity of sabbatical leave at the Australian National University and the Canberra College of Advanced Education; thanks are due to these organizations and to the University of Durham for helping to make this possible. The other author (TAA) was simultaneously involved in collaborative work with Ordnance Survey and wishes to acknowledge the value of discussions with staff of that organization.

REFERENCES

ADAMS, T. A. and RHIND, D. W. (forthcoming). The projected characteristics of a national digital topographic data bank.

ANON. (1977). Heinrich Wild 1877–51. Zum hundersten Geburtstag des Schöpfers neuzeitlicher geodätischer Instrumente. *Astronomisch-Geodatische Arbeiten in der Schweiz*, 31, 39 pp.

ASHKENAZI, V., CROSS, P. A., DAVIS, M. J. K., and PROCTOR, D. W. (1971). The readjustment of the retriangulation of Great Britain and its relationship to the European terrestrial and satellite networks. *Proc., Conf., of Commonwealth Survey Officers*, Cambridge.

ATKINSON, K. (ed.) (1977). *Photogrammetry and Surveying: a selection of papers by E. H. Thompson 1970–1976*. Photogram. Soc., London.

BELL, J. F. (1978) The development of the Ordnance Survey 1:25 000 scale derived digital map. *Cartographic Journal*, 15. 1: 7–13.

BICKMORE, D. P. (1968). Maps for the computer age. *Geogrl. Magazine*, 61. 3: 221–7.

BICKMORE, D. P. (1971). Experimental maps of the Bideford area. *Proc., Conf. Commonw. Survey Offrs, Paper El*.

BICKMORE, D. P. and BOYLE, R. (1964). An automated system of cartography. *Tech. Symp. int. Cartogr. Ass., Edinburgh*.

BLACHUT, T. and VAN WIJK, M. (1978). Results of the international orthophoto experiment 1972–76, in *International Archives of Photogrammetry*. Proceedings of the International Society for Photogrammetry XIII Congress, Helsinki, 1976, Commission II Reports, paper 3, 32 pp.

BOARD, C. (ed.) (1976. *Bibliography of works on cartographic communication*. ICA Commission V.

BOARD, C. (1977). Maps and Mapping. *Progress in Human Geography*. 12: 288–95.

BOARD, C. and TAYLOR, R. M. (1977). Perception and maps: human factors in map design and interpretation. *Trans. Inst. Brit. Geogr*. NS 2. 1: 19–36.

BOMFORD, A. G. (1979). The role of a national mapping organisation. *Proc. Conf. of Commonwealth Survey Officers*, Cambridge.

BRANDES, D. (1976) The present state of perceptual research in cartography. *Cartographic Journal*, 13: 172–6.

CALVOCORESSES, A. P. (1979). Proposed parameters for Mapsat. *Photogr. Eng. and Remote Sensing*, 45. 4: 501–6.

CLARKE, J. (1966). Morphometry from maps, in Dury, G. H. (ed.). *Essays in Geomorphology*, 235–74. London.

COBB, M. C. (1971). Changing map scales by automation. *Geogrl. Magazine*, 43: 786–9.

COOPER, D. (1979). The IBG and cartographical education in Britain. *Area*, 11. 2: 171–2.

COPPOCK, J. T. (1975). Maps by line printers, in Davis, J. C. and McCullagh, M. J. (ed.), *Display and Analysis of Spatial Data*. Wiley, London.

DALE, P. F. (1977). The land surveyor and land information management. Paper given to FIG conference, Sweden.

DAVEY, P., HARRIS, J. F., and PRESTON, G. (1978). Automatic boundary digitising for the Department of the Environment. *Working Paper, Oxford University Department of Nuclear Physics, Image Analysis Group*, 91/78.

DEBENHAM, F. (1936). *Map making*. Blackie and Son, London.

DELL FOSTER, H. and FOLCHI, W. L. (1977). The RS 300-11 digital stereo compiler. *Photogr. Eng. and Rem. Sensing*, 43. 11: 1437–43.

198   *Geography Yesterday and Tomorrow*

DOYLE, F. J. (1978a).   The next decade of satellite remote sensing. *Photogr. Eng. and Remote Sensing*, 44. 2: 155–64.

DOYLE, F. J. (1978b).   Contribution to panel discussion on digital terrain models. *Photogr. Eng. and Remote Sensing*, 44. 12: 1490–3.

DREWITT, B. (1973).   The changing profile of the map user in Great Britain. *Cartographic Journal*, 10. 1: 42–8.

DUBUISSON, B. L. Y. (1977).   Why analytical plotters? *Photogr. Eng. and Rem. Sensing*, 43. 11: 1363–6.

DUFFIELD, B. S. and COPPOCK, J. T. (1975).   The delineation of recreational landscapes: the role of a computer-based information system. *Trans. Inst. Brit. Geogr.* 66. 141–8.

ECU (1978).   *Land use mapping by local authorities in Britain.* Architectural Press, London.

EVANS, I. S. (1970).   The implementation of an automated cartography system, in Cutbill, J. L. (ed.), *Data processing in biology and geology.* Systematics Assocn. Special Volume 3. 39–55: Academic Press, London.

FISCHER, W. A., HEMPHILL, W. R., and KOVER, A. (1976).   Progress in remote sensing (1972–76). *Photogrammetria*, 32. 2: 33–72.

FRIEDMAN, S. J. (1977).   The OMI analytical stereoplotter model AP/C4. *Photogr. Eng. and Rem. Sensing*, 43.11: 1419–25.

GAITS, G. (1969).   Thematic mapping by computer. *Cartographic Journal*, 6.1: 50–68.

GARDINER-HILL, R. C. (1972).   The development of digital maps. *Ordnance Survey Prof. Paper*, NS, No. 23.

GEO. ABS. (1979).   *A matter of degree:* a directory of geography courses 1979/80. Geo. Abstracts, Norwich.

HARDY, G. (1974).   The Ordnance Survey 1 : 50 000 map series. *Geogrl. J.* 140. 274–83.

HARLEY, J. B. (1974).   Changing the Minister's mind: a guide to the Ordnance Survey review. *Area*, 3. 211–19.

HARLEY, J. B. (1975).   *Ordnance Survey maps: a descriptive manual.* HMSO, London.

HELAVA, U. V. (1977).   The analytical plotter—its future. *Photogr. Eng. and Rem. Sensing*, 43.11: 1361–2.

HILL, A. R. (1973).   Military user performance with orthophotomaps. *Report to US Engineer Topographic Laboratories.*

HMSO (1979a).   *Report of the Ordnance Survey Review Committee.* London.

HMSO (1979b).   *People in Britain—a census atlas.* London.

HOBBIE, D. (1977).   C-100 Planicomp, the analytical stereoplotting system from Carl Zeiss. *Photogr. Eng. and Rem. Sensing*, 43. 11: 1377–90.

HOLDEN, G. J. F. (1979).   *Submission to the Applied Science/Engineering Committee into Technological Change*, School of Surveying, University of New South Wales.

HOWMAN, C. and WOODSFORD, P. A. (1978).   The Laserscan FASTRAK automatic digitising system. Paper given to the 9th International Conference on Cartography, Maryland, USA.

INGHILLERI, G. (1977).   The DS-type Galileo analytical plotters. *Photogr. Eng. and Rem. Sensing*, 43. 11: 1427–35.

JACKSON, M. J., CARTER, P., GARDNER, W. G., and SMITH, T. F. (forthcoming).   Urban land use mapping from remotely sensed data. *Photgr. Eng. and Rem. Sensing.*

JOHNSTON, R. J. (1978).   *Multivariate statistical analysis in geography.* Longman.

KEATES, J. S. (1973).   *Cartographic design and production.* Longman.

KELLY, R. E., McCONNELL, P. R. H., and MILDENBERGER, S. J. (1977).   The Gestalt photomapping system. *Photogr. Eng. and Rem. Sensing*, 43. 11: 1407–17.

KONECNY, G. (1977).   Software aspects of analytical plotters. *Photogr. Eng. and Rem. Sensing*, 43. 11: 1363–6.

LAMBERT, B. P. (1971). Production of small and medium scale maps with the aid of orthophotography. *Proc. Comf. Commonwealth Surv. Offrs.* Part 1, 149–57.

LANG, T. (1971) *Computer programs for mapping.* Architectural Press, London.

LYONS, H. G. (1931). Land surveying in early times. *Proc. Conf., Empire Survey Officers.*

MALING, D. H. (1974). *Coordinate systems and map projections.* Phillips, London.

MARGERISON, T. (1977). *Computers and the renaissance of cartography.* Natural Env. Res. Council, London.

MARLES, A. (1971). The place of map miniaturisation in map production. *Proc. Conf. Commonweath Survey Offrs.*

MOLINEUX, A. (1978). Data bank at work. *Geogrl. Magazine,* 50. 754–5 (see also 736–53).

PERS (1978). Proceedings of the Digital Terrain Model Symposium, published in *Photogram. Eng. and Remote Sensing,* 44. 12: 1481–1586.

PETRIE, G. (1970 and 1971). Some considerations regarding mapping from earth satellites. *Photogram. Record,* 6. 36: 590–624 and 7. 37: 55–6.

PETRIE, G. (1977). Orthophotomaps. *Trans. Inst. Brit. Geogr.,* NS 2. 1: 49–70.

RHIND, D. W. (1971). The production of a multi-colour geological map by automated means. *Nachr. ausden Karten und Vermessungswesen* Heft Nr 52. 47–52.

RHIND, D. W. (1974). An introduction to the digitising and editing of mapped data, in Dale, P. F. (ed.), *Automation in Cartography,* Br. Cartogr. Soc. Spec. Publ. 1.

RHIND, D. W. (1976). Towards universal, intelligent and usable automated cartographic systems. *ITC Jl.* 3: 515–45.

RHIND, D. W. (1977). Computer-aided cartography. *Trans. Inst. Brit. Geog.* NS 2. 1: 71–97.

RHIND, D. W. (forthcoming,1). Geographical information systems in Britain, in Wrigley, N. and Bennett, R. (ed.), *Quantitative geography in Britain.* Routledge and Kegan Paul.

RHIND, D. W. (forthcoming, 2). Spatial Data Analysis: the evolution and progress of a master's degree course. *Jl. Geog. H. Ed.*

RHIND, D. W., EVANS, I. S., and VISVALINGAM, M. (1980). Making a national atlas of population by computers, *Cartographic Journal,* 17, 1.

RHIND, D. W. and HUDSON, R. (1980). *Land Use.* Methuen.

RHIND, D. W. and TREWMAN, T. (1975). Automatic cartography and urban data banks—some lessons from the UK. *International Year Book of Cartography,* 15: 143–57.

RHIND, D. W., VISVALINGAM, M., PERRY, B., and EVANS, I. S. (1976). Mapping people by laser beam. *Geogrl. Magazine,* 49. 3: 148–52.

ROBBINS, A. R. (1977). Geodetic astronomy in the next decade. *Survey Review,* 185.

ROBINSON, A. H. and PETCHENNIK, B. B. (1976). *The nature of maps.* Chicago Univ. Press.

ROBINSON, A. H., MORRISON, J. L., and MUEHRCKE, P. C. (1977). Cartography 1950–2000. *Trans. Inst. Brit. Geogr.,* NS, 2. 1: 3–18.

ROY. SOC. (1978). *The future role of the Ordnance Survey. The Royal Society submission to the Ordnance Survey Review Committee.* London.

SEYMOUR, R. H. (1977). The US-1 universal stereoplotter. *Photogr. Eng. and Rem. Sensing,* 43. 11: 1399–1406.

SMITH, W. P. (1978). Ordnance Survey AD 2000. *Geogrl. Magazine,* 50. 6: 361–4.

SOUTHARD, R. B. (1978). Development of a digital cartographic capability in the National Mapping Program. Paper given to the 9th International Conference on Cartography, Maryland, USA.

TAYLOR, P. (1976). An interpretation of the quantification debate in British geography. *Trans. Inst. Brit. Geogr.,* NS 1: 129–42.

THOMPSON, C. N. (1978).   Digital mapping in the Ordnance Survey 1968–1978. Paper given to the ISP Commission IV Inter-Congress Symposium, Ottawa.

THOMPSON, C. N. (1979).   The need for a large scale topographic data base. *Proc., Conf. Commonw. Survey Offrs., Cambridge.*

THOMPSON, C. N. and WOODSFORD, P. A. (1979).   The Ordnance Survey topographic data base concept for the 1980s. Paper given to the Second UN Regional Cartographic Conference for the Americas, Mexico City.

THOMPSON, E. H. (1958).   The prospect for British photogrammetry. *Photogram. Record*, 2. 11: 355–62.

THOMPSON, E. H. (1971).   The development of photogrammetry in the seventies. *Photogram. Record*, 7. 38: 195–200.

THOMPSON, M. M. and MIKHAIL, E. M. (1976).   Automation in photogrammetry: recent developments and applications (1972–1976). *Photogrammetria*, 32. 4: 111–45.

UNITED NATIONS (1976).   *Status of world mapping ( World Cartography 14)*. Dept. of Economic and Social Affairs, UN, New York, 102 pp.

VIGNERON, C. (1977).   Traster 77: Matra analytical stereoplotter. *Photogr. Eng. and Rem. Sensing*, 43. 11: 1391–7.

WALLIS, H. (ed.) (1976).   *Map Making to 1900: an historical glossary of cartographic innovations and their diffusion*. Royal Society, London.

WALLIS, H., COPPOCK, J. T., and THOMAS, D. (1975).   Ordnance Survey policy review *Area*, 7. 1: 21–2.

WAUGH, T. C. and TAYLOR, D. R. F. (1976).   GIMMS/An example of an operational system for computer cartography. *Canadian Cartographer*, 13. 2: 158–66.

WELCH, R. and LO, C. P. (1977).   Height measurements from satellite images. *Photogr. Eng. and Remote Sensing*, 43. 10: 1233–41.

WOLF, P. R. (1974).   *Elements of photogrammetry*. McGraw Hill.

ZIMMERMANN, F. (1978).   A mapping system for the White House. Paper given to Harvard University Conference on Computer Graphics, Cambridge, Mass., July.

# 8 Theory in Human Geography: A Review Essay

ALAN WILSON    *School of Geography, University of Leeds*

## 1. WHAT IS THEORY?

THIS volume commemorates the founding of the Royal Geographical Society in 1830. It is interesting to begin with the observation that von Thunen's *Isolated State* was first published in 1826. Theory has a long history in geography, though it has not always been practised by people who would call themselves geographers.

What is theory? It is sometimes important to explore questions of this kind with the deepest philosophical rigour. I shall take the simpler view here, however, that it is easy to recognize good theory when we see it. A theory is a representation of our understanding of some entity, and its workings, which we are interested in. The philosophical problems arise because of the difficult relationship between theories and truth. Sometimes, theories are patently obviously not true, but they offer insights and perhaps a step towards something better. Theories are not 'facts', but generate predictions which can be compared with data. Above all, perhaps, in theorizing, we seek ever *deeper* understanding of the elements of our discipline. Whatever has been achieved by a particular time, more difficult theoretical problems usually turn out to lie around the corner. In achieving some degree of understanding, our eyes are opened more clearly to what we do not understand.

We can also discuss what theory is not! In particular, it is important to note at the outset that theory is not the same as 'technique'. Theory is not to be identified with 'quantification', for example, important though that may be in a number of respects.

There is a wide range of forms within which theory can be expressed: as much understanding of the workings of a city may be obtained from a novel as from a mathematical model, and the kinds of understanding represented in this range will usually be complementary rather than in competition. Notwithstanding this observation, the view of this essay will, on the whole, be a more restrictive one. This arises from the way we choose to define disciplines, and geographers are not usually so imperialist as to enter directly the preserves of literature, and also because a review of a field as broad as this is inevitably stamped with personal blinkers and idiosyncrasies.

It is impossible to explore geographical theory without deciding what is geographical in relation to other disciplines and this preliminary topic is dealt with in the next section. This is followed by a review, in broad terms, of some of the major general features of geographical theory in section 3. Finally, a review of future prospects is presented in section 4.

## 2. HUMAN GEOGRAPHY AND RELATED DISCIPLINES

Human geography is concerned with people, their activities, and their spatial distributions; and with organizations of all kinds, their activities—the production of goods and services in the broadest sense—and their spatial distributions. The discipline is subdivided in very many ways which, inevitably, overlap. The two broadest divisions are into social and economic geography, reflecting the (at times artificial) distinction between people and organizations made at the outset. Another way of slicing the subject is into urban and rural geography. Or more detailed subdivisions are sometimes used: the geographies of resources, agriculture, transport, and so on. If a long time period is involved (or perhaps a particular method?), we think of historical geography. Synthesis makes us think of regional geography in the old sense—of which more later. A more modern concern with techniques produces fields like spatial analysis. Current concerns, as we will see later, relate to welfare or radical geography.

What is a discipline or a sub-discipline anyway? It is clear from the previous paragraph that whatever geography is, it will overlap with many other disciplines. The definition of a discipline in part involves substantive issues and in part social convention. At its broadest, and given its history, we might argue that geography is concerned with places and with spatial relationships. But then so are many other disciplines. This difficulty is resolved by recognizing that there is substantial overlap between disciplines, and by noting that it is a matter of convention as to what the group of people who call themselves geographers actually do. This has important consequences for the discussion of theory in geography. Firstly, such discussion cannot take place without regard to the close links with a number of other disciplines. Secondly, what is distinctively geographical theory will be part of the social convention. And thirdly, there is an important organizational corollary of the second point for the future of geography: when new and important questions arise, if geographers do not take them on, then someone else will.

At the level of general theory, it can be argued that the geographer, in his or her concern with spatial analysis, is working with the concepts of location—both of structure and of activities—and flows; with spatial networks; and with spatial dynamics. These concepts provide the building blocks for most geographical theory whatever the sub-discipline from the kind of list discussed earlier. It may also be argued, whether from the perspective of systems theory or of Marxist-structuralism, that only a unified social science is possible, and I return to this issue in the final section.

The other disciplines which are relevant obviously include economics, sociology, politics, social administration, history, and even something as far removed as civil engineering. There are also methodological disciplines like philosophy and mathematics. This list is to be borne in mind as our explorations proceed.

### 3. A REVIEW OF SOME MAJOR DEVELOPMENTS AND IDEAS

#### 3.1. *First breakthroughs*

The early development of theory in human geography illustrates the point made in the previous section about the importance of recognizing conventions. There were major developments from at least the time of von Thunen (1826). Perhaps the most important subsequent early locational theorists were Weber (1909), Christaller (1933), and Lösch (1940). There were important contributions from economists (for example, Haig 1926, and Hotelling 1929) and from sociologists (such as Burgess 1927, and Hoyt 1939). The study of flows and interactions, using the gravity model and the intervening opportunities model for example, also has a long history in the work of authors like Carey (1858), Ravenstein (1885), Reilly (1931), Stouffer (1940), Stewart (1941), and Zipf (1946). But most of this work was not recognized, or worked with, in the then profession of geographers until the 1950s (cf. Gould 1979, for a discussion of this point and for a personal history of subsequent developments). Geographers at the time were more concerned with simpler accounts of 'areal differentiation' and most of the current theoretical basis until, say, twenty-five years ago, was a concern with taxonomy and elementary verbal theorizing. Gould makes an exception of some work in historical geography, which is interesting in that it has a basis in another discipline.

Thus, some very important theoretical foundations were laid during the first 125 years of the period we are considering, but they were not recognized and built on within geography until much more recently. Stoddart (1965) commented bluntly that 'the subject has managed to isolate itself from virtually every major development in the field of scientific thought since 1859.' What was the first general breakthrough which forced this position to change? This is not the place to chart it in terms of the variety of people involved in a very difficult battle. Many others do this—see for example many of the articles in Hudson (1979), Chisholm (1975), and Johnston (1978, 1979). A good basic reference is Haggett, Cliff, and Frey (1977). The essence of the change was a shift to a concern with generality and away from a belief in uniqueness. Bunge (1979) puts the matter succinctly: 'If Hartshorne was right, not only in his insistence that locations were only unique, but humans were also only unique, then neither locations nor societies could be predicted. Then location theory, as well as scientific socialism, was impossible.' Shortly afterwards, in the same article, he makes simply the point which is perhaps all too obvious to an outsider, but was clearly not so at the time: 'Locations and people are clearly general. But they are also unique.' It does not seem difficult to accept the idea that locations and people have some general properties and some which are unique to them. Theory is in a sense concerned with both: primarily with generality, but also with the way in which uniqueness generates exceptions.

The shift was ultimately seen as one from regional geography to systematic geography and it can be argued—as I shall below—that this has had some unfortunate consequences also.

## 3.2. *Some general considerations: problem types, scale, representation*

Before charting some of the main ideas and movements which have shaped geographical theory in more recent times, it is useful to discuss some general concepts which provide a useful background.

First, we consider Weaver's (1958) categorization of problem types (which can also be taken as a classification of system types). He noted three kinds of system. Those which could be described by a very small number of variables, such as two or three or four, which he called simple systems. They could be modelled by the methods of traditional mathematics and formed the main subjects of science up to the late nineteenth century. Then there are systems of disorganized complexity: these need many variables to describe them, thousands or millions, but their components interact together only weakly. It was the discoveries of workers like Boltzmann and Gibbs in the late-nineteenth and early-twentieth centuries which enabled general methods to be made available for modelling these kinds of systems. In physics, these were the methods of statistical mechanics: entropy maximizing methods. More broadly, they can be seen as statistical averaging methods. They do not offer a description of system behaviour in detail, but do enable more aggregate questions to be answered, and these may be important ones in many disciplines. Finally, he recognized systems of organized complexity. These also involve large numbers of components and variables to describe them but where there is some strong interaction between the components. Because of these new complexities, no general methods have as yet been discovered for modelling such systems, and Weaver described these as characteristic of the problems of modern science.

Weaver offered his categories in the context of arguing that research funds should be increasingly devoted to the biological sciences, rather than the physical, but it is easy to see that his analysis has implications for the social sciences such as human geography. This will become evident in the discussion below. We can see the simpler methods of analysis discussed in the preceding subsection as dealing with geographical systems as simple systems and the newer methods as part of attempts to handle greater complexity.

The next preliminary point concerns scale, or level of resolution, at which a system is viewed. The finer the scale, the more detail is observed and the more complex a system is likely to seem; and vice versa. This means that Weaver's categories, and the corresponding methods of analysis, have to be applied not simply to systems *per se*, but to systems viewed at particular scales. There is nothing odd about this. Indeed, some disciplines are defined by scale: microbiology, biology, and ecology, for example. The useful idea to absorb, and this is particularly important for geography which operates over a very wide variety of scales, is that there are usually interesting results to be obtained at each scale. While reductionism, for example, might be a useful driving force in research, it does not follow that all useful results can only be obtained at the finest scale—which in geography is the 'behaviourist' view.

The third point to bear in mind is that even when scale is appropriately chosen, there are likely to be various ways in which the system can be represented. A particularly important example of this within geography is the way in which space is represented. Most of the early theorists mentioned above treated space continuously, and this often forces restrictive assumptions—for example in urban location theory that all employment is at the centre of the city. By treating space as a set of discrete units, there is some loss of resolution, but other restrictive assumptions of that type can be avoided. Much skill in theoretical geography has to be devoted to these kinds of decisions which at first may appear relatively trivial but which are often of crucial importance.

With this backcloth, we can now proceed to review some of the main general ideas which have had an impact on theoretical geography in recent years.

### 3.3. *Quantitative methods: statistics and model building*

One of the sources of 'new geography' has been the so-called quantitative revolution (Burton 1963). This has sometimes been misleading and has lead to the identification of theoretical geography with quantification and to some misdirected effort. None the less, the impact of quantification was important, and much theoretical geography has been and is stated in statistical or mathematical terms. So a discussion of the range of techniques involved is a necessary beginning. We should also note the almost obvious point that the timing of the development of quantitative methods on a large scale was almost certainly associated with the development of computers in the 1950s.

It is important to recognize at the outset the distinction between the approaches of statistics and of mathematics. Broadly speaking, the former are inductive and the latter deductive; the first trying to cull generalities from data, the second trying to invent mathematical statements of theories and then testing these against data, modifying them, and so on. So almost always in the quantification of a discipline, statistical analysis come first. This is true of geography at least in terms of the scale of the contributions of the different styles even though the examples cited of early geographical theory to some extent contradict the argument. It should also be emphasized that the dichotomy is not as sharp as stated here. A successful piece of statistical analysis may lead to a mathematical model; conversely, a model can only be effectively tested using the methods of statistics, and some of the methods of model building, such as entropy maximzing, are closely related to statistical methods, such as those based on Bayes's Theorem. Further, there are plenty of circumstances when models have to represent probabilistic phenomena and then the two styles come very close since the roots of probability theory might be considered to lie jointly in statistics and mathematics (as, for example, in Curry 1964).

The first consequence of quantification has been that many geographers have become competent in a wide range of statistical techniques, from the construction of linear models of varying degrees of complexity, through factor analysis in its various forms, to statistical control theory. The array of literature is impressive—for typical examples, see King (1979), Berry (1972), Cliff and Ord (1973), Rees (1979), and Bennett (1979).

The second consequence was the adoption by geographers of a range of elementary mathematical models of which the most popular was undoubtedly the gravity model. When such a model is used in its most elementary form, it in effect represents a separate model for each flow in a geographical system and can be considered as an application of Newtonian, Weaver-I methods for simple systems. Similar comments could be applied to population models of the Malthusian exponential growth or the Verhultz logistic growth variety, or to economic models of the economic base type. Location theory was firmly rooted in the authors listed in section 3.1 above and the underlying mathematics (if one neglects the geometrical complexity—in which field geographers have contributed much distinguished work—see Dacey 1976 and Beavon 1977, for recent reviews) was either elementary or not completely worked out—take the rank size rule and the Löschian economic landscape as representing the different ends of this spectrum.

A new range of mathematical modelling problems were arising out of theory and it is probably no exaggeration to assert that the root of the difficulties lay in the so-called aggregation problem—well-known in economics, less explicitly recognized in geography. The theory of locational or transport behaviour was based on the theory of consumers' behaviour or the theory of the firm. The phenomena which were measured, and which were to be modelled, were often, indeed usually for geography, at a coarser scale. How does one aggregate the results of a micro-scale analysis to a meso (Haggett (1965) usefully recognized that the geographer's scale was often between the economist's micro- and macro-) or a macro-scale? The recognition of this problem is, with hindsight of course, an understanding that the systems being modelled are complex—Weaver-II or Weaver-III. We can look at these general ideas to help us to interpret the forms of development through which theoretical geography, as represented by mathematical modelling, passed.

First, consider the systems which may be taken as Weaver-II. These are the major subsystems which represent the behaviour of people: residential location, job choice, use of services, transport flows, the population model itself. Weaver would lead us to expect various forms of statistical averaging procedure to be used to lift us from the micro scale to the meso, and this is indeed how we can interpret much of the work which has been done. Linear programming, entropy maximizing, non-linear programming, Markov models, non-linear account-based models have all been used, often as alternative methods for modelling the same system. More recently, it has proved possible to tackle the economic aggregation question more directly using the methods of random utility theory. Probabilistic methods, involving simulation, have also been used so that high-dimensional arrays can be avoided, and may become more popular. Some of the work has been done in disciplines adjacent to geography, but the results have now been absorbed by geographical model builders.

The large set of complicated and important results referred to briefly above can perhaps best be charted by an introduction to some of the relevant literature. Each reference cited here may be used as the basis for building a much more extensive bibliography.

The modern foundations of location theory, building on the authors cited in section 3.1, are typified by Alonso (1964) and Muth (1969) in relation to urban residential structure and by Isard (1956) for the economic side. Their work is still in a continuous space representation and operational models have been generated by working with discrete space. Herbert and Stevens (1960) showed how to use linear programming to develop Alonso's scheme and Harris (1962) began the long task of estimating some of the utility functions involved. As Weaver might have predicted if he had examined the geographical field in 1958, there have also been extensive applications of entropy maximizing methods (Wilson 1970). This can also be seen as a form of mathematical programming and now fits into a wider framework. Nijkamp (1972) shows how non-linear programming can be applied to economic location. An alternative basis is random utility theory—as presented for example by Williams (1977) building on work by McFadden (1974) and Cochrane (1975) among others. The aggregate forms of these models can be presented in mathematical programming form, based on the idea of maximizing locational surplus (Coelho and Williams 1977). Indeed, the recognition that many models can be cast in mathematical programming form has produced some rich results. Evans (1973) showed that the transportation problem of linear programming was a special case of the corresponding entropy maximizing formulation and this was extended in relation to dual variables and its range of applications by Wilson and Senior (1974). This work showed that entropy maximizing methods, and later random utility methods, could be seen as incorporating 'dispersion' to represent imperfections in economic models in contrast to the perfect market assumptions of linear programming models. It is also clear that such models can be embedded within broader planning optimization frameworks and this further extends their usefulness (Coelho and Wilson 1976; Coelho, Williams, and Wilson 1978). The whole class of models is usefully reviewed in two articles by Senior (1973), 1974) and in two books (Wilson 1974; Batty 1976) among a field of literature which is now becoming prolific.

Most of the models referred to involve large arrays of variables. These can be seen more formally as accounts, and another Weaver-II method is the application of Markov techniques (see, for example, Ginsberg 1973). It is also possible to build account-based models on a broader foundation, seeking to make the best use of the often-imperfect data which are available (Rees and Wilson 1977). As models and theories become more detailed, however, the arrays become very large. One escape from this problem is to use simulation methods, following the work of Orcutt *et al.* (1961). Wilson and Pownall (1976) showed how these methods could be applied in a geographical context and such models are now being actively developed on a larger scale (Clarke, Keys, and Williams 1979).

What of Weaver-III systems, when we face organized complexity? For the over-all economic model, input–output and related mathematical programming methods offer a model which is a Weaver-II approximation, and geographers have helped to develop versions of these in which space is explicit (for example, Cripps, Macgill, and Wilson 1974). Otherwise, we have to seek new methods for handling complexity and non-linearities in this type of system. An important and well-defined subsystem with these characteristics is any which involves network flows and much progress has been made with models which build on and extend the shortest-path-through-a-network algorithm (see Haggett and Chorley 1969 for a broad review, and Wilson 1979 for a review in the context of congestion and transportation). Other systems, in which non-linear dynamic behaviour is important, have to be represented directly by differential or difference equations. An important branch of work in this field, contributed from the direction of engineering, was Forrester's (1969) *Urban dynamics*. This has been an important stimulus in geography, but has also been heavily criticized and not taken up directly. Because of their slow rate of change, it is probably more difficult to model structure—the location of buildings and transport infrastructure—than to model the location of economic activity and the newer methods of dynamic modelling show signs of promise here (see Harris and Wilson 1978 for an application to urban retail structure). Progress is often piecemeal, but recent developments in mathematics are beginning to offer general results under the heading of bifurcation theory (cf. Thom 1975; Hirsch and Smale 1974).

Other approaches to the study of process, and hence dynamics, have a longer history in geography. Of particular importance is Hagerstrand's (1953, 1968) work on the diffusion of innovations which has developed into a wider concern with time geography (Hagerstrand 1970 and many of the articles in the three volumes edited by Carlstein, Parkes, and Thrift 1978). In effect, much of this work is dynamical analysis at the micro scale. It offers important insights and concepts—the notion of space-time prisms for example—and may ultimately be integrated with other aspects of dynamical analysis.

The Weaver characterization of geographical systems helps us to see why, on the whole, more progress has been made with modelling and theory in population and social geography (in the broadest sense) than in economic geography: the systems of disorganized complexity are easier to handle than systems of organized complexity.

## 3.4 *Systems analysis*

Another source of, or model for, the new geography is often said to be systems analysis ( and two important recent surveys are those of Chapman 1977 and Bennett and Chorley 1978). I shall argue here that systems analysis has five roles, and, in briefly articulating these roles, an implicit definition emerges (see Wilson 1977 for a more detailed account).

Firstly, straightforward help is provided in the analysis of complexity. Orderly definition of state variables at particular levels of resolution in particular representations is encouraged. Diagrams can be constructed showing the main relationships. All this then provides a foundation for developing theory and models.

Secondly, the systems analyst focuses on the interdependence of the elements of the system. This can help in the definition of subsystems (and indeed sub-disciplines). This focus on the coupling of elements is then intimately connected to the notion of whole-system behaviour: the behaviour of the whole system is likely to be very different from anything which could be predicted from a knowledge of any of the individual components. This is sometimes called systemic behaviour.

Thirdly, it provides a potential mapping of possible methods of analysis or modelling against system types. At this level, a concern with generality is developing and it builds in an obvious way on the argument of the previous subsection. If general system characteristics can be identified, perhaps as an elaboration of Weaver's scheme, and if analytical methods can also be categorized, then when a new problem turns up, it should be possible to identify the system type and then to state a list of the alternative methods for modelling. This can form the basis of analogy between systems in different disciplines. For example, the work of authors like Rescigno and Richardson (1967) in ecology can now be applied in urban dynamics.

The fourth point seeks an even higher level of generality: the development of general systems theory. At its most ambitious, this involves the building of models of a completely abstract kind to which any real situation can then be fitted. At a less ambitious level, this might involve the building of a number of geographical models, say concerned with location theory, interaction theory, network theory, and dynamics, which would each have a wide range of applicability.

Fifthly, when effective systems models can be built, they are usually seen as having a potential role in planning or problem-solving, and this has always been one of the main motivations of systems analysts. But more of this in the next subsection.

The list, in an obvious sense, represents increasing complexity and ambition. I would argue that the concepts of systems analysis are useful to geographers from the first stage onwards. So you do not have to believe in the possibility, at this time, of being able to represent systemic behaviour in a city, or of being able to develop general theory, for some aspects of the techniques to be useful. Perhaps the most important role is very general and not at all specific; that is, to remind us continually that systems and subsystems are very rarely isolated from each other and that this should never be forgotten in any theory-building exercise.

### 3.5. *Planning, welfare, behaviour, radicalism*

It is tempting to divide into separate subsections the four further themes identified in the title above; but they are so closely linked that it seems better to deal with them together. Smith (1977) talks about geography's 'second revolution' as being 'radical' or 'concerned with social relevance'. From the early sixties onwards, and earlier in the United States, social science has been seen as relevant to public policy, and geographers have played their part in this, though not as prominent a part as economists. This has had an impact on geographical theory through the demand for 'relevance' and has had a direct influence on what is often argued to be the best scale of analysis for this purpose—the welfare of the individual, and therefore a concern with behaviour. The radical approach is also concerned with individuals and welfare, partly in the sense of contributing new methods of analysis based on Marxian theory and partly with an argument for radical or revolutionary change. What has been the impact of these ideas on the development of geographical theory?

Planning involves the control by the state of various instruments of public policy for some kind of public good to be determined by the political process. So the first impact was to force the theorist, particularly the modeller, to distinguish those variables which were potentially controllable. Various settings of these variables, now treated as exogenous to the models, could then be inserted as inputs to the models so that the impact of alternative plans could be assessed. It also forced the construction of evaluation measures to facilitate this assessment, usually some variant of cost–benefit analysis. Perhaps the best-known examples of this kind of work are represented in the transportation studies which have now been carried out in most large British towns and cities. (See Hall 1975 for a general survey and Wilson 1974 for a technical one; SELNEC Transportation Study 1972 offers a specific illustration, and Scott and Roweis 1977 a critique of the whole approach.)

The measures of welfare in these studies were based on meso-scale spatial interaction models, and were usually further aggregated to a macro-scale before being used. This, and more general arguments, lead to the criticism of these techniques as being insufficiently concerned with individual behaviour (cf. Hagerstrand 1970, for a plea from a different quarter). To some extent, these criticisms have been met directly with the development of so-called disaggregate models based on random utility theory with an economic foundation, or, looking in the direction of psychology, on the work of Luce (1959). But there was also criticism of the representation of individual behaviour in such models, and one school of thought developed a concern with perception. This is perhaps best known in the work on mental maps (as in Gould and White 1974) but has also been approached in a more technical way.

The radical critique is more fundamental and is best characterized by Harvey's (1973) *Social justice and the city.* King (1979) and Peet (1979) offer recent opinion and reviews. There are many strands to the argument. First, a warning against environmental and spatial determinism in geographical theory. The argument is that the social relations created by particular modes of production are the main determinants of change and development in society, and that spatial processes should be seen as a consequence of this. The driving force is class conflict and struggle. This leads to a concern with the study of processes driving such phenomena as urbanization which are different from those of traditional models with their neo-classical economic foundations. It can be argued, in this sense, that this new perspective offers a new approach to dynamics through a focus on process which is more fundamental than hitherto. But there is an as yet unsolved aggregation difficulty here for geographical theory. It is also argued that the liberal approach to change through planning within the capitalist system is always likely to be ineffective (cf. Sayer's (1976) critique of the use of many standard models in this context) and that ultimately, the only effective *theory* will be that which is the basis of new practice. Some authors, notably Scott (1976), have attempted to combine this style of theorizing with the methods of quantitative modelling in geography, but generally, this theory has not reached such a degree of articulation. It should also be emphasized that the type of theoretical work arising from this basis is more likely to be of the style of historical geography rather than leading to the modification and development of new models (cf. Thompson 1978).

## 4. REFLECTIONS AND PROSPECTS

A concluding section in an essay of this nature has to start with an apology. While I hope that I have been able to present some themes which create some useful perspectives for understanding progress in theoretical geography, I am very conscious at this stage of many omissions. In some ways, I hope that the examples I have used to illustrate the argument cover the main ground, but it is also clear that in the application of theory, I have omitted many important areas of study. I shall mention two only and ask in advance for forgiveness in relation to others. Firstly, geography is at the start of what is likely to be a rapidly expanding field in the study of resources. There is a long and distinguished history in relation to water and land and this will expand in the future to include energy. Secondly, there has been much applied work which has also contributed to theory in the study of development, particularly in relation to the third world.

There are also omissions of themes. Gould (1979) in his reflections on the last twenty years notes an increasing concern of geographers with philosophy (for example, Harvey 1969; Olsson 1975; and Gale and Olsson 1979). He also notes advances in cartography, particularly singling out Tobler's work dating from the time of his doctoral dissertation (Tobler 1961). And finally, he identifies new mathematical advances in the study of complexity which have not been incorporated in any of the themes developed above, particularly the notions of q-analysis developed by Atkin (1977).

It is none the less clear from the ground which has been covered that there is now a very rich basis of ideas in theoretical geography. Many of these have been contributed by practitioners in other disciplines, notably economics, but also sociology, history, and methodological disciplines such as mathematics. It could be argued that if we believe in the ideas of system analysis, we should believe in a unified social science, or perhaps even more generally, a unified systems science: all things are connected. However, there will continue to be a division of labour and the problems of spatial analysis and the study of processes in space will continue to provide the geographers with some difficult and interesting problems, though there will continue to be contributions, even in these specific fields, from elsewhere. The level of development achieved, and the connections to a variety of other disciplines, will make life rather difficult for the geographer of the future in that he or she will have to acquire skills from a wide variety of sources; and these skills are continually becoming a more complex. The demands of the quantitative revolution of the 1960s are nothing compared to what faces the graduate student of today—whether approaching theory from the viewpoint of the mathematical modeller or from that of Marx.

So the prospects are exciting and the tasks difficult. Ideally, it would be good to think that geography could continue to be a synthesizing discipline and that a particular theoretical concern would be to put together effective theories for whole-systems of interest to geographers. It has been argued elsewhere that this could lead to a new kind of regional geography (Wilson, Rees, and Leigh 1977), but this is proving to be a very difficult task. There are also broad issues—like the long-run theory of urban development—which perhaps need contributions from both modellers and Marxists, where the bulk of the progress with a very large problem has yet to be made.

The rate of progress in the last twenty-five years has been particularly remarkable. There is no sign of the pace slowing down and a corresponding review to this in ten years' time should make exciting reading to those of us who are struggling with today's problems.

# REFERENCES

ALONSO, W. 1964. *Location and land use*. Cambridge Mass.

ATKIN, R. 1977. *Combinatorial connectivities in social systems*. Basle.

BATTY, M. 1976. *Urban modelling: algorithms, calibration, prediction*. Cambridge.

BEAVON, K. S. O. 1977. *Central place theory: a re-interpretation*. London.

BENNETT, R. J. 1979. *Spatial time series*. London.

BENNETT, R. J. and CHORLEY, R. J. 1978. *Environmental systems: philosophy, analysis and control*. London.

BERRY, B. J. L. (ed.) 1972. *City classification handbook*. New York.

BUNGE, W. 1979. Perspective on Theoretical Geography. *Annals Association of American Geographers*, 69: 169–74.

BURGESS, E. W. 1927. The determination of gradients in the growth of the city. *Publications American Sociological Society*, 21: 178–84.

BURTON, I. 1963. The quantative revolution and theoretical geography. *Canadian Geographer*, 7: 151–62.

CAREY, H. C. 1858. *Principles of social science*. Philadelphia.

CARLSTEIN, T., PARKES, D. N., and THRIFT, N. J. (ed.) 1978. *Timing space and spacing time*. 3 volumes. London.

CHAPMAN, G. T. 1977. *Human and environmental systems*. London.

CHISHOLM, M. 1971. In search of a basis for location theory: micro-economics or welfare economics. *Progress in Geography:*, 3: 111–33.

CHISHOLM, M. 1975. *Human geography: evolution or revolution*. Harmondsworth.

CHRISTALLER, W. 1933. *Die centralen orte in Süddeutschland*. Jena; English translation by C. W. Baskin. *Central places in southern Germany*. Englewood Cliffs.

CLARKE, M., KEYS, P., and WILLIAMS, H. C. W. L. 1979. Household dynamics and economic forecasting: a micro-simulation approach. Paper presented at the Regional Science Association, London, August 1979.

CLIFF, A. D. and ORD, J. K. 1973. *Spatial autocorrelation*. London.

COCHRANE, R. A. 1975. A possible economic basis for the gravity model. *Journal of Transport Economics and Policy*, 9: 34–9.

COELHO, J. D. and WILLIAMS, H. C. W. L. 1977. On the design of land use plans through location surplus maximisation. To be published in *Papers, Regional Science Association* (Working Paper 202, School of Geography, University of Leeds).

COELHO, J. D., WILLIAMS, H. C. W. L. and WILSON, A. G. 1978. Entropy maximising submodels within overall mathematical programming frameworks: a correction. *Geographical Analysis*, 10: 195–201.

COELHO, J. D. and WILSON, A. G. 1976. The optimum size and location of shopping centres. *Regional Studies*, 10: 413–21.

CRIPPS, E. L., MACGILL, S. M. and WILSON, A. G. 1974. Energy and materials flows in the urban space economy. *Transportation Research*, 8: 293–305.

CURRY, L. 1964. The random spatial economy: an exploration in settlement theory. *Annals Association of American Geographers*, 54: 138–46.

DACEY, M. 1976. *An introduction to the mathematical theory of central places*. Vol. 1. Evanston, USA.

EVANS, S. P. 1973. A relationship between the gravity model for trip distribution and the transportation problem in linear programming. *Transportation Research*, 7: 39–61.

FORRESTER, J. W. 1969. *Urban dynamics*. Cambridge, Mass.

GALE, S. and OLSSON, G. (ed.) 1979. *Philosophy in geography*. Dordrecht.

GINSBERG, R. B. 1973. Stochastic models of residential and geographic mobility for heterogeneous populations. *Environment and Planning*, 5: 113–24.

GOULD, P. 1979. Geography 1957–1977: The Augean period. *Annals Association of American Geographers*, 69: 139–51.

GOULD, P. and WHITE, R. 1974. *Mental maps*. Harmondsworth.

HAGERSTRAND, T. 1953. Innovations for loppet ur korologisk synpunkt. Lund.

HAGERSTRAND, T. 1968. *The diffusion of innovation*. Chicago.

HAGERSTRAND, T. 1970. What about people in regional science? *Papers Regional Science Association*, 24: 7–21.

HAGGETT, P. 1965. *Locational analysis in human geography*. London.

HAGGETT, P. and CHORLEY, R. J. 1969. *Network analysis in Geography*. London.

HAGGETT, P., CLIFF, A. D., and FREY, A. 1977. *Locational analysis in human geography*. Second Edition. London.

HAIG, R. M. 1926. Towards an understanding of the metropolis. *Quarterly Journal of Economics*, 40: 179–206 and 402–34.

HALL, P. 1975. *Urban and regional planning*. Harmondsworth.

HARRIS, B. 1962. Linear programming and the projection of land uses. Paper 20. Penn-Jersey Transportation Study, Philadelphia.

HARRIS, B. and WILSON, A. G. 1978. Equilibrium values and dynamics of attractiveness terms in production-constrained spatial-interaction models. *Environment and Planning, A*, 10: 371–88.

HARVEY, D. 1969. *Explanation in geography*. London.

HARVEY, D. 1973. *Social justice and the city*. London.

HERBERT, D. J. and STEVENS, B. H. 1960. A model for the distribution of residential activity in urban areas. *Journal of Regional Science*, 2: 21–36.

HIRSCH, M. W. and SMALE, S. 1974. *Differential equations, dynamical systems and linear algebra*. New York.

HOTELLING, H. 1929. Stability in competition. *Economic Journal*, 39: 41–57.

HOYT, H. 1939. *The structure and growth of residential neighborhoods in American cities*. Federal Housing Administration, Washington DC.

HUDSON, J. C. (ed.) 1979. *Annals Association of American Geographer*, 69: 1–185, Washington DC.

ISARD, W. 1956. *Location and space-economy*. Cambridge, Mass.

JOHNSTON, R. J. 1978. Paradigms and revolutions or evolution? *Progress in Human Geography*, 2: 189–206.

JOHNSTON, R. J. 1979. *The making of modern geography: a history of Anglo-American human geography since 1945*. London.

KING, L. J. 1979. The seventies: disillussionment and consolidation. *Annals Association of American Geographer*, 69: 155–7.

LÖSCH, A. 1940. *Die raumliche ordnung der wirtschaft*. Jena; English translation by W. H. Woglam. *The economics of location*. New Haven.

LUCE, R. D. 1959. *Individual choice behavior*. New York.

MCFADDEN, D. 1974. Conditional logit analysis of qualitative choice behaviour, in P. Zarembka (ed.), *Frontiers in econometrics*. New York.

MUTH, R. F. 1969. *Cities and housing*. Chicago.

NIJKAMP, P. 1972. *Planning of industrial complexes by means of geometric programming*. Rotterdam.

OLSSON, G. 1975. *Birds in egg*. Department of Geography, University of Michigan, Ann Arbor.

ORCUTT, G. H., GREENBERGER, M., KORBEL, J. and RIVELEN, A. M. 1961. *Microanalysis of socio-economic systems: a simulation study*. New York.

PEET, R. 1979. Societal contradiction and Marxist geography. *Annals Association of American Geographers*, 69: 164–9.

RAVENSTEIN, E. G. 1885.   The laws of immigration. *Journal of the Royal Statistical Society*, 4: 165–285 and 241–305.
REES, P. H. 1979.   *Residential patterns in American cities*. Research Paper 189. Department of Geography, University of Chicago.
REES, P. H. and WILSON, A. G. 1977.   *Spatial population analysis*. London.
REILLY, W. J. 1931.   *The law of retail gravitation*. New York.
RESCIGNO, A. and RICHARDSON, I. 1967.   Struggle for life I: two species. *Bulletin of Mathematical Biophysics*, 29: 377–88.
SAYER, R. A. 1976.   *A critique of urban modelling: from regional science to urban and regional political economy*. Oxford.
SCOTT, A. J. 1976.   Land use and commodity production. *Regional Science and Urban Economics*, 6: 147–60.
SCOTT, A. J. and ROWEIS, S. 1977.   Urban planning in theory and practice: a reappraisal. *Environment and Planning*, A, 9: 1097–119.
SELNEC TRANSPORTATION STUDY 1972.   *A broad plan for 1984*. Town Hall, Manchester.
SENIOR, M. L. 1973.   Approaches to residential location modelling 1: urban ecological and spatial interaction models. *Environment and Planning*, 5: 165–97.
SENIOR, M. L. 1974.   Approaches to residential location modelling 2: urban economic models and some recent developments. *Environment and Planning*, A, 6: 369–409.
SMITH, D. M. 1977.   *Human geography: a welfare approach*. London.
STEWART, J. Q. 1941.   An inverse distance variation for certain social influences. *Science*, 93: 89–90.
STODDART, D. R. 1965.   Geography and the ecological approach: the ecosystem as a geographical principle and method. *Geography*, 50: 242–51.
STOUFFER, S. A. 1940.   Intervening opportunities: a theory relating mobility and distance. *American Sociological Review*, 5: 845–67.
THOM, R. 1975.   *Structural stability and morphogenesis*. Reading, Mass.
THOMPSON, E. P. 1978.   *The poverty of theory*. London.
VON THUNEN, J. H. 1826.   *Der isolierte staat in beziehung auf land wirtschaft und national ökonomie*. Hamburg. English translation: *The Isolated State*, by C. M. Wartenburg with an Introduction by P. Hall. Oxford, 1966.
TOBLER, W. 1961.   Map transformations of geographic space. Ph.D. dissertation, University of Washington, Seattle.
WEAVER, W. 1958.   A quarter century in the natural sciences. *Annual Report*. The Rockefeller Foundation, New York, 7–122.
WEBER, A. 1909.   *Uber den standort der industrien*. Tubingen; English translation by C. J. Friedrich, Chicago.
WILLIAMS, H. C. W. L. 1977.   On the formation of travel demand models and economic evaluation measures of user benefit. *Environment and Planning*, A, 9: 285–344.
WILSON, A. G. 1970.   *Entropy in urban and regional modelling*. London.
WILSON, A. G. 1974.   *Urban and regional models in geography and planning*. Chichester.
WILSON, A. G. 1977.   Recent developments in urban and regional modelling: towards an articulation of systems theoretical foundations. In *Proceedings, Volume 1, Giornate di Lavoro*. AIRO, Parma.
WILSON, A. G. 1979.   Equilibrium and transport system dynamics, in Hensher, D. and Stopher, P. (ed.), *Behavioural travel modelling*. London.
WILSON, A. G. and POWNALL, C. E. 1976.   A new representation of the urban system for modelling and for the study of micro-level interdependence. *Area*, 8: 246–54.
WILSON, A. G., REES, P. H., and LEIGH, C. M. (ed.) 1977.   *Models of cities and regions*. Chichester.

WILSON, A. G. and SENIOR, M. L. 1974.  Some relationships between entropy max-imising models, mathematical programming models and their duals. *Journal of Regional Science*, 14: 207–15.

ZIPF, G. K. 1946.  The $P_1P_2/D$ hypothesis: on the intercity movement of persons. *American Sociological Review*, 11: 677–86.

# 9 Land-Use Survey Today and Tomorrow

ALICE COLEMAN *King's College, London; Director of the Second Land Utilisation Survey of Britain*

## THE FOUNDING FATHER

PROFESSOR Sir (Laurence) Dudley Stamp, the father of land-use survey, one of the world's great geographers, and the only academic geographer to date to have received the honour of the Presidency of the Royal Geographical Society, lectured to the Society on land-use topics on a number of occasions. In 1931 he explained the work of the young Land Utilisation Survey of Britain and compared it for the first time to the Domesday Survey. In 1940, in a lecture on 'Fertility, Productivity and Classification of Land in Britain' he showed how land-use maps assisted in the urgent wartime need to increase home agricultural production. He also described a Types of Farming Map, his first derivative from the land-use maps and outlined his second, a Land Classification Map showing productive potential; he also stressed the need for soil and vegetation surveys. All these ideas have proved to be fertile, and have been further developed by others. By 1942 his concern for land misuse and potential improvement led him to participate in the Society's discussion on 'The Geographical Aspects of Regional Planning', and to serve as vice-chairman of the Scott Committee on Land Utilisation in Rural Areas; he lectured on the latter's report in the following year. Once again his ideas were followed up, this time in the Town and Country Planning Act of 1947, which introduced a comprehensive system of land-use control. In that year he presented to the Society a study of wartime changes in British agriculture, and was also looking abroad to the international scene, and to the establishment, at S. Van Valkenburg's suggestion, of the Commission on a World Land Use Survey by the International Geographical Union. This, too, bore copious fruit in a whole series of national land-use maps, summarized in Table 9.1 (A. Clark 1976). Subsequently, in 1961, Stamp took the chair at the Royal Geographical Society at a meeting which introduced the Second Land Utilisation Survey, and had the satisfaction of seeing his own great endeavour handed on to another generation of British geographers.

## DATA COLLECTION

A felicitous combination of circumstances makes Britain peculiarly suited to data collection by ground survey. In the first place there exists a complete cover of maps at the scale of 1 : 10 560 or 1 : 10 000 showing plot boundaries, so that most of the field mapping consists of inserting categories into prepared spaces without the need for topographic survey *ab initio*. Secondly, Britain is a small, densely populated country, well served by roads and footpaths, and characterized by a strong tradition of voluntary service, so that both the First and Second Land Utilisation Surveys have been largely carried out by volunteer teams. The bringing to fruition of these enterprises has been difficult and demanding, but not impossible.

Most countries lack one or more of these advantages. Canada, for example, does not have any maps with field boundaries, and land-use work has to be based on air photographs. Hare's vegetation mapping of the Ungava Peninsula was followed in 1964 by manuscript land-use maps at 1 : 25 000, covering the whole settled area of the country. In other countries the desirability of working from the image rather than from the terrain has been promoted by factors such as poor definition of land uses on the ground as well as on the map, lack of road access or extensiveness of coverage.

It is the development of imagery for data collection that has received the greatest concentration of effort in the whole field of land-use studies during the last thirty years. From black and white photography to the use of colour to enlarge the number of identifiable categories; from the waveband of light to multispectral imagery to penetrate cloud, provide sharper topographic detail, and extend the data beyond land use into areas such as pollution; from the use of aircraft to the use of satellites to extend the range and increase the speed; from the limitations of the image obtained to the advantages of computerized enhancement; from visual interpretation of the image to computer-based identification; all these technical advances take land-use survey out of the hands of ground surveyors. Ground survey is now rare in the land-use surveys of the world. Approximately forty of the map series listed in Table 9.1 have been based upon imagery.

A second development in data collection has been the plotting of land-use statistics. This is a feature of East European map series, inspired by the leadership of J. Kostrowicki of the Polish Academy of Sciences. Like the development of imagery, it arose out of a constraint that precluded ground survey. In this case the constraint was a military one. If, for example, pastoral valleys were to be differentiated as marshy and non-marshy grassland, information for enemy tanks would be provided; such differentiation was thus prohibited on security grounds. The plotting of statistics according to administrative units minimizes locational information but compensates by increasing quantitative information.

A third development is what may be termed the logging of data. This is a response to computer opportunities, which permit the recording of multiple information items for each plot, or hereditament. Items of interest to local authorities, such as rateable value, ownership, or occupancy, can be logged under a hereditament identification number, and land uses can be logged in much greater detail than can be contained within the plot's area on the map. The different uses of the various floors within a building can all be logged individually and an enormously complex land-use classification is made possible. The National Land Classification of the Department of the Environment (1975), and variants of it devised by local authorities, run into many hundreds of categories. Ever-growing complexity would seem to be the dominant characteristic in a world where new land uses are constantly coming into existence.

A fourth set of data-collection developments has arisen in an attempt to reduce the magnitude and cost of land-use surveys. It falls into two classes.

TABLE 9.1

*A Sample of Land-Use Map Series*

| Country | | Scale | No. of Categories | No. of Colours | Morphogram |
|---|---|---|---|---|---|
| Antigua | 1: | 145 221 approx. | 8 | 2 | No |
| Australia | 1: | 63 360 | 12 | Red overprint | Yes |
| Austria | 1: | 500 000 | 7 | 5 | No |
| Azores | 1: | 50 000 approx. | 12 | Black and white | |
| Bahamas | 1: | 25 000 | 6 | Black and white | No |
| Belize | 1: | 100 000 | 29 | 5 | No |
| Botswana | 1: | 3 000 000 | 7 | Black and white | No |
| Botswana | 1: | 1 609 350 approx. | 4 | Black and white | No |
| Canada | 1: | 250 000 | 22 | 8 | No |
| Canada | 1: | 50 000 | 26 | 9 | Yes |
| Ceylon | 1: | 63 360 | 6 | Black and white | Yes |
| (see also Sri Lanka) | 1: | 253 440 | 32 | 4 | Yes |
| China | 1: | 600 000 | 11 | 6 | No |
| Christmas Island | 1: | 25 000 | 1 | 1 | Yes |
| Cook Islands | 1: | 7 290 | 21 | 6 | Yes |
| Cyprus | 1: | 50 000 | 11 | Black, white, & brown | No |
| Czechoslovakia | 1: | 40 000 | 76 | 10 | No |
| Czechoslovakia | 1: | 500 000 | 6 | 4 | No |
| Dominica | 1: | 25 000 | 7 | Black, white, & brown | No |
| Ethiopia | 1: | 50 000 | 14 | Black and white | Yes |
| Europe | 1: | 2 500 000 | 24 | 9 | |
| Fiji | 1: | 50 000 | 24 | 5 | No |
| Fiji | 1: | 250 000 | 13 | 6 | No |
| Gambia | 1: | 25 000 | 16 | 6 | No |
| Germany | 1: | 50 000 | 17 | 4 | Yes |
| Guatemala | 1: | 1 900 000 | 8 | Black and white | No |
| Gilbert & Ellice Islands | 1: | | | | |
| Hong Kong | 1: | 80 000 | 8 | 8 | No |
| Hungary | 1: | 25 000 | 98 | 10 | No |
| Indonesia | 1: | 40 000 | 14 | Black and white | Yes |
| Iran | 1: | 27 000 | 15 | 4 | Yes |
| Iraq | 1: | 1 000 000 | 10 | 8 | No |
| Israel | 1: | 1 000 000 | 13 | 4 | No |
| Italy | 1: | 200 000 | 21 | 8 | No |
| Italy | 1: | 25 000 | 9 | 5 | |
| Japan | 1: | 30 000 | 10 | 8 | Yes |
| Japan | 1: | 50 000 | 15 | 15 | Yes |
| Japan | 1: | 200 000 | | 10 | No |
| Japan | 1: | 800 000 | 2 | 4 | No |
| Japan | 1: | 2 500 000 | 9 | 12 | No |
| Java | 1: | 50 000 | 22 | 4 | Yes |
| Kenya | 1: | 250 000 | 170 approx. | | No |
| Lebanon | 1: | 200 000 | 17 | 6 | No |
| Lesotho | 1: | 447 040 approx. | 1 | 1 | No |
| The Maghreb | 1: | 5 000 000 | 10 | 3 | No |
| Malagasy Republic | | | | | |
| Malaya | 1: | 760 320 | 8 | 6 | No |
| Mauritius | 1: | 389 490 approx. | 7 | Black, white, red, & grey | No |

| Country | Scale | No. of Categories | No. of Colours | Morphogram |
|---|---|---|---|---|
| Nepal | 1 : 125 000 | 9 | 4 | No |
| Netherlands | 1 : 325 000 | 8 | 8 | No |
| New Zealand | 1 : 63 360 | 18 | 4 | Yes |
| Niger | | | | |
| Nigeria | 1 : 25 000 | 3 | 3 | Yes |
| Nigeria | 1 : 50 000 | 12 | Black, white, & green | Yes |
| Nigeria | 1 : 1 000 000 | 23 | 5 | No |
| Nigeria | 1 : 2 000 000 | 6 | 4 | No |
| Norway | 1 : 100 000 | 6 | 2 | No |
| Norway | 1 : 250 000 | 4 | 2 | No |
| Pakistan-Bangladesh | 1 : 18 933 approx. | 6 | 4 | Yes |
| Philippines | 1 : 5 000 | 9 | | Yes |
| Poland | 1 : 25 000 | 67 | 8 | Yes |
| Poland | 1 : 200 000 | 32 | 7 | No |
| Rhodesia | 1 : 1 000 000 | 22 | 6 | No |
| Sabah | 1 : 50 000 | 30 | — | Yes |
| Sarawak & Brunei | 1 : 250 000 | 8 | 4 | No |
| Senegal | 1 : 1 000 000 | 21 | 2 | No |
| Sierra Leone | 1 : 16 000 | 13 | 4 | Yes |
| Sierra Leone | 1 : 40 000 | 9 | 3 | No |
| Singapore | 1 : 6 336 | 42 | 8 | Yes |
| Solomon Islands | 1 : 150 000 | 9 | 4 | No |
| South Arabia | 1 : 20 000 | 8 | 3 | Yes |
| South Africa | 1 : 125 000 | 11 | 8 | No |
| Spain | 1 : 1 000 000 | 22 | 8 | No |
| Sri Lanka | 1 : 1 000 000 | 7 | 4 | No |
| St. Helena | | | | |
| Sudan | 1 : 4 000 000 | 29 | Black and white | No |
| Sweden | 1 : 10 000 | 2 | 2 | Yes |
| Taiwan | 1 : 25 700 | 6 | Black and white | No |
| Tanzania | 1' 750 000 | 12 | 2 | No |
| Tanzania | 1 : 25 000 | 8 | Black, white, & green | No |
| Thailand | 1 : 50 000 | 4 | 3 | |
| Tobago | 1 : 63 360 | 12 | Black and white | |
| Togo | | | | |
| Trinidad | 1 : 150 000 | 19 | 6 | No |
| Uganda | 1 : 1 000 000 | 9 | Black and white | No |
| Uganda | 1 : 1 500 000 | 5 | 4 | No |
| UK First Survey | 1 : 63 360 | 7 | 7 | Yes |
| UK Second Survey | 1 : 25 000 | 70 | 11 | Yes |
| UK Fairey/DOE | 1 : 50 000 | 5 | Black and white | No |
| USA | 1 : 250 000 | 37 | Black and white | Yes |
| Zambia | 1 : 750 000 | 54 | 6 | No |

This table was compiled by I. J. Feaver. It is not a comprehensive list, which would be very difficult to assemble. By 1971 under Prof. Radó of Budapest a bibliography of 750 items representing 73 countries had been compiled but these included vegetation maps, soil maps, etc. as well as land-use maps. Cartactual of Hungary is currently attempting to update the bibliography and has compiled a European Land Use map noted in the list of references.

Thanks are expressed to map librarians and others at King's College, the London School of Economics, the Royal Geographical Society, the British Library Map Room, the Land Resources Division, the Directorate of Overseas Surveys Library, Fairey Surveys Limited, Hunting Surveys and Consultants Limited, and Geographical Publications Limited.

One extols the value of problem-based as opposed to inventory-based mapping, and restricted the task to selected categories and selected areas (J. W. Birch 1968; V. C. Robertson and R. F. Stoner 1970). This concept has also been represented as being intellectually superior in that it permits vigorous hypothesis testing, a view expressed, for example, by J. T. Coppock who established a data bank dealing with recreational and other problems, having earlier specialized in agricultural land use (1964, 1979). The other concept is the substitution of sample mapping for comprehensive coverage, e.g. field transects, sample lines or points on air photographers, or block samples (J. G. Osborne, 1942; P. Haggett, 1963; D. Thomas, 1970; G. C. Dickinson and M. G. Shaw, 1977(b)). Dickinson and Shaw see the future of land-use data collection as lying in efficiency and simplicity unlike the apparent official trend towards greater complexity.

In this diversity of techniques, the original Stamp method has come to occupy a minority position, and is often regarded as a somewhat amateurish poor relation. Looked at dispassionately, however, there are still grounds for considering it the best approach wherever it is practicable, and these will be illustrated in relation to the Second Land Utilisation Survey of Britain.

The Stamp method may be defined as a comprehensive, inventory-based, ground survey which records each land use as a plot of true shape, i.e. a morphogram.

Ground survey has the great advantage of affording a free choice of land-use categories. Many categories are readily identifiable in the field but not identifiable at all on imagery documents. The average number of categories portrayed by the known imagery-based surveys listed in Table 9.1 is only 13, whereas the ground-based Second Land Utilisation Survey shows 70 on its published maps and 250 on its field maps, in addition to over 13 000 permutations of species abundance in vegetation communities. In 'Land Use Studies by Remote Sensing' (W. G. Collins and J. L. Van Genderen 1976) a great deal was said about imagery techniques but very little about the land uses that could actually be mapped from them. The most specific achievements claimed for Landsat imagery were the differentiation of coniferous from deciduous woods in winter and the deduction of wheat cultivation from ploughed fields in areas known, from external evidence, to produce wheat. It was stated that greater, but unspecified, detail could be obtained by American workers.

The smallest resolvable area on satellite imagery (i.e. pixel size) is about $79 \times 56$ m., which is too large to reveal many individual uses (C. T. Paludan 1976). This will doubtless be reduced in the future so that the number of categories will be increased, but ground survey is likely to retain a big advantage for a long time. For example, there may be no visible evidence of the conversion of a church into a toy factory except a brass plate, and whereas this gives full information to the field surveyor it completely eludes detection from either satellite or aircraft. Instances of this type could be multiplied.

The inventory-based nature of the Stamp method has advantages not shared by the problem-based approach. The two should be regarded as complementary and not mutually exclusive. Thus, while problem-based survey may ensure that all the relevant categories are mapped, inventory-based survey can detect the existence of problems in the first place. For example, it was not until after the Second Land Utilisation Survey had demonstrated the existence of a waste land problem, that concern for a supposed shortage of building land was replaced by a spate of surveys of vacant land. Similarly, local authority surveys, focusing on urban problems, failed to notice the speed of encroachment upon rural land. Inventories have value in making problems visible; problem-based surveys can then be initiated in a relevant context.

The morphogram nature of the Stamp method is an important safeguard for accuracy. The lesser degree of detail discernible from imagery is nearly always a temptation to desert pure categories and fall back on mixtures. For example, the five-category map produced by Fairey for the Department of the Environment (T. F. Smith *et al.* 1977) shows predominantly residential and predominantly industrial categories over minimum areas of 5 hectares. Since there is much less industry than housing in Britain it is likely that less industry will be concealed in the residential plots than vice versa, and the figures derived from these maps cannot be expected to be accurate. The disadvantage is magnified by the inclusion of derelict land in the industrial category. It is hoped that note will be taken of this sort of defect to ensure that a proper morphogram basis is incorporated in land-use surveys of the future.

Another advantage of the morphogram principle is the immediate feedback to the surveyor. If he meets a new category that is not covered by the classification, he has to leave a blank on his map and seek guidance on how to include it. He does not have the option of fudging it by inclusion in some kind of mixed category.

Finally, the Stamp method produces a map on which each land use is clearly defined in a single operation. It does not need the double process of 'first catch your imagery and then cook it', and is therefore substantially cheaper. A financial comparison can now be made between the Department of the Environment's recent map referred to above, and the Second Land Utilisation Survey's 1976–8 mapping of three counties with full-time workers. The official survey had five times the funding to cover less than two and a half times the area, and to map only 2 per cent of the number of categories.

### AREA MEASUREMENT

A first stage in the analysis of land-use data is its quantification. P. Haggett (1963) believed that random line sampling was the quickest and most accurate method but M. R. Sampford (1962) demonstrated that standard errors are smaller with systematic than with random methods. D. Thomas (1970) used systematic line sampling for his study of the London green belt, agreeing with Osborne (1942) that random line sampling may be only one-half to one-quarter as efficient as systematic line sampling of the same intensity.

The present writer investigated a score of methods in a project for the Ministry of Land and Natural Resources (1965–6), in order to assess their relative accuracy, speed, equipment cost, labour cost, and suitability to land-use material. The estimated time requirement for measuring the 6500 maps of the Second Land Utilisation Survey ranged from man-millennia for geometrical methods to nineteen years for systematic point sampling. The latter, which was exhaustively tested on a 'pattern map' devised and accurately measured by J. J. Mackay of the Ordnance Survey also proved to be the easiest method to computerize, so that the eventual time taken, including thorough checking, was eight man years.

In surveys which have to digitize boundaries in order to produce their maps from imagery, it may not involve a great additional cost to programme the computer to calculate the areas of plots from the digital data. This is done by the Canada Land Inventory and CSIRO in Australia for example. The Experimental Cartography Unit of NERC is also primarily concerned with digitizing and the automated reproduction of maps. For example, it has reproduced category separations showing the spatial patterns produced by individual types of land use (D. Bickmore and A. Molineux 1978).

In addition to the direct measurement of maps and imagery, there has been another important field of endeavour pioneered by R. H. Best. Starting from evidence that Stamp's measurement of agricultural land was more accurate than that of the Ministry of Agriculture (1962), he has assiduously advanced the case for better official statistics which would add up to the total national area, leaving no land unaccounted for. This state of perfection has not yet been reached, but meanwhile Best has been plugging gaps in the figures by carrying out sample measurements of his own, and calibrating the results in a series of improved methods. At present Best's calculations give figures that tally with those of the Second Land Utilisation Survey (private communication), although his interpretation of the measurement is quite a different one.

### ANALYSIS AND INTERPRETATION

There is no doubt that the main drive in land-use survey has been in the field of data collection, and whereas recent developments in area measurement have reached the point of matching the need, the stage of analysis and interpretation continues to lag behind. J. Kostrowicki (1964) summed this up with the comment that we still need to find a land-use typology.

This lack is the more surprising in that planning passes beyond the stage of analysis and interpretation to the recommendation of land-use changes. Increasingly, however, it has moved away from the analysis of land use towards the analysis of population trends or socio-economic factors which it uses to predict quantitative needs for houses, work places, school places, and other facilities. Driven by an overriding ethic of newness and spaciousness, the structure plans are in the business of moving people. As between counties and districts no less than five million more people in England and Wales alone are to be resettled on greenfield sites, and the total figure is probably much larger if the intentions to rehouse within districts could also be discerned.

It is of interest to calculate how much enforced land-use change would take place by 1991 if all the structure plans are duly implemented. Since over 10 per cent of the population is to be moved, at least 10 per cent of the urban area will be affected, and rehousing at more generous density standards means that two to four times as much greenfield land will be taken up by building sites. If the practices of the 1970s continue to prevail, new building sites will sprawl through farmland, subjecting it to intolerable urban pressures and creating as much, or even twice as much, new wasteland as new settlement. The total change could easily equal or exceed the area equivalent to all the existing urban land, with all the expense, dislocation, resource consumption, and sterilization of productive soils that this implies. It is not surprising that a fresh look is being taken at the land-use analysis, or lack of it, which underlies these massive proposals, and consequently current land-use analysis can be seen as crystallizing into two schools, respectively land-budgeting and conservationist (A. Hillis 1979).

The land-budgeting approach is essentially quantitative in nature and concerned to allocate the land surface optimally among desirable land uses. Underpinned by the work of notable quantifiers at Wye College, it seeks to justify the drive for more spacious urbanization and the resulting loss of farmland. It invokes arguments such as an anticipated perpetual intensification of agriculture to produce ever more food from less land, and the capacity of house gardens to produce food of greater financial value than the farmland taken for the housing. A number of papers have been published in the *Geographical Journal*, beginning with G. P. Wibberley's 'Some Aspects of Problem Rural Areas in Britain' (1954). R. H. Best's 'Recent Changes and Future Prospects of Land Use in England and Wales' (1965) was an important paper dealing with data sources and comparing rates of urban growth with calculations of how much agricultural land Britain could afford to lose. It was concluded that the economy was unlikely to suffer in any foreseeable future, although aesthetic, amenity, and recreational aspects needed to be given a greater weighting. Other publications by the Wye College school of thinking include J. Wyllie (1954), J. T. Ward (1956), B. R. Davidson and G. P. Wibberley (1956), R. Pahl (1965), R. Gasson (1966 and 1967), A. M. Edwards and G. P. Wibberley (1971), and A. W. Rogers (1978). Many papers were written in collaboration with Best or Wibberley.

The conservationist school writes largely in non-geographical publications, in this case environmental journals and books. However, the Royal Geographical Society was the venue for the first national report of the Second Land Utilisation Survey, in a paper entitled 'Is Planning Really Necessary?' (A. Coleman 1976). This introduced a quite different ethical stance from the land-budgeting school. In place of the allocation principle of urbanizing as much productive land as it was thought could be spared, a case was made for an insurance principle of more economical urbanization and maximum conservation of renewable resource land. Since the needs of the future cannot be foreseen, it is important to keep productive options open.

The insurance principle, which was similar to the now rather overlooked ideas of Stamp, Willatts, and the early planners, was prompted by the quantitative results of the Second Land Utilisation Survey. Improved farmland was shown to be shrinking more rapidly than the land-budgeting school believes and heading for total extinction within 200 years unless a policy reversal takes place. The discrepancy between land-budget figures and conservationist figures is accounted for by the widespread land wastage that has accompanied the massive population relocations of the last three decades, especially the growth of derelict land in the inner city and rough wasteland

FIG. 9.1. Land Uses in each scape and fringe by counties, simplified to show only the three supercategories: settlement, improved farmland and vegetation. The three scapes are clearly dominated by one supercategory only, with the other two subordinate or absent. Marginal fringe is characterized by subordinate settlement and codominant farmland and vegetation. Although vegetation could exceed farmland in theory, and often does so in smaller areas, farmland is more extensive in practice in most counties measured to date. Rurban fringe shows settlement co-dominant with rural land uses. Although rurban vegetation need not necessarily occupy a smaller area than farmland, it does so in all the measured counties. It should be noted that area measurement does not feature in the delineation of the scapes and fringes. They are identified purely by pattern recognition techniques, and measured subsequently. The emerging quantitative consistency of characteristics makes this system of distinctive land-use patterns a useful tool in discussing land-use policies.

in the rurban fringe. Some of the most active land-use surveys at present are the vacant-land surveys being carried out by the planning authorities.

The discovery of land wastage results from another difference between the land-budgeting and the conservationist approach. The latter uses spatial patterns as well as quantitative values because it is based on map evidence. These patterns are beginning to supply the missing land-use typology noted by Jerzy Kostrowicki. They consist of five basic types, townscape, farmscape, wildscape, rurban fringe, and marginal fringe, together with a number of subdivisions. A precise pattern-recognition technique has been developed to locate scape and fringe boundaries objectively and area measurements have been made of the land uses within the boundaries (Fig. 9.1).

It is of particular interest to the scientific development of land-use analysis that two alternative approaches to the Second Land Utilisation Survey's data can now be compared. Each has been reported in the *Geographical Journal*. In the 'Containment of Urban England' (1974) P. Hall used a grid square method, arguing that if any square contained less than 20 per cent settlement it was rural and if it contained more than 40 per cent settlement it was urban. The remaining squares with 20 to 40 per cent were not widespread, and hence it was concluded that rural–urban fringes were rare and urban containment had been successful. However, 'Is Planning Really Necessary?' (Coleman 1976), used the same data to come to the opposite conclusion. Rurban fringe was identified on a pattern-recognition basis and proved to be very extensive. When measured it was found to include areas where 35 to 65 per cent was rural land, fragmented by 35 to 65 per cent settlement, the great bulk of this range lying within what Hall described as fully urbanized. It is from this sort of disagreement that land-use analysis is becoming more precise.

## LAND-USE SURVEY IN THE FUTURE

Land-use survey has followed the pattern of topographic, hydrographic, geological, and soil surveys in being pioneered by private individuals and being taken up by government agencies. Unlike its predecessors, however, land use survey was not entrusted to a unified professional survey organization, but was split up among a multiplicity of planning authorities, who had no interest in maintaining any over-all consistency of classification, or in publishing maps that would allow the public to monitor the progress of land-use planning. Rural areas were not surveyed at all, and if urban areas were surveyed, the lack of analytical methods led to their eclipse as a planning tool in favour of population projections, etc. The first and foremost need for the future is the transfer of survey responsibility to a single professional body which can supply planners and the public alike with a comprehensive cover of well-designed and informative maps.

Over a decade ago W. G. V. Balchin suggested that the Ordnance Survey should be charged with producing a decennial census of land use to accompany the decennial census of population. This would obviously satisfy the requirements of professionalism and public availability of the results. It would also be the most economical approach, as the OS already has surveyors on the ground and maps going through the press. Land use could be added to existing processes without the expense of a special infrastructure of its own.

The use of ground survey would allow certain important criteria to be satisfied.

It is important, for example, that categories should be compatible with those of the Second Land Utilisation Survey so that trends of land-use change can be calculated. New categories should be subdivisions within the existing ones and should not cut across them in ways that would invalidate the comparison. It is also desirable that the classification should be cartographically hierarchical, so that different degrees of detail can be furnished at different map scales to suit the needs of different workers. The hierarchy of the National Classification was devised for logging and not mapping, and is not, in fact, mappable. An alternative hierarchy, which has evolved during practical survey, consists of three levels: curtilage, colour, and convention. The name 'curtilage' was suggested by Dickinson and Shaw in 1977 and has been developed as a practical technique in the Second Land Utilisation Survey's maps of Surrey (1977), Buckinghamshire (1978), and Tower Hamlets (1979). It consists of a labelled line drawn round a complex land use such as a hospital, school, or estate of flats. Colour is then used to differentiate the separate uses within the curtilage, such as buildings, tended open space, or vehicle ways, while conventions are employed for subtypes within the colour, e.g. an orange wash for roads and orange cross-hatching for vehicle parks. The objective is to produce a map which can be clearly read at all three levels, and where patterns of use can be appreciated as well as individual categories.

A comprehensive, inventory-based map produced at ten-yearly intervals could satisfy the need for greater land-use detail, but could also satisfy the need for greater simplification since the curtilage level could be used alone for certain purposes, or the curtilage and colour levels combined for other purposes. Additionally, the map could provide information on the need for problem-based surveys of small areas or individual uses to be carried out at intermediate dates.

The future will also certainly hold further developments in the field of imagery production, and the identification of a larger range of land uses from improved images. The world has sunk an enormous investment into this type of technology and its momentum will carry it forward into great advances. It is not impossible, however, that its potential for land-use advancement may be exceeded by its potential for other purposes.

The most vital need for the future, however, is better analysis of land-use surveys to give more perceptive insights into problems and opportunities as a basis for more effective and imaginative policies and action. Better scientific testing of proposed solutions in advance of their mass application will have to replace the great fashionable surges of the last three decades, which have been followed all too often by mass failure, as in the case of high-rise flats or inner city improvements, for example. The direction of constructive changes are difficult to foresee, but one feature seems certain. The public is no longer content to leave land-use planning to a professional few. They are no longer content with ineffective participation exercises. They are eager for insights and explanations which are comprehensible to the layman, and which ring true. This is a splendid future opportunity for geographers to offer their skills and understanding for the common good.

# REFERENCES

BEST, R. H. and WARD, J. T. 1956. *The Garden Controversy*. Studies in Rural Land Use, Report No. 2, Wye College, Kent.

BEST, R. H. and COPPOCK, J. T. 1962. *The Changing Use of Land in Britain*. London.

BEST, R. H. 1965. Recent Changes and Future Prospects of Land Use in England and Wales. *Geogrl. J.* 131: 1–12.

BICKMORE, D. and MOLINEUX, A. 1978. *Land Use Mapping by Local Authorities in Britain*. Experimental Cartography Unit. NERC.

BIRCH, J. W. 1968. *Rural Land-Use: A Central Theme in Geography*. Inst. Brit. Geographers, Special Publication No. 1.

CLARK, A. 1976. The World Land Use Survey. World Land Use Survey. *Geographica Helvetica*, pp. 27–8. Kümmerley and Frey.

COLEMAN, A. 1961. The Second Land Use Survey: Progress and Prospect. *Geogrl. J.* 127: 168–86.

COLEMAN, A. 1964. Some Cartographic Aspects of the Second Series Land Use Maps. *Geogl. J.* 130: 167–70.

COLEMAN, A. 1970. The Conservation of Wildscape: A Quest for Facts. *Geogrl. J.* 136: 199–205.

COLEMAN, A. 1976. Is Planning Really Necessary? *Geogrl. J.* 142: 411–37.

COLEMAN, A. 1977. Land-Use Planning: Success or Failure? *Architects' Journal*, 165. 3: 91–134.

COLEMAN, A. 1978. Land-Use Planning in a world of Shrinking Resources. *Journal of Long Range Planning*, 11: 47–52.

COLLINS, W. G. and VAN GENDEREN, J. L. 1976. *Land Use Studies by Remote Sensing*. Remote Sensing Society, University of Aston.

COPPOCK, J. T. 1964. *An Agricultural Atlas of England and Wales*. London.

COPPOCK, J. T. 1979. *Planning Data Management Service: Service Description*. University of Edinburgh.

COX, I. H. (ed.) 1970. *New Possibilities and Techniques for Land Use and Related Surveys*. World Land Use Survey, Occasional Papers, No. 9, Geographical Publications, Ltd.

DAVIDSON, B. R. and WIBBERLEY, G. P. 1956. *The Agricultural Significance of the Hills*. Studies in Rural Land Use, Report No. 3, Wye College, Kent.

DICKINSON, G. C. and SHAW, M. G. 1977(a). *Monitoring Land Use Change*. Planning Research Applications Group, Centre for Environmental Studies.

DICKINSON, G. C. and SHAW, M. G. (1977(b)). What is 'Land Use'? *Area*, 9. 1: 38–42.

EDWARDS, A. M. and WIBBERLEY, G. P. 1971. *An Agricultural Land Budget for Britain, 1965–2000*. Studies in Rural Land Use, Report No. 10, Wye College, Kent.

GASSON, R. 1967. *The Influence of Urbanisation on Farm Ownership and Practice*. Studies in Rural Land Use, Report No. 7, Wye College, Kent.

HAGGETT, P. 1963. Regional and Local Components in Land Use Sampling: A Case Study from the Brazilian Triangulo. *Erdkunde*, 17: 108–14.

HALL, P. 1974. The Containment of Rural England. *Geogrl. J.* 140: 386–418.

HARE, F. K. 1964. New Light from Labrador-Ungava. *Annals of the Association of American Geographers*, 54: 459–76.

HILLIS, A. 1979. Investigation of Transfers of Land, Mainly Agricultural to Other Uses in Northern Ireland Since 1945. Unpublished Seminar, Coleraine.

HMSO 1975. *National Land Use Classification*. London.

KOSTROWICKI, J. 1964.   *The Polish Detailed Survey of Land Utilization: Methods and Techniques of Research*. Dokumentacja Geograficzna, 2. Warsaw.

OSBORNE, J. G. 1942.   Sampling errors of systematic and random surveys of cover-type areas. *Journal of the American Statistical Association*, 37: 256–64.

PAHL, R. 1965.   *Urbs in Rure*. London School of Economics.

PALUDAN, C. T. 1976.   Land Use Surveys Based on Remote Sensing from High Altitudes. World Land Use Survey Report. *Geographica Helvetica*, 31: 1–32. Kümmerley and Frey.

ROBERTSON, V. C. and STONER, R. F. 1970.   Land Use Survey: A Case for Reducing the Costs. World Land Use Survey, Occasional Papers, No. 9, pp. 3–15. Geographical Publications, Ltd.

ROGERS, A. W. (ed.) 1978.   *Urban Growth, Farmland Losses and Planning*. Inst. Brit. Geographers. London.

SAMPFORD, M. R. 1962.   *An Introduction to Sampling Theory with Applications to Agriculture*. Edinburgh.

SCOTT, L. 1942.   *Report of the Committee on Land Utilisation in Rural Areas*. HMSO. Cmd 6378.

SMITH, T. F., Van Genderen, J. L., and HOLLAND, E. W. 1977.   A Land Use Survey of Developed Areas of England and Wales. *Cartographic Journal*, 14. 1: 23–30.

STAMP, L. D. 1931.   The Land Utilisation Survey of Britain. *Geogrl. J*. 78: 40–53.

STAMP, L. D. 1940.   Fertility, Productivity and Classification of Land in Britain. *Geogrl. J*. 96: 389–413.

STAMP, L. D. 1943.   The Scott Report. *Geogrl. J*. 101: 16–30.

STAMP, L. D. 1947.   Wartime Changes in British Agriculture. *Geogrl. J*. 109: 39–57.

STAMP, L. D. Third edition 1962.   *The Land of Britain: Its Use and Misuse*. London.

THOMAS, D. 1970.   *London's Green Belt*. London.

TIMAR, E. 1971.   *International Bibliography of Land Use Maps*. Hungarian Agricultural Museum, Budapest.

WARD, J. T.   Studies in Rural Land Use. Wye College, Kent.

WIBBERLEY, G. P. 1954.   Some Aspects of Problem Rural Areas in Britain. *Geogrl. J*. 120: 43–61.

WILLATTS, E. C. 1933.   Changes in Land Utilization in the Southwest of the London Basin. *Geogrl. J*. 82: 515–28.

WYLLIE, J. 1954.   *Land Requirements for the Production of Human Food*. Studies in Rural Land Use, Report No. 1, Wye College, Kent.

# 10 Historical Geography in 1980

HUGH PRINCE *University College London*

HISTORICAL geography is not so much a discrete branch of knowledge, but some would say more a way of life. Most of its practitioners are independent-minded individualists whose interests and outlooks vary widely, indeed, so widely that no statement can adequately summarize the views of all or even a majority. Outside observers are perplexed and sometimes dismayed by different scholars pursuing divergent aims, employing different methods, disagreeing about the boundaries of their fields of inquiry. But debate, argument, and revision lie at the heart of our work.

This survey claims to be no more than a personal view of work in progress, prefaced by some fleeting impressions of the recent history of the subject. From many diverse strands of thought, I have selected three topics for discussion: interpreting sources of evidence, working with other disciplines, and debating ideologies. These three topics were discussed on several occasions during a symposium held at Cambridge in July 1979 by the International Geographical Union's working group on historical changes in spatial organization.

## A RECENT HISTORY OF HISTORICAL GEOGRAPHY

In 1940, Carl Sauer delivered a presidential address to the Association of American Geographers, 'Foreward to historical geography', at a moment of crisis. Historical geographers together with geomorphologists had been struggling and were failing to justify their positions within departments adhering to the dictum enunciated by Immanuel Kant, carried into geography by Alfred Hettner, that defined geography as a study of areas and areal differentiation, whilst the study of work, thought, and change were assigned to the province of history (Hartshorne, 1939). Carl Sauer declared historical geography a part of culture history and he invited geographers to contribute to the task of tracing the development of cultural differences. 'We deal not with Culture, but with cultures,' he wrote (1941, p. 24), 'except in so far as we delude ourselves into thinking the world made over in our own image.' Sauer and his disciples have greatly extended our understanding of cultural distinctiveness and cultural change.

In 1949, Roger Dion inaugurated a course at the Collège de France by describing *la géographie humaine rétrospective* as 'an archaeological study in the broadest sense of the word, a search for reasons which have determined the choice of sites for man's works' (Dion, 1949, p. 8). Across the Channel, H. C. Darby (1951) lectured to the Royal Geographical Society on the changing English landscape, dealing in turn with themes of change: clearing the woods, draining the marshes, reclaiming the heaths, landscape gardening, industrialization, and population migration. A stock-taking account of American geography in 1954 contained Andrew Clark's reasoned appeal for a genetic approach, expressing the need to examine geographical change as a means of understanding geographies of present or past periods. Guided by these examples, historical geographers in many countries broke through the thin surface of Kant's static, two-dimensional representation of the world and began to explore transformations of societies, economies, and environments.

While other geographers took part in or were victims of the quantitative revolution and hurriedly renounced the orthodoxy of areal differentiation in order to embrace a newer orthodoxy of spatial science, many historical geographers were enjoying a new-found freedom to investigate historical questions (Baker, Butlin, Phillips, and Prince, 1969). They were collaborating closely with historians and archaeologists, reading widely and deeply, visiting archives, digging and mapping relict features, discovering fresh sources of inspiration in art, literature, and the history of science, pondering upon the meaning of these finds. During the 1960s, spatial scientists focused upon static patterns. The statistical techniques required to operationalize diachronic models are much more complicated than those used in synchronic models and the power to predict diminishes rapidly with each step through time (Langton, 1972; Prince, 1978, p. 17). Torsten Hägerstrand's interest in time geography arises from his study, published in 1947, of the life-paths of villagers in Åsby, where parish records allowed him to trace the births, marriages, deaths, and movements of all inhabitants from 1840 to 1944. Thirty years later, Hägerstrand (1975) reviewed the life histories of individuals in relation to their geographical environments. In 1978, the editors of *Timing Space and Spacing Time* expressed a hope that further explorations 'will help us all, theoreticians, empiricists or visionaries, to make better sense of time and so improve our understanding of geographical patterns and processes'. (Carlstein, Parkes, and Thrift, 1978, vol. i, p. 5.)

Historical geographers drawing statistical generalizations from historical sources have sensitively repaired and rectified irregularities and deficiencies in the data (Overton, 1977; Adrian, 1977; Smith, 1978; Wrigley, 1966, 1977; Finlay, 1978; Langton and Laxton, 1978). Leslie Curry, one of the few spatial scientists to construct dynamic models to account for change through time (Curry, 1966) regarded the historical method as essentially empirical and ideographic, concerned with understanding unique events. To most analytical minds, discourse among historical geographers appears open-ended, unsystematic, and inconclusive, incapable of yielding valid generalizations. For their part, historical geographers have made few contributions to spatial analysis. Even in the context of Swedish theoretical studies, the simulation model of colonization designed by Eric Bylund (1960) is exceptionally austere and abstract. John Jakle (1971, p. 1096) concluded that 'simulation work has increased the capabilities of geography for describing dynamic spatial distributions, but this has not, for the effort, brought the discipline any closer to understanding the real rationales underlying human spatial behavior'. In their empirical research, historical geographers are engaged in examining the records of an infinite variety of individual and social activities, and are beginning to account for some of the ways in which behaviour differs from place to place and from epoch to epoch.

## INTERPRETING SOURCES OF EVIDENCE

While a few spatial scientists have endeavoured to elaborate models for examining time series, historical geographers have pursued the search for new sources of evidence and have critically evaluated old sources. They have proceeded in the faith articulated by Andrew Clark (1975, p. 16) that 'nothing is more primary to historical geographers than complete, relevant information, whether we find it in field, library or archive'. Of these sources of information, archives are the fullest and most rewarding. Herman Friis (1975, p. 156) describes the National Archives in Washington as 'a treasure trove so vast, so pulsating with opportunity, that no one can comprehend the infinite potentialities of the resources'. Those who have gone to the repositories cherish a hope that a hundred years hence the materials for historical geography 'will not be as we know them now' (Darby, 1960, p. 156). Year by year, collections of documents and maps will be catalogued, edited, abstracted for data banks, and interpreted.

The stock of archival material is continually changing. Unknown quantities of documents are destroyed each year by fire, flood, by the ravages of insects and fungi, and, wantonly, by being thrown away. At the same time, remarkable new discoveries are made. Finds of historic maps are frequently reported in *Imago Mundi* and in the pages of the *Geographical Journal*. During the past twenty-five years the *Journal* has noted discoveries of medieval topographical maps (Price, 1955), early sixteenth-century field maps (Yates, 1960, 1961), three manuscript maps by Christopher Saxton (Batho, 1959; Evans, 1972), county maps of Essex and of Cornwall by John Norden (Emmison and Skelton, 1957; Ravenhill, 1970), and a wealth of cartographic material from the eighteenth and nineteenth centuries, including Thomas Milne's land-use map of the London area in 1800 (Bull, 1956), pioneer maps of health and disease (Gilbert, 1958), and the 1837 maps of Henry Harness (Robinson, 1955). The *Journal of Historical Geography* has announced an atlas of settlement in Finland compiled from sixteenth-century evidence (Mead, 1975) and has critically appraised a collection of sixteenth-century official reports from the Spanish colony of Yucatan (Edwards, 1975). It would be tedious to list the volumes of topographical writings that have been brought to light in recent years, but the scholarly publications of the Hakluyt Society, covering one small part of the field, indicate the enormous quantities of material awaiting transcription and editing. Among professional journals, the *Geographical Journal* has reported the discovery of travellers' accounts from all continents, has identified source materials for seventeenth-century county histories, for Defoe's 'Tour', for eighteenth-century geographical compendia, and has collected biographical notes on the lives of geographers, explorers, and antiquaries.

Studies devoted to single sources of evidence emulate the patient and meticulous scholarship of the Domesday geography of England, whose seventh and last volume appeared in 1977. 'We all felt the fascination of the Domesday text,' writes H. C. Darby (1977, p. 376). 'It seemed as if F. W. Maitland's words of 1897 had come true: many men from all over England had come within King William's spell, had bowed themselves to him, and, in the language of the feudal lawyers, had become that man's men.' The formidable task of abstracting and interpreting the changing social structure of Victorian households, streets, and neighbourhoods from the enumerators' books of the population censuses from 1841 onwards was begun by Richard Lawton in 1955. Nearly twenty years later, Lawton (1972, p. 218) reported that following 'these early and unsophisticated efforts, the use of enumerators' books in empirical studies in urban history has increased enormously'. Further advances are announced in *The Census and Social Structure* (Lawton, 1978) and Richard Dennis (1979, p. 125) refers to local studies by 'a host of investigators'. Both Domesday Book and the nineteenth-century population censuses are exceptional in providing vast amounts of information for a large part of the country for particular years.

Other extensive data sources illuminate our knowledge of past geographies. Robin Glasscock's (1975) careful editing and ingenious reconstitution of the 1334 Lay Subsidy returns throws fresh light on medieval England before the onset of plague and economic decline. John Patten (1978) documents the revival of pre-industrial towns from 1500 to 1700. The collection and abstraction of parliamentary enclosure awards, begun by W. E. Tate and completed by Michael Turner (1978), a re-interpretation of common field enclosures (Yelling, 1977), a transcription of the 1801 crop returns begun by H. C. K. Henderson (1952), an abstraction and tabulation of the Gazette Corn Returns by Lucy Adrian (1977), and mapping of land-use, crop, and livestock data from the Tithe Surveys by Roger Kain (1979) are providing detailed bases for an agricultural geography of England and Wales from 1750 to 1850. Studies of maps and mapmakers, notably J. H. Andrews's (1975) account of the work of the Ordnance Survey in Ireland, J. B. Harley and Yolande O'Donoghue's (1975–7) studies of the Ordnance Survey in England, J. B. Harley's (1966; 1974) and Paul Laxton's (1976) contributions to the history of map publishing, have expanded our vision of Georgian Britain. 'An appreciation of the historical context of source materials and of the expertise necessary to use them properly' (Baker, Hamshere, and Langton, 1970, p. 22) is exemplified by geographical interpretations of more than twenty different bodies of historical evidence.

Some workers have concentrated upon the problems of reading and interpreting a single source to the neglect of other material, and their views have been criticized in the light of evidence drawn from different sources. In studying agriculture in Wales during the Napoleonic wars, David Thomas (1963) illustrated the partiality of his sources by presenting three separate pictures based on the 1801 returns, the county reports to the Board of Agriculture, and estate and enclosure maps. Upon most questions, sources do not appear in isolation and most writers attempt to compare and to integrate material from different sources. Occasionally, a researcher has the good fortune to find not only differing accounts of the same subject but critical assessments of the accuracy of scientific observations written by contemporaries. Discussing an eighteenth-century dispute over the prospects for settlement in Rupert's Land, D. W. Moodie (1976, p. 293) remarks: 'In any period, reality is little more than consensus which, even in science, is not absolute and can be strikingly ephemeral in nature.'

In recent years, important advances have been made by combining field evidence with air photographic, cartographic, and documentary evidence. Harry Thorpe (1965, 1975), Brian Roberts (1968, 1977), and Glanville Jones (1972, 1976) have greatly extended our knowledge of changing forms of rural settlements by bringing early maps and documents into field studies. M. W. Beresford and John Hurst (1971) have confirmed that over 2300 medieval villages were deserted. W. R. Mead and others (1954, 1965) have traced ridge and furrow over thousands of midland fields. C. J. Bond (1979) has recorded hundreds of moated sites. In Norfolk, thousands of disused marl pits have been identified (Prince, 1964), and from Herefordshire to Huntingdonshire hundreds of miles of hedges have been surveyed and dated (Hooper, 1971; Pollard, Hooper, and Moore, 1974). M. W. Beresford and J. K. S. St Joseph (1958) illustrate these and other themes in *Medieval England: an aerial survey*. Neil Cossons and Kenneth Hudson (1971) and R. A. Buchanan (1972) have begun to survey and list the relics of the industrial revolution. A combination of literary, statistical, and archival evidence is being brought to bear on the geography of eighteenth- and nineteenth-century industrialization.

The interpretation of primary evidence is fraught with difficulties. The documents upon which our knowledge of the past is founded can only be understood in the light of the usages of the times in which they were written. We must first translate the information we receive, often couched in dead languages, conveying messages alien to our ways of thinking. We must attempt to understand the personalities of and to assess the reliability of the writers and later custodians of records. With reason and imagination, we must infer the circumstances in which statements were recorded. To trace processes of change, it is necessary to examine the behaviour of individuals and of small groups, asking what people intended to do and how they achieved their objectives (Guelke, 1974). Records of decisions and of reasons for taking decisions provide material for case-studies, but because important considerations were often kept off the record, individual cases cannot be read at their face value, nor may they be aggregated and generalized.

On the basis of a few proposed land-use changes or development plans, discussed and minuted at meetings, negotiated by correspondence, written into contracts, or decided in court, a scholar is left to conjecture the motives, tastes, preferences, and prejudices of developers and planners. Such evidence may be supplemented by observation of relict features in the landscape and by a critical examination of personal histories. French historians, Marc Bloch (1954), Lucien Febvre (Burke, 1973), Fernand Braudel (1972–3; Prince, 1975), Emmanuel Le Roy Ladurie (1974, 1975, 1976), Georges Duby (1961) have made use of biography and psycho-analysis in reconstructing the past. Following their lead, British historians have shown an interest in a biographical approach, especially in tracing the lives of *petit peuple* and of rebels. Eric Hobsbawm and George Rudé (1969), Richard Cobb (1975, 1976, 1978; Prince, 1977, 1979), and E. P. Thompson (1963, 1975) have drawn eloquent testimony from reluctant witnesses and from witnesses whom government officials wished to silence. Stuart Macdonald (1979, p. 5) has elucidated the reactions of eighteenth-century working farmers, 'individuals with muddy boots striving to make money from the land', to George Culley's efforts to improve agricultural techniques in a Northumberland dale. It is a study in resistance to change as much as an account of the acceptance of innovations. In these investigations, the first priority has been 'to allow people to speak for themselves and to give as much licence as possible to individual behaviour and to popular habit.' (Cobb, 1970, p. xix).

The terms on which individuals belong to groups are more precisely defined and enforced than are personal goals and habits, and people interact in historical events more frequently as members of groups than as private individuals. The religious beliefs, political opinions, class consciousness, economic aims, and cultural values of churches, schools, clubs, regiments, businesses, unions, prisons, hospitals, and other institutions are often sharply differentiated (Harris, 1978). Knowledge of past social geographies may be gained through what Lawrence Stone (1971) calls prosopography, 'the investigation of the common background characteristics of a group of actors in history by means of a collective study of their lives'. Mark Billinge (1980) has examined the social and economic backgrounds and attitudes of members of literary and scientific societies in English industrial cities in the nineteenth century. Harold Fox (1979) has similarly examined the membership of nineteenth-century farmers' associations, assessing their importance as transmitters of information about new farming methods. It is not only measurable characteristics such as age, sex, occupation, and income that are significant in membership of a group, but attitudes towards work, marriage, education, health, wealth, poverty, charity, privilege, authority, segregation, beauty, and utility. Attitudes of groups have changed from time to time and have caused violent conflicts. They have also differed widely from place to place, leading to regional separatism, to clashes between city and country, or secession of colony from capital.

A map or document from the past has a life of its own. It was at one time an instrument and a witness to change. An eighteenth-century surveyor drawing an enclosure map was laying out a new a pattern of fields that perhaps had not undergone changes of such magnitude for a thousand years. At a unique point in time, a plan or document was drawn or written by a living individual within a complex daily routine of activities and relaxations, and in deciphering the writing, we may with difficulty recall some of the discourse and conflict that lies behind each sentence. A document is often filed together with reports and commentaries that enhance its significance. A shelf-full of reports and papers surrounding the census returns shed light on nineteenth-century society (Wrigley, 1972). Records of library sales, personal diaries, and letters tell us about map-users and the purposes for which maps were used in eighteenth-century North America (Harley, 1976) and from Hampshire, around 1800, related collections of company papers tell us how stage coaches were run and how much traffic they carried (Freeman, 1975). We can match the promises of publishers' prospectuses or newspaper advertisements with lists of subscribers of passengers, with reports from directors to shareholders, and ultimately, we may learn upon what journeys map-users or coach travellers and their promoters embarked.

Not the smallest reward for research on primary sources is an increased knowledge of the use of words. Differing meanings of words in different sources may reflect changes in social organizations or farming practices or introductions of new ideas from foreign countries. Ian Adams's (1976) glossary of agrarian landscape terms brings together and interprets a large number of such variants. Sources may be read and reread in an infinite variety of ways. Brian Harley (1979) asserts that 'the most fundamental contribution of historical geography may yet lie—whatever its broader philosophical or social purposes—in the maintenance of scholarly values, in the reading and understanding of evidence—a precious gift—and in an undogmatic approach to past societies and landscapes in their manifold interactions'.

## WORKING WITH OTHER DISCIPLINES

Sharing the use of the same source materials and the study of the same landscape features, historical geographers have worked alongside historians, archaeologists, and physical geographers. Archival research has frequently supplied a vital piece of evidence for the elucidation of the origin of a landform. An explanation for the present form of the Norfolk Broads has been revised in the light of C. T. Smith's discovery of thirteenth- and fourteenth-century records of the digging of millions of cubic feet of turf (Lambert, Jennings, and others, 1960). Another RGS Research Memoir (Salway and others, 1970) has examined changes brought about by Roman colonization in the Fens. Other studies (Sherlock, 1922; Brown, 1970) indicate that many landscape features, once thought to have been formed by natural processes, are in fact man-made. Knowledge of climatic and vegetational changes has been put on a secure footing by careful studies of documentary, cartographic, and pictorial evidence (Manley, 1952; Lamb, 1972, 1977; Ladurie, 1971; Parry, 1978; Pennington, 1969; Rackham, 1976). Physical geographers have learned to pose questions in ways that may be answered by historical evidence.

Both geographers and historians have speculated about the influence of geography upon history, have delineated geographies of the past, have traced themes of geographical change, and have investigated historical elements in geographies of the present. Co-operating in practical projects, geographers have played active parts in the Medieval Village Research Group, in the Moated Sites Research Group, in the Society for Medieval Archaeology, in the Association for Industrial Archaeology, in the Planning History Group, in the Historic Landscapes Steering Group (Brandon and Millman, 1978), and in many local archaeological and historical societies. They have also contributed to the work of the Cambridge Group for the History of Population and Social Structure and the English Place-Name Society. They have presented research papers to the Economic History Society, to the Social History Society, to the British Agricultural History Society, to the Garden History Society, to the Urban History Group, and to the Past and Present Society. In 1975, the Social Science Research Council convened a joint seminar of historical geographers and economic and social historians to discuss common problems in the study of nineteenth-century Britain. In 1978, another SSRC joint seminar considered the internal structure of nineteenth-century cities. Links between historical geographers, historians, and archaeologists are growing stronger, and in this review of recent advances in historical geography I have repeatedly acknowledged the work of historians. Indeed, it is scarcely possible to separate the contributions of geographers from those of historians to the study of medieval and modern cities, agrarian questions (Baker and Butlin, 1973), and population history. R. A. Dodgshon and R. A. Butlin's (1978) *An Historical Geography of England and Wales* illustrates more fully than previous books that links between historical geography and social and economic history are firmly established. The studies of Langton (1978), Pawson (1978), Cromar (1977), and Gregory (1978) on industrial changes and working conditions are enriched by exchanges of ideas among scholars in different disciplines.

In studying the development of religious, scientific, and intellectual ideas, historical geographers have remained aloof from cultural geographers and have not maintained contacts with social anthropologists (Billinge, 1976). Carl Sauer offered a lead in this direction and Clarence Glacken, Paul Wheatley, Marvin Mikesell, David Lowenthal, Yi-Fu Tuan, and J. B. Jackson have kept open channels of communication with social scientists, mostly in the United States. The *Geographical Review* and *Landscape* have served as vehicles for discussion of cultural and historical topics, but no similar journal has appeared in Britain, and the borderlands between historical geography, landscape history, and social anthropology have been crossed by few intrepid explorers. H. J. Fleure's *A Natural History of Man in Britain* (1951) focused upon people and the ways they lived in changing habitats. The notion that, through long histories of occupance, people impress a personality upon a region is illustrated in Estyn Evans's study of Ireland (1973). Jane Zaring (1977) has depicted from romantic literature the distinctive characteristics of Welsh scenes, Ian Adams (1978) has identified Scottish traits in the making of towns, and Robin Butlin (1977) has looked at changes in the Irish town. E. W. Gilbert's (1954) study of the patronage and promotion of Brighton is a rare example of an urban geography that relates building and population growth to the aspirations of people who designed the place and the tastes of those who took up residence.

In the borderlands between historical geography and art history, the question of preferences for different kinds of scenery and of changing fashions in landscape appreciation have received attention (Lowenthal and Prince, 1965, 1969; Appleton, 1975). In particular, changing styles in park and garden design and in the building of ornamental villas have been chronicled (Prince, 1967; Slater, 1978; Brandon, 1979). Drawing upon literary sources, David Lowenthal (1968, 1975, 1976) has traced changing attitudes towards the past in the United States. Larry Ford (1978) has looked at the preservation of historic townscapes in Bath, Chester, and Norwich, Roger Kain (1975) has surveyed conservation in France, and Robert Newcomb (1979) has compared preservation in Denmark with practices in Britain and the United States. Both refined and popular tastes in architecture and planning are now being studied. The part played by a large land-owner in designing a landscape and giving it a historic flavour is exemplified in Heather Fuller's (1976) account of Lord Yarborough's property. For an earlier period, Robin Donkin (1978) has documented the changes imposed by a strictly disciplined monastic order in England and Wales. Vernacular buildings have been set in their regional contexts by Gwyn Meirion-Jones (1971, 1973, 1977) and by June Sheppard (1966). In New Zealand, the activities of politicians, administrators, and settlers in building and laying out farms are recounted by J. M. Powell (1971). In India, the connections between cultures, social power, and an alien environment are examined in Anthony King's (1976) penetrating study of British colonial urban settlements. Returning colonial officials brought back to Cheltenham and Torquay bungalows and verandahs (King, 1973a,b). In studies of aesthetic geography, the past flows into the present, genuine relics mingle with sham ruins, exotic styles are assimilated into native architecture, and preservation merges into restoration.

A few historical geographers have reached out towards literature and art, whilst literary scholars and art historians have reconstructed past geographies. H. C. Darby (1948), John Paterson (1965, 1976), E. W. Gilbert (1972), Douglas Pocock (1978), and Yi-Fu Tuan (1978) have surveyed the relations between literature and geography. Through John Clare's poetic experience of enclosure in his Northamptonshire village, John Barrell (1972) provides insights into eighteenth-century ideas of landscape. J. R. Watson (1970) looks at picturesque landscapes and lakeland scenes through the eyes of English romantic poets. Paintings, no less than literature, offer symbolic as well as realistic representations of past landscapes (Heathcote, 1972). Constable's landscape imagery and cloud studies (Rees, 1976; Thornes, 1979) and Ruskin's ideas about nature and culture (Cosgrove, 1979) are especially illuminating. The spread of art forms across the world has yet to be examined.

Until recently, historical geographers belonged to tightly knit national groups, divided by barriers of language and held apart by separate national academic systems. *Progress in Historical Geography* (Baker, 1972) readily stereotyped a French school, a German-language area, and American historical geography. In Scandinavia, past and present were subsumed under systematic branches of the subject, in Latin America and Africa geographies of the present and future took priority over the past, and in the USSR historical geography was not recognized. During the past decade, however, international co-operation has increased. A colloquium on the evolution of rural landscapes in Europe, held at Nancy in 1957 (Juillard *et al.*, 1957), has been followed by nine subsequent conferences in different countries. Their proceedings (Kielczewska–Zaleska, 1978; Flatres, 1979) suggest cross-cultural connections in the development of field systems, settlement patterns, and farming practices. Attempts to translate terms from one language to another have widened the scope of investigations into agrarian institutions. European urban historians have held similar discussions. The preparation of a series of historic towns atlases is but one project upon which geographers are collaborating with historians in different European countries (Conzen, 1976). In 1976, the International Geographical Union set up a Working Group on historical changes in spatial organization to promote international co-operation in the study of historical geography (Tanioka and Annenkov, 1979). In addi-'tion to informing members about activities of historical geographers around the world, the Working Group arranges symposia on themes of international concern. Another series of discussions, initiated in 1975 by the Historical Geography Research Group and the Canadian Association of Geographers, has now enlarged its constituency to include Americans, Australians, and New Zealanders. A third colloquium in this series, devoted to rural, urban, and physical environments in frontier transition, was held at Los Angeles in August 1979 (Vicero, 1979).

These joint efforts are surmounting real and imagined barriers, but many scholars seem isolated and out of touch with workers in other countries. The *Journal of Historical Geography* remains largely a preserve for English-speaking geographers. Contributors from other countries have included David Hannerberg, Mats Widgren, Ursula Ewald, and, because she wrote about lost villages in Germany, I should add Anngret Simms. Rare citations of foreign works in articles and the small number of foreign books reviewed suggest that British and American historical geographers are not fully integrated in an international community of scholars (Denecke, 1978). With regard to social geography in general and landscape studies in particular, Robert Geipel (1978, p. 155) notes that 'relatively little information has reached the English-speaking academic community' from Germany since 1939. British historical geographers hear virtually nothing from fellow workers in Japan or other non-European language areas.

On the credit side, Jean Gottmann (1955), Wreford Watson (1979), and other scholars move easily back and forth across the Atlantic. During the past ten years, British historical geographers have made important contributions to the historical geography of European and Middle Eastern regions (Bater, 1976; Baker, 1979; Brice, 1979; Carter, 1972, 1977; Clout, 1977; Delano-Smith, 1979; French, 1970; Jones, 1977; Lambert, 1971; Mayhew, 1973; Mead, 1953; Ogden, 1977; Pounds, 1973; Shaw, 1977; Sutton, 1971). Others have worked on the historical geography of the Caribbean and Latin America (Blakemore and Smith, 1971; Clarke, 1975; Donkin, 1979; Galloway, 1975; Harris, 1965; Lowenthal, 1972; Newson, 1976; Robinson, 1979; Watts, 1966). K. B. Dickson (1969) and A. J. Christopher (1976) have written on contrasting regions in Africa. The cumulative effect of this work is to lay firm foundations for cultural comparisons among various realms at different periods.

## DEBATING IDEOLOGIES

Not since Lucien Febvre (1925) exposed the assumptions underlying environmental determinism have the philosophical bases of our discipline been questioned as deeply and as rigorously as during the past decade. Debate was precipitated by the claim some spatial scientists made—not unlike the claim environmental determinists made fifty years earlier—that spatial patterns were governed by unchanging, universal morphological laws (Harris, 1971). While most historical geographers would agree that men do not make history entirely under conditions of their own choosing, few would accept that economic and political events are subject to forces beyond human control. The reluctance or inability of some spatial scientists to document their hypotheses cast further doubt on the validity of their arguments (Baker, 1976; Gregory, 1976; Guelke, 1971, 1974, 1975, 1976). At best, geometric models only account for spatial arrangements of phenomena.

They add little to an understanding of processes bringing about those patterns. Walther Christaller (1933) examined a hierarchy of urban centres in medieval southern Germany and related the spacing of centres to what he called a marketing principle, in which higher order places supplied all the goods supplied by lower order places plus a number of higher order goods and services. Ignoring the historical improbability of fortified towns, episcopal seats, and imperial free cities in territorially-fragmented, feudal Swabia and Franconia functioning like small towns and cities on a flat surface in twentieth century, capitalist Iowa, Barry Garner (1967, p. 313) asserted that 'regularities of this sort may well exist in many areas despite differences in population density, regional economy and social and economic history'. Postulating causes for observed results is open to serious objections (Langton, 1972; Baker, 1975) and some historical geographers prefer to 'deduce statements about spatial form from a knowledge of generating processes' (Baker, 1976, p. 172). Explanations in historical geography are valid for particular periods and for specific cultural milieus. Writing between 1775 and 1810, Adam Smith sketched a model for the location of manufacturing appropriate at a time when canal-building, mechanization of cotton-spinning, smelting of iron by coke, and harnessing steam-power were beginning to revolutionize industries in Britain, and Alfred Weber (1909) outlined a least-cost theory of industrial location when German industries were triumphing in competition over other European powers. The classical model for agricultural location drawn by J. H. von Thünen in 1826 was based on an econometric analysis of his own estates in Mecklenburg in the period before and after the Napoleonic wars. Central place theory and classical location theories make sense only within the context of a competitive market economy (Brook, 1976a,b); they offer no help in explaining how markets expanded in area and changed in character as international corporations and state-controlled boards took over smaller enterprises. History refutes claims to permanent, universal laws of spatial organization.

In questioning assumptions behind different theories, historical geographers have brought into the open fundamental conflicts in their own political, economic, and social attitudes. Interpretation of past events is coloured by different observers' present-day value judgements. Whether the state of Britain at the beginning of the nineteenth century is viewed in terms of increased productivity or increased suffering is implicit in alternative pictures of 'an age of improvement' or 'a bleak age'. H. C. Darby (1973) describes the period of extensive reclamation of waste land, parliamentary enclosure, and expansion of industrial production as 'the age of the improver', but not all changes were benign. It was also a period of social upheaval, political insecurity, and a time of hardship for labourers in towns and villages. From 1350 to 1450, another period in which some people gained while others lost, has been interpreted 'both as a time of economic decline and as the golden age of English peasantry' (Postan, 1975, p. 158). A scholar who regards land improvement as an index of technological progress, a record of increasing mastery over the means of production, will view geographical change in a different light from a writer who focuses attention upon men and women at their places of work, at home, at play, and in their social environments. Those who ask how millions of acres of new land were brought into cultivation will write one version of historical geography, whilst those who ask why the task was ventured upon or who followed the plough will write another. F. J. Turner's (1894, p. 199) explanation of American history in terms of 'the existence of an area of free land, its continuous recession and the advance of American settlements westward', says little about the personal motives, social pressures, individual and group behaviour of migrants, speculators, and politicians. Recent studies have traced the transference of cultural attributes from the old world to the new and have gained fresh insights into the importance of ethnic identity and community in settling and exploiting new lands (Lemon, 1972; Adams, 1976; Handcock, 1976; Rice, 1977; Harris and Guelke, 1977; McQuillan, 1978; Ostergren, 1979; Gibson, 1978).

The orientation of recent studies has shifted from the outward appearance of landscapes to the social relations, activities, and ideas of farmers, manufacturers, builders, and others engaged in modifying their environments. Some writers (Ross, 1970; Allen, 1975; Hemming, 1970, 1978) have attempted vicariously to relive the past, to experience in imagination former environments and historic events. Some (Heathcote, 1965; Blouet and Lawson, 1975; Jakle, 1977; Powell, 1970, 1977; Jackson, 1978) have exercised the wisdom of hindsight to go back step by step from recorded achievements, tracing the antecedents of ideals and traditions, probing histories of heresies and misunderstandings, uncovering latent fears and desires from intimations in the documents. Studies of the cosmo-magical origins of ancient cities (Wheatley, 1969) and the influence of dietary taboos upon agriculture (Simoons, 1961) point the way towards vast unexplored fields of inquiry into the relations between religious beliefs, aesthetic ideals, building, cultivation, and destructive exploitation of land. These questions may be examined at an individual level, biographically, and sociologically, through studies of the behaviour of tribes, age groups, social classes, village communities, and other groups.

In this essay, I have asserted the primacy of individuals as agents of geographical change, but not all individuals have been equally free, willing, or active participants in the making of history. Slaves have been forced to till fields, build roads, row galleys, women have been forced to raise children, young men have been conscripted to fight wars, defeated civilians have been driven from their homes, criminals and dissidents have been imprisoned or exiled, labourers have been compelled to accept starvation wages, tenants have been evicted by landlords. Millions have been forced to serve as involuntary actors in history and countless other millions have been helpless, passive victims of floods, fires, famines, epidemics, wars, civil disorders, economic slumps, and inflation. Neither individuals nor even large powerful groups have been able to shape the future as they would have wished. The same individuals have not remained week after week, year after year, out of work, homeless, or hungry. The faces in the dole queues have changed but continual efforts by governments and international agencies have failed to shorten the endless lines of unemployed, vagrants, and undernourished. Individuals have attributed their failure to obtain jobs, shelter, or food to bad luck or to local and personal circumstances, whereas governments have been concerned with long-term, large-scale malfunctioning in world-wide economic systems. People do not practise what they preach, do not fulfil their ambitions, are pursued by misfortunes and disasters.

Many historical geographers have investigated impersonal environmental changes affecting the use of land and resources, such as silting of harbours, desertification, soil erosion, deforestation, exhaustion of mines, but few have studied the complex histories of cultural and institutional development. Beneath a recital of day-to-day incidents, Fernand Braudel (1972–3) discerns and delineates deeper levels of causation underlying economic decline, social disintegration, and political oppression as well as long-term causes for intellectual enlightenment, technological progress, and regional reorganization. In attempting to describe and explain the forces behind changes of long duration, historical geographers may draw upon and combine materialist modes of explanation, whether classical or Marxist, with idealist modes of explanation, whether humanist or structuralist (Baker, 1979). Among large, persistent problems awaiting investigation are geographical changes in imperialism, slavery, feudalism, capitalism, racial conflict, alienation among immigrants, cultural absorbtion, warfare, rebellion, territoriality, and nostalgia. In approaching these themes, historical geographers must avoid positivist pitfalls of disaggregating and analysing relations between separate elements and they must also steer clear of phenomenological swamps of intuitive and sympathetic understanding (Baker, 1978, p. 500; Gregory, 1978a, b). Ideological issues occupy a more important place in searching for causes of large-scale changes over long periods of time than they do in detailed inquiries into local events. As access to sources of evidence widens and co-operation with scholars in neighbouring disciplines and in other countries becomes closer, debate, argument, and revision increase and intensify.

### ACKNOWLEDGEMENT

I wish to thank Alan Baker, Robin Butlin, Steve Daniels, Richard Dennis, Derek Gregory, Brian Harley, and David Lowenthal for their help in offering criticisms and comments on an earlier draft.

# REFERENCES

ADAMS, I. H. 1976. *Agrarian landscape terms: a glossary for historical geography.* I.B.G. Special Pubn. 9.

ADAMS, I. H. 1976. The changing face of Scotland and the role of emigration, 1760–90. In ed. B. S. Osborne 2–12.

ADAMS, I. H. 1978. *The making of urban Scotland.* London.

ADRIAN, L. 1977. The nineteenth century Gazette Corn Returns from East Anglian markets. *J. Hist. Geogr.* 3: 217–36.

ALLEN, J. L. 1975. *Passage through the garden: Lewis and Clark and the image of the American northwest.* Urbana, Illinois.

ANDREWS, J. H. 1975. *A paper landscape: the Ordnance Survey in nineteenth-century Ireland.* Oxford.

APPLETON, J. H. 1975. *The experience of landscape.* London.

BAKER, A. R. H. (ed.) 1972. *Progress in historical geography.* Newton Abbot.

BAKER, A. R. H. 1975. *Historical geography and geographical change.* Basingstoke.

BAKER, A. R. H. 1976. The limits of inference in historical geography. In ed. B. S. Osborne, 169–82.

BAKER, A. R. H. 1977. Historical geography. *Progress in Human Geography* 1: 465–74.

BAKER, A. R. H. 1978. Historical geography: understanding and experiencing the past. *Progress in Human Geography* 2: 495–504.

BAKER, A. R. H. 1979. Rural settlements and early development of agricultural syndicalism: the case of Loir-et-Cher during the second half of the 19th century. In ed. P. Flatres, 267–75.

BAKER, A. R. H. 1979. Historical geography: a new beginning? *Progress in Human Geography* 3: 560–70.

BAKER, A. R. H. and BUTLIN, R. A. (eds.) 1973. *Studies of field systems in the British Isles.* Cambridge.

BAKER, A. R. H., BUTLIN, R. A., PHILLIPS, A. D. M., and PRINCE, H. C. 1969. The future of the past. *Area* 4: 46–51.

BAKER, A. R. H., HAMSHERE, J. D., and LANGTON, J. (eds.) 1970. *Geographical interpretations of historical sources.* Newton Abbot.

BARRELL, J. 1972. *The idea of landscape and the sense of place, 1730–1840: an approach to the poetry of John Clare.* Cambridge.

BATER, J. H. 1976. *St Petersburg: industrialisation and change.* London.

BATHO, G. R. 1959. Two newly discovered manuscript maps by Christopher Saxton. *Geogrl. J.* 125: 70–4.

BERESFORD, M. W. and HURST, J. G. (eds.) 1971. *Deserted medieval villages.* London.

BERESFORD, M. W. and ST JOSEPH, J. K. S. 1958. *Medieval England: an aerial survey.* Cambridge.

BILLINGE, M. 1976. The other neighbors. *Historical Geography* 6: 16–18.

BILLINGE, M. 1980. Towards a cultural geography of industrialization in late Georgian and early Victorian Britain. In eds. A. R. H. Baker and D. J. Gregory, *Explorations in historical geography.* Folkestone.

BLAKEMORE, H. and SMITH, C. T. (eds.) 1971. *Latin America: geographical perspectives.* London.

BLOCH, M. 1954. *The historian's craft.* Manchester.

BLOUET, B. W. and LAWSON, M. P. (eds.) 1975. *Images of the plains: the role of human nature in settlement.* Lincoln, Nebraska.

BOND, C. J. (ed.) 1979. *Moated Site Research Group, Report 6*. Woodstock, Oxon.

BRANDON, P. F. 1979. The diffusion of designed landscapes in south-east England. In eds. H. S. A. Fox and R. A. Butlin, 165–87.

BRANDON, P. F. and MILLMAN, R. N. (eds.) 1978. *Historic landscapes: identification, recording and management*. Department of Geography, Polytechnic of North London.

BRAUDEL, F. 1972, 1973. *The Mediterranean and the Mediterranean world in the age of Philip II*. 2 vols. London.

BRICE, W. 1978. *The environmental history of the Near and Middle East since the last ice age*. London.

BROOK, A. 1976a. Spatial systems in American history. *Area* 8: 47–52.

BROOK, A. 1976b. Industry, history and location theory. *J. Hist. Geogr.* 2: 165–68.

BROWN, E. H. 1970. Man shapes the earth. *Geogrl. J.* 136: 74–84.

BUCHANAN, R. A. 1972. *Industrial archaeology in Britain*. Harmondsworth.

BULL, G. B. G. 1956. Thomas Milne's land utilization map of the London area in 1800. *Geogrl. J.* 122: 25–30.

BURKE, P. 1973. *A new kind of history from the writing of Febvre*. London.

BUTLIN, R. A. 1977. *The development of the Irish town*. London.

BYLUND, E. 1960. Theoretical considerations regarding the distribution of settlement in inner north Sweden. *Geografiska Annaler* 52: 225–31.

CARLSTEIN, T., PARKES, D., and THRIFT, N. (eds.) 1978. *Timing space and spacing time*. 3 vols. London.

CARTER, F. W. 1972. *Dubrovnik: a classic city-state*. London.

CARTER, F. W. (ed.) 1977. *An historical geography of the Balkans*. London.

CHRISTALLER, W. 1933. *Die zentralen Orte in Süddeutschland*. Jena.

CHRISTOPHER, A. J. 1977. *Southern Africa*. Folkestone.

CLARK, A. H. 1954. Historical geography. In eds. P. E. James and C. F. Jones, *American geography: inventory and prospect*. Syracuse, NY 70–105.

CLARK, A. H. 1975. First things first. In ed. R. E. Ehrenberg, 9–21.

CLARKE, C. G. 1975. *Kingston, Jamaica: urban development and social change, 1692–1962*. Berkeley, California.

CLOUT, H. D. (ed.) 1977. *Themes in the historical geography of France*. London.

COBB, R. C. 1970. *The police and the people: French popular protest, 1789–1820*. Oxford.

COBB, R. C. 1975. *Paris and its provinces, 1792–1802*. Oxford.

COBB, R. C. 1975. *A sense of place*. London.

COBB, R. C. 1976. *Tour de France*. London.

COBB, R. C. 1978. *Death in Paris, 1795–1801*. Oxford.

CONZEN, M. R. G. 1976. A note on the historic towns atlas. *J. Hist. Geogr.* 2: 361–2.

COSGROVE, D. E. 1979. John Ruskin and the geographical imagination. *Geog. Rev.* 69: 43–62.

COSSONS, N. and HUDSON, K. (eds.) 1971. *Industrial archaeologists' guide, 1971–73*. Newton Abbot.

CROMAR, P. 1977. The coal industry on Tyneside, 1771–1800 oligopoly and spatial change. *Econ. Geog.* 53: 79–94.

CURRY, L. 1966. Chance and landscape. In ed. J. W. House, *Northern geographical essays*. Newcastle upon Tyne, 40–55.

DARBY, H. C. 1948. The regional geography of Thomas Hardy's Wessex. *Geog. Rev.* 38: 426–43.

DARBY, H. C. 1951. The changing English landscape. *Geogrl. J.* 117: 377–98.

DARBY, H. C. 1954.   On the relations of history and geography. *Trans. Inst. Brit. Geographers.* 19: 1–11.

DARBY, H. C. 1960.   An historical geography of England: twenty years after. *Geogrl. J.* 126: 147–59.

DARBY, H. C. (ed.) 1973.   *A new historical geography of England.* Cambridge.

DARBY, H. C. 1977.   *Domesday England.* Cambridge.

DELANO-SMITH, C. 1979.   *Western Mediterranean Europe.* London.

DENECKE, D. 1978.   Journal of Historical Geography. *Berichte zur deutschen Landeskunde* 52: 150–52.

DENNIS, R. J. (ed.) 1979.   The Victorian city. *Trans. Inst. Brit. Geographers.* N.S. 4: 125–319.

DICKSON, K. B. 1969.   *A historical geography of Ghana.* Cambridge.

DION, R. 1949.   La géographie humaine rétrospective. *Cahiers Internationaux de Sociologie* 6: 3–27.

DODGSHON, R. A. and BUTLIN, R. A. (eds.) 1978.   *A historical geography of England and Wales.* London.

DONKIN, R. A. 1978.   *The Cistercians: studies in the geography of medieval England and Wales.* Toronto.

DONKIN, R. A. 1979.   *Agricultural terracing in the aboriginal new world.* Tucson, Arizona.

DUBY, G. 1961.   L'histoire des mentalités. In ed. C. Samaran, *L'histoire et ses méthodes.* Paris.

EDWARDS, C. R. 1975.   The Relaciones de Yucatán as sources for historical geography. *J. Hist. Geogr.* 1: 245–58.

EHRENBERG, R. E. (ed.) 1975.   *Pattern and process: research in historical geography.* Washington, DC.

EMMISON, F. G. and SKELTON, R. A. 1957.   'The description of Essex' by John Norden, 1594. *Geogrl. J.* 123: 37–41.

EVANS, E. E. 1973.   *The personality of Ireland.* Cambridge.

EVANS, I. M. 1972.   A newly discovered manuscript estate plan by Christopher Saxton relating to Faversham, Kent. *Geogrl. J.* 138: 480–86.

FEBVRE, L. 1925.   *A geographical introduction to history.* London.

FINLAY, R. A. P. 1978.   The accuracy of the London parish registers, 1580–1653. *Population Studies* 32: 95–112.

FLATRES, P. (ed.) 1979.   *Paysages ruraux européen: travaux de la conférence européenne, Rennes–Quimper 26–30 Sept. 1977.* Rennes.

FLEURE, H. J. 1951.   *A natural history of man in Britain.* London.

FORD, L. R. 1978.   Continuity and change in historic cities: Bath, Chester and Norwich. *Geog. Rev.* 68: 253–73.

FOX, H. S. A. 1979.   Local farmers' associations and the circulation of agricultural information in nineteenth century England. In eds. H. S. A. Fox and R. A. Butlin, 43–63.

FOX, H. S. A. and BUTLIN, R. A. (eds.) 1979.   *Change in the countryside: essays on rural England, 1500–1900.* I.B.G. Special Pubn. 10.

FREEMAN, M. J. 1975.   The stage-coach system of southern Hampshire, 1775–1851. *J. Hist. Geogr.* 1: 259–81.

FRENCH, R. A. 1970.   The three-field system of sixteenth century Lithuania. *Agric. Hist. Rev.* 18: 106–25.

FRIIS, H. 1975.   Original and published sources in research in historical geography, a comparison. In ed. R. E. Ehrenberg, 133–59.

FULLER, H. A. 1976.   Landownership and the Lindsey landscape. *Annals Assoc. Amer. Geog.* 66: 14–24.

GALLOWAY, J. H. 1975.   Northeast Brazil 1700–50: the agricultural crisis re-examined. *J. Hist. Geogr.* 1: 21–38.

GARNER, B. J. 1967. Models of urban geography and settlement location. In eds. R. J. Chorley and P. Haggett, *Models in geography*. London, 303–60.

GEIPEL, R. 1978. The landscape indicators school in German geography. In eds. D. Ley and M. S. Samuels, 155–72.

GIBSON, J. R. (ed.) 1978. *European settlement and development in North America: essays on geographical change in honour and memory of Andrew Hill Clark*. Toronto.

GILBERT, E. W. 1954 *Brighton, old ocean's bauble*. London.

GILBERT, E. W. 1958. Pioneer maps of health and disease in England. *Geogrl. J.* 124: 172–83.

GILBERT, E. W. 1972. British regional novelists and geography in *British pioneers in Geography*. Newton Abbot, 116–127.

GLASSCOCK, R. E. 1975. *The Lay Subsidy of 1334*. British Academy, London.

GOTTMANN, J. 1955. *Virginia at mid-century*. New York.

GREGORY, D. J. 1976. Rethinking historical geography. *Area* 8: 295–99.

GREGORY, D. J. 1978a. The discourse of the past: phenomenology, structuralism and historical geography. *J. Hist. Geogr.* 4: 161–73.

GREGORY, D. J. 1978b. *Ideology, science and human geography*. London.

GUELKE, L. 1971. Problems of scientific explanation in geography. *Canadian Geographer* 15: 38–53.

GUELKE, L. 1974. An idealist alternative in human geography. *Annals Assoc. Amer. Geog.* 64: 193–202.

GUELKE, L. 1975. On rethinking historical geography. *Area* 7: 135–38.

GUELKE, L. 1976. The philosophy of idealism. *Annals Assoc. Amer. Geog.* 66: 168–9.

HÄGERSTRAND, T. 1947. En landsbygdsbefolknings förflyttmingsröelser. *Svensk Geografisk Årsbok* 23: 114–42.

HÄGERSTRAND, T. 1975. Survival and arena: on the life-history of individuals in relation to their geographical environment. *Monadnock* 49: 9–29.

HANDCOCK, W. G. 1976. Spatial patterns in a trans-Atlantic migration field: the British Isles and Newfoundland during the eighteenth and nineteenth centuries. In ed. B. S. Osborne, 13–40.

HARLEY, J. B. 1966. The re-mapping of England 1750–1800. *Imago Mundi* 19: 17–24.

HARLEY, J. B. 1976. The map user in eighteenth-century North America: some preliminary observations. In ed. B. S. Osborne, 47–69.

HARLEY, J. B. 1979. Historical geography and its evidence. Cited with permission from a paper read at a symposium on research methods in historical geography. Cambridge, 26 July 1979.

HARLEY, J. B. and LAXTON P. (eds.) 1974. *A survey of the county palatine of Chester, P. P. Burdett 1777*. Hist. Soc. Lancs and Ches., Liverpool.

HARLEY, J. B. and O'DONOGHUE, Y. (eds.) 1975–77. *The old series Ordnance Survey maps of England and Wales*. Lympne Castle.

HARRIS, D. R. 1965. *Plants, animals and man in the outer Leeward Islands, West Indies*. University of California Pubn. in Geog. 18, Berkeley.

HARRIS, R. C. 1971. Theory and synthesis in historical geography. *Canadian Geographer* 15: 157–72.

HARRIS, R. C. 1978. The historical mind and the practice of geography. In eds. D. Ley and M. S. Samuels, 123–37.

HARRIS, R. C. and GUELKE, L. 1977. Land and society in early Canada and South Africa. *J. Hist. Geogr.* 3: 135–53.

HARRISON, M. J., MEAD, W. R., and PANNETT, D. U. 1965. A midland ridge-and-furrow map. *Geogrl. J.* 131: 366–69.

HARTSHORNE, R. 1939. *The nature of geography: a critical survey of current thought in the light of the past*. Annals Assoc. Amer. Geog. 29, Lancaster, Penn.

HEATHCOTE, R. L. 1965.    *Back of Bourke: a study of land appraisal and settlement in semi-arid Australia.* Melbourne.
HEATHCOTE, R. L. 1972.    The artist as geographer: landscape paintings as a source for geographical research. *Proc. Royal Geog. Soc. Australia, South Australian Branch* 73: 1–21.
HEMMING, J. 1970.    *The conquest of the Incas.* London.
HEMMING, J. 1978.    *Red gold: the conquest of the Brazilian indians.* London.
HENDERSON, H. C. K. 1952.    Agriculture in England and Wales in 1801. *Geogrl. J.* 118: 338–45.
HOBSBAWM, E. J. and RUDÉ, G. 1969.    *Captain Swing.* London.
HOOPER, M. D. 1971.    *Hedges and local history.* Standing Conference for Local History. London.
JACKSON, R. 1978.    Mormon perception and settlement. *Annals Assoc. Amer. Geog.* 68: 317–34.
JAKLE, J. A. 1971.    Time, space and the geographic past: a prospectus for historical geography. *Amer. Hist. Rev.* 76: 1084–1103.
JAKLE, J. A. 1977.    *Images of the Ohio valley: a historical geography of travel, 1740 to 1860.* Oxford.
JONES, G. R. J. 1972.    Post-Roman Wales. In ed. H. P. R. Finberg, *The Agrarian History of England and Wales,* vol. i, pt. i Cambridge, 281–382.
JONES, G. R. J., 1976.    Multiple estates and early settlement. In ed. P. H. Sawyer, *Medieval settlement: continuity and change* London 15–40.
JONES, M., 1977.    *Finland: daughter of the sea.* Folkestone.
JUILLARD, E., MEYNIER, A., PLANHOL X. DE, and SAUTTER, G. (eds.) 1957.    Structures agraires et paysages ruraux. *Annales de l'Est* 17: 1–188.
KAIN, R. J. P. 1975.    Urban conservation in France. *Town and Country Planning* 43: 428–32.
KAIN, R. J. P. 1979.    Compiling an atlas of agriculture in England and Wales from the Tithe surveys. *Geogrl. J.* 145: 225–35.
KIELCZEWSKA-ZALESKA, M. (ed.) 1978.    *Rural landscape and settlement evolution in Europe: proceedings of the conference, Warsaw, September 1975.* Warsaw.
KING, A. D. 1973a.    The bungalow: the development and diffusion of a house-type. *Architectural Assoc. Quart.* 5,3: 5–26.
KING, A. D. 1973b.    Social process and urban form: the bungalow as an indicator of social trends. *Architectural Assoc. Quart.* 5, 4: 4–21.
KING, A. D. 1976.    *Colonial urban development: culture, social power and environment.* London.
LADURIE, E. LE ROY 1971.    *Times of feast, times of famine.* London.
LADURIE, E. LE ROY 1974.    *The peasants of Languedoc.* Urbana, Illinois.
LADURIE, E. LE ROY 1975.    Du social au mental, une analyse ethnographique. In H. Neveux, J. Jacquart, and E. Le Roy Ladurie, *L'histoire de la France rurale. 2 L'age classique des paysans 1340–1789.* Paris.
LADURIE, E. LE ROY 1978.    *Montaillou, Cathars and Catholics in a French village, 1294–1324.* London.
LAMB, H. H. 1972, 1977.    *Climate: present, past and future.* 2 vols. London.
LAMBERT, A. M. 1971.    *The making of the Dutch landscape.* London.
LAMBERT, J. M., JENNINGS, J. N., SMITH, C. T., GREEN, C., and HUTCHINSON, J. N. 1960.    *The making of the Broads.* R. G. S. Research Series 3, London.
LANGTON, J. 1972.    Potentialities and problems of adopting a systems approach to the study of change in human geography. *Progress in Geog.* 4: 126–79.
LANGTON, J. 1978.    *Geographical change in the industrial revolution: the south west Lancashire mining industry, 1590–1799.* Cambridge.

LANGTON, J. and LAXTON, P. 1978. Parish registers and urban structure, the example of late eighteenth-century Liverpool. *Urban History Yearbook* 74–84.

LAWTON, R. 1955. The population of Liverpool in the mid-nineteenth century. *Trans. Hist. Soc. Lancs. and Cheshire* 107: 89–120.

LAWTON, R. 1972. An age of great cities. *Town Planning Review* 43: 199–224.

LAWTON, R. (ed.) 1978. *The census and social structure.* London.

LAXTON, P. 1976. The geodetic and topographical evaluation of English county maps, 1740–1840. *Cartographic Journ.* 13: 37–54.

LEMON, J. T. 1972. *The best poor man's country: a geographical study of early southeastern Pennsylvania.* Baltimore.

LEY, D. and SAMUELS, M. S. (eds.) 1978. *Humanistic geography: prospects and problems.* Chicago.

LOWENTHAL, D., 1968. The American scene. *Geog. Rev.* 58: 61–88.

LOWENTHAL, D. 1972. *West Indian societies.* Oxford.

LOWENTHAL, D. 1975. Past time, present place: landscape and memory. *Geog. Rev.* 65: 1–36.

LOWENTHAL, D. 1976. The place of the past in the American landscape. In eds. D. Lowenthal and M. J. Bowden, 89–117.

LOWENTHAL, D. and BOWDEN, M. J. (eds.) 1976. *Geographies of the mind: essays in historical geosophy in honor of John Kirtland Wright.* Oxford.

LOWENTHAL, D. and PRINCE, H. C. 1965. English landscape tastes. *Geog. Rev.* 55: 186–222.

LOWENTHAL, D. and PRINCE, H. C. 1969. English façades. *Architectural Assoc. Quart.* 3: 50–64.

MACDONALD, S. 1979. The role of the individual in agricultural change: the example of George Culley of Fenton, Northumberland. In eds. H. S. A. Fox and R. A. Butlin, 5–21.

McQUILLAN, D. A. 1978. Farm size and work ethic: measuring the success of immigrant farmers on the American grasslands, 1875–1925. *Hist. Geogr.* 4: 57–76.

MANLEY, G. 1952. *Climate and the British scene.* London.

MAYHEW, A. 1973. *Rural settlement and farming in Germany.* London.

MEAD, W. R. 1953. *Farming in Finland.* London.

MEAD, W. R. 1954. Ridge and furrow in Buckinghamshire. *Geogrl. J.* 120: 34–42.

MEAD, W. R. 1975. An atlas of settlement in sixteenth-century Finland. *J. Hist. Geogr.* 1: 17–20.

MEIRION-JONES, G. I. 1971. The use of the Hearth Tax Returns and vernacular architecture in settlement studies. *Trans. Inst. Brit. Geographers.* 53: 133–60.

MEIRION-JONES, G. I. 1973. The long-house in Brittany: a provisional assessment. *Post-Medieval Archaeology* 7: 1–19.

MEIRION-JONES, G. I. 1977. Vernacular architecture and the peasant house. In ed. H. D. Clout, 343–406.

MOODIE, D. W. 1976. Science and reality: Arthur Dobbs and the eighteenth-century geography of Rupert's Land. *J. Hist. Geogr.* 2: 293–309.

NEWCOMB, R. M. 1979. *Planning the past: historical landscape resources and recreation.* Folkestone.

NEWSON, L. A., 1976. *Aboriginal and Spanish colonial Trinidad.* London.

OGDEN, P. E., 1977. *Foreigners in Paris: residential segregation in the nineteenth and twentieth centuries.* Department of Geography, Queen Mary College, London.

OSBORNE, B. S. (ed.) 1976. *The settlement of Canada: origins and transfer. Proceedings of the 1975 British–Canadian Symposium on historical geography.* Kingston, Ontario.

OSTERGREN, R. 1979. A community transplanted: the formative experience of a Swedish immigrant community in the upper Middle West. *J. Hist. Geogr.* 5: 189–212.

OVERTON, M. 1977.   Computer analysis of an inconsistent data source: the case of probate inventories. *J. Hist. Geogr.* 3: 317–26.

PARRY, M. L. 1978.   *Climatic change, agriculture and settlement.* Folkestone.

PATERSON, J. H. 1965.   The novelist and his region: Scotland through the eyes of Sir Walter Scott. *Scot. Geog. Mag.* 81: 146–52.

PATERSON, J. H. 1976.   The poet and the metropolis. In eds. J. W. Watson and T. O.'Riordan, *The American environment: perceptions and policies,* New York, 93–108.

PATTEN, J. H. C. 1978.   *English towns 1500–1700.* Folkestone.

PAWSON, E. 1978.   *The early industrial revolution: Britain in the eighteenth century.* London.

PENNINGTON, W. 1969.   *The history of British vegetation.* London.

POCOCK, D. C. D. 1978.   *The novelist and the north.* Department of Geography, University of Durham.

POLLARD, E., HOOPER, M. D., and MOORE, N. W. 1974.   *Hedges.* London.

POSTAN, M. M. 1975.   *The medieval economy and society.* Harmondsworth.

POUNDS, N. J. G. 1973.   *An historical geography of Europe, 450 B.C.–A.D. 1330.* Cambridge.

POWELL, J. M. 1970.   *The public lands of Australia Felix: settlement and land appraisal in Victoria, 1834–91.* Melbourne.

POWELL, J. M. 1971.   White collars and moleskin trousers. *New Zealand Geog.* 27: 151–71.

POWELL, J. M. 1977.   *Mirrors of the new world: images and image-makers in the settlement process.* Folkestone.

PRICE, D. J. 1955.   Medieval land surveying and topographical maps. *Geogrl. J.* 121: 1–10.

PRINCE, H. C. 1964.   The origin of pits and depressions in Norfolk. *Geography* 49: 15–32.

PRINCE, H. C. 1967.   *Parks in England.* Shalfleet.

PRINCE, H. C. 1975.   Fernand Braudel and total history. *J. Hist. Geogr.* 1: 103–6.

PRINCE, H. C. 1977.   Richard Cobb: a spy in revolutionary France. *J. Hist. Geogr.* 3: 363–72.

PRINCE, H. C. 1978.   Time and historical geography. In eds. T. Carlstein, D. Parkes, and N. Thrift, 1: 17–37.

PRINCE, H. C. 1979.   Richard Cobb: on intimate terms with the past. *J. Hist. Geogr.* 5: 327–36.

RACKHAM, O. 1976.   *Trees and Woodland in the British landscape.* London.

RAVENHILL, W. D. 1970.   The newly-discovered manuscript maps of Cornwall by John Norden. *Geogrl. J.* 136: 593–95.

REES, R., 1976.   John Constable and the art of geography. *Geog. Rev.* 66: 59–72.

RICE, J. G. 1977.   The role of culture and community in frontier prairie farming. *J. Hist. Geogr.* 3: 155–75.

ROBERTS, B. K. 1968.   A study of medieval colonisation in the Forest of Arden, Warwickshire. *Agric. Hist. Rev.* 16: 101–13.

ROBERTS, B. K. 1977.   *Rural settlement in Britain.* Folkestone.

ROBINSON, A. H. 1955.   The 1837 maps of Henry Drury Harness. *Geogrl. J.* 121: 440–50.

ROBINSON, D. J. (ed.) 1979.   *Social fabric and spatial structure in colonial Latin America.* Syracuse, NY.

ROSS, E. 1970.   *Beyond the river and the bay: some observations on the state of the Canadian northwest in 1811.* Toronto.

SALWAY, P., HALLAM, S. J., and BROMWICH, J. L'A 1970.   *The Fenland in Roman times: a study of a major area of peasant colonization.* R.G.S. Research Memoir 5, London.

SAUER, C. O. 1941. Foreward to historical geography. *Annals Assoc. Amer. Geog.* 31: 1–24.

SHAW, D. J. B., 1977. Urbanism and economic development in a pre-industrial context: the case of southern Russia. *J. Hist. Geogr.* 3: 107–22.

SHEPPARD, J. A. 1966. Vernacular buildings in England and Wales: a survey of recent work by architects, archaeologists and social historians. *Trans. Inst. Brit. Geographers.* 40, 21–37.

SHERLOCK, R. L. 1922. *Man as a geological agent, an account of his action on inanimate nature.* London.

SIMOONS, F. J. 1961. *Eat not this flesh: food avoidance in the old world.* Madison, Wisconsin.

SLATER, T. R. 1978. Family, society and the ornamental villa on the fringes of English country towns. *J. Hist. Geogr.* 4: 129–44.

SMITH, R. M. 1978. Population and its geography in England 1500–1730. In eds. R. A. Dodgshon and R. A. Butlin, 199–237.

STONE, L. 1971. Prosopography. *Daedalus* 100: 46–79.

SUTTON, K. 1971. The reduction of wasteland in the Sologne: nineteenth-century French regional improvement. *Trans. Inst. Brit. Geographers.* 52: 129–44.

TANIOKA, T. and ANNENKOV, V. 1979. Historical changes in spatial organisation. *I.G.U. Bulletin* 30: 129–34.

THOMAS, D. 1963. *Agriculture in Wales during the Napoleonic wars, a study in the geographical interpretation of historical sources.* Cardiff.

THOMPSON, E. P. 1963. *The making of the English working class.* London.

THOMPSON, E. P. 1975. *Whigs and hunters: the origin of the Black Act.* London.

THORNES, J. E. 1979. The meteorological understanding behind John Constable's atmospheric art. *Burlington Magazine* 121: 697–704.

THORPE, H. 1965. The lord and the landscape, illustrated through the changing fortunes of an English parish, Wormleighton. *Trans. and Proc. Birmingham. Archeol. Soc.* 80: 38–77.

THORPE, H. 1975. Air, ground, document. In ed. D. R. Wilson, *Aerial reconnaissance for archaeology.* Council for Brit. Archaeology Research Report 12: 14 1–53.

TURNER, F. J. 1894. The significance of the frontier in American history. *Amer. Hist. Assoc. Annual Report for 1893*, 199–227.

TURNER, M. (ed.) 1978. *W. E. Tate: a domesday of English enclosure acts and awards.* Reading.

VICERO, R. D. (ed.) 1979. *Proceedings of CUKANZUS '79: rural, urban and physical environments in frontier transition.* Los Angeles, California.

WATSON, J. R. 1970. *Picturesque landscape and English romantic poetry.* London.

WATSON, J. W. 1979. *Social geography of the United States.* London.

WATTS, D. 1966. *Man's influence on the vegetation of Barbados, 1527 to 1800.* Department of Geography, University of Hull.

WEBER, A. 1909. *Über den Standort der Industrien.* Tübingen.

WHEATLEY, P. 1969. *City as symbol.* London.

WRIGLEY, E. A. (ed.) 1966. *An introduction to English historical demography.* London.

WRIGLEY, E. A. (ed.) 1972. *Nineteenth century society.* Cambridge.

WRIGLEY, E. A. 1977. Births and baptisms: the use of Anglican baptism registers as a source of information about the numbers of births in England before the beginnings of civil registration. *Population Studies* 31: 281–312.

YATES, E. M. 1960. Map of Ashbourne, Derbyshire. *Geogrl. J.* 126: 479–81.

YATES, E. M. 1961. Blackpool, A. D. 1533. *Geogrl. J.* 127: 83–5.

YELLING, J. A. 1977. *Common field and enclosure in England 1450–1850.* London.

Yi-Fu Tuan 1978. Literature and geography: implications for geographical research. In eds. D. Ley and M. S. Samuels, 194–206.

Zaring, J. 1977. The romantic face of Wales. *Annals Assoc. Amer. Geog.* 67: 397–418.

# 11  Social Geography

EMRYS JONES  *London School of Economics*

DURING the last two decades the term 'social geography' has become an accepted one to denote a limited field of interest within the very broad area of human geography. Consequently, many courses in universities and polytechnics in Britain now carry this title; articles and books on the topic form a class of literature reviewed in *Geo Abstracts*; it merits a study group within the Institute of British Geographers; the latest edition of the *Encyclopaedia of Social Sciences* has an article on social geography; and the first texts have appeared which incorporate the name in their titles. The use and general acceptance of the term does not necessarily imply a commonly accepted understanding of the concept, but rather it reflects an upsurge in interest in certain topics which were previously subsumed, in a very diffuse way, in human geography. It also reflects the beginning of an enquiry whether new concepts are emerging which justify the specificity of the title. The separate identity of the field is nebulous and attempts at definition hesitant, and this is a good reason why we should remind ourselves of the antecedents of studies we now call collectively 'social geography' and understand their origin in human geography as a whole.

The term 'social geography' is not new. In a very useful review of the historical derivation of the term, Professor Dunbar (1978) reminds us that it appeared at least as long ago as 1884 in France, and that it was used several times by Demolins and others in the Le Play school in the two decades following. It was largely equated with human geography as a whole, as it was by Vidal de la Blache, and some writers interpreted the field in the traditional environmentalist way as the study of the action of the environment on society.

A much more contemporary-sounding approach was made by G. W. Hoke (1907), when he described social geography as 'the distribution in space of social phenomena' and 'the description of the sequence and relative significance of those factors, the result of whose influences is the localisation in space of the series of phenomena chosen for investigation'. This might well be summed up in a simple phrase: patterns and processes in society.

Hoke seems to have gone unnoticed by subsequent writers on social geography, and, in general, scholars did not see the necessity of specifying a restricted field of interest: they were human geographers in the widest sense of the term. The influence of Le Play was felt strongly in Britain in the early years of this century, particularly through the writing and teaching of Fleure, Geddes, and Roxby, and it was embedded in a holistic approach to human geography; though Roxby (1930) saw social geography as a branch of human geography, together with racial, political, and economic. Gilbert and Steel (1945) did not depart radically from this view, and in 1946 Fitzgerald could still say 'If it were possible to replace our nomenclature *ab initio*, I feel certain that I would recommend replacing "human geography" with "social geography".'

Dunbar is right in accusing those who use the term 'social geography' of being 'unconcerned with their intellectual roots', but this is largely because of the lack of differentiation up to the mid-century, and the holistic way in which human geography was taught. It was not until Watson's paper of 1957 that some of the issues were clarified by attempting to show the relationship between certain geographical topics and the burgeoning field of sociology. But meanwhile one of the first strands in the emergence of social geography was being woven within a traditional context in the Aberystwyth school of geography. Here the anthropological link was very explicit, the honours degree being a combination of geography and anthropology. Fleure himself was a Fellow of the Royal Society by virtue of his anthropological research; his successor, C. Daryl Forde, subsequently took a Chair of anthropology at University College London: the department produced two current holders of Chairs in anthropology. The French influence was strong, via Le Play and Vidal. The best products of this school—exemplified in the works of Estyn Evans (1942, 1951)—are outstanding achievements in studying society within an environmental and spatial setting and with a deep understanding of the cultural and historical background.

In the 1950s, and following the pattern set by Alwyn Rees (1951) in his penetrating work on a Welsh countryside community, a series of studies emerged from Aberystwyth which dealt with contemporary society (Jenkins *et al.* 1960). There was, by this time, a more explicitly sociological influence, though these researches might still be thought of as being more concerned with the traditional than with the emerging aspects of social processes. Just as Evans was concerned with peasantry and with continuity in rural life, so the Welsh series emphasized the traditional in rural communities, as did Williams (1956) in his work on Cumberland and Devon.

Abruptly the rural interest faded and almost disappeared in the 1960s with a sudden interest in urban society, and it was this field which established itself as the basis of contemporary social geographical studies. There were two reasons for this; one was a growing awareness of American literature in this field, based on the ecological school of Chicago, and the other was that the urban interest coincided with the introduction of a new methodology in geography generally.

There had been no dearth of interest in urban geography, but the emphasis had been on historically based morphological and comparative studies, although Dickinson (1957) had introduced British geographers both to Christaller—paving the way to studies of urban fields and the vast amount of work on urban systems—and to the Chicago school. For social geographers the future lay in social differentiation within the town or city, and here the American influence was paramount. To many, human geography was very closely akin to human ecology (indeed Barrows, as far back as 1923, had sought to equate them). Parks's work on natural areas in the city, Burgess's zonal model, and Hoyt's work on sectors suggested frameworks for the study of social groups in the city which were eagerly accepted (Timms 1971). Watson's study of Toronto was a pioneer work by a geographer, as opposed to a sociologist, but as we have seen, he was acutely aware of the sociological implications, and although his aim was the identification of 'regions', it was 'by associations of social phenomena' (Watson 1957).

This kind of study demanded much more detailed data than were normally available to British students. American sociologists had long had the advantage of urban data published by census tract. A British sociologist, Ruth Glass (1948), used enumeration district data in a pioneer work on social groups in Middlesborough, and Mogey (1956), a geographer student of Evans turned sociologist, used tracts in studying social groups in Oxford. Belfast was the first city in which demographic and social data were mapped by enumeration districts, revealing patterns which demanded a new level of explanation; and this study also applied the classical ecological models—if only to reveal their inadequacies (Jones 1960).

This disaggregation of data could develop further only by virtue of the second fillip, multivariate analysis and the use of the computer. This particular nudge also came from outside geography, this time from a statistician and a sociologist in a seminal book on British towns (Moser and Scott 1961). It was followed by a spate of urban studies in this country, closely paralleling similar work in the United States and incorporating a whole body of research on social area analysis from that country.

The 'scientific revolution' in British geography found a fruitful field in social geography. The volume of empirical work on social areas in cities, allied to work on segregation and ethnic distributions, has been very considerable (Jones and Eyles 1977). Techniques have become increasingly sophisticated both in mapping and analysing data. But fundamentally quantification has not taken us much further than social patterns, and the dearth of theory has been obvious (Eyles 1974). In acknowledging the scientific—albeit social-scientific bias, social geographers have seen themselves as basically classificatory and model-oriented: we have been slow to appreciate that this takes us very little further than description in the first place, and there has been a reluctance to move from pattern to process. It is in the latter that we seek explanations, and as a rule we have had to borrow our explanatory models from other social sciences. This was admirably demonstrated by Pahl (1967)—the classic examples being the urban ecological models—and is as true now as when he wrote his paper on models in social geography.

Social area analysis and models of areal differentiation accorded well with the positivist current of thought of the 1960s, and there was a tacit acceptance of mechanistic models based on causality and determinism. The ecological approach combined a simplistic biological model with a distance-decay economic model.

Empirical work certainly raced ahead of theoretical. Robson (1975) refers to the 'fossilisation of urban theory ... The Victorian city is still much to the forefront ... and at the back of our minds is still a romantic view of Chicago, the city which ... has become the seed-bed of theory, the norm, the point of reference, the source of urban fact and urban fiction.' This was partly because, for a time, fascination with the new methodology seemed to allow room for little else, and the manipulation of masses of data accorded well with 'scientific' methods and macro-generalizations.

Even before the spate of social-urban studies, a warning had been sounded that the classical ecological models, which regarded people and social groups as pawns to be moved by ecological or economic laws, ignored the richness of human behaviour and those elements in social value-systems which were non-economic. Firey's (1947) work on Boston was virtually ignored: it was not referred to in British literature until 1960 (Jones 1960), and consequently distributions of social groups were still studied as if they bore no relationship to the meaning those groups gave to the territory they occupied. But there was a massive reaction—which some have seen as a 'behavioural counter-revolution' (Davies 1972)—and this constitutes a further strand in social geography today.

This 'new' approach was strongly influenced by phenomenological arguments. The objectivity of the classical scientific approach was being questioned and more and more place given to the subjective. Again the antecedents are readily to hand. Kirk first published his ideas in 1951, but it was his *Geography* article of 1963 that made the impact in Britain. By accepting the idea of a 'perceived' world, as far as social geographers were concerned the old, 'real' objective world seemed to dissolve into a flux. The geographer's solid world, delineated and mapped as it was gradually brought into view by travellers and explorers, had been the touchstone of the subject. Now we were realizing that people's behaviour is not necessarily directly related to this objective world, but to the world as they see it. There are as many worlds as there are people. Traditionally 'the kind of space that geographers are concerned with is earth space . . . [which] can be indefinitely sub-divided into segments of various sizes' (James 1972). Space, which we divided into pigeonholes according to intent or whim, was a container of things—something 'out there' which was the constant in the geographers' universe. To the social geographer, however, space becomes an attribute of human behaviour, something we create by what we do and how we think. Location is no more than a reference point in an abstract model: social geography deals with places, and places are invested with values. Incidentally, these values may often seem to be the very opposite of the rationality which location theory and all normative approaches assume.

The surge of interest in the phenomenological approach was by no means unrelated to the efforts of the early generation of geographers to see areal differentiation in social terms. Vidal de la Blache's interest was in patterns of living (*genres de vie*) in which society and culture were the main variables (Buttimer 1968). *Genres de vie* are essentially projections of society. This was the heart of the *pays*, why there was a constant search for the 'personality of a region'. Evans's work on Ulster was in terms of the human interpretation of the land: Bowen (1959) looked for social variables in describing part of Wales as *le pays de Galles*. This approach is not unrelated to Bobek's work in Germany in which large culture regions are broken down into regions of social interaction (Hajdu 1968).

The French connection has been given a contemporary and renewed vigour in the work of Buttimer (1968, 1969, 1978), both in her interpretation of the philosophy of the French school, particularly that of Vidal de la Blache, Chombart de Lauwe, and Sorre, and in her own contribution to the idea of social space. This allows for an objective as well as a subjective view of space, the former reflecting social frameworks, structure, and organization conditioned by ecological and cultural factors, the latter the view of particular social groups. Chombart incorporated the notion of the use and perception of social space into a hierarchy of activity systems.

Both the philosophical aspects of the phenomenological approach and their implications have been taken further by writers like Tuan (1974) and by the exploration of mental maps. The idea is to see the world through the eyes and minds of specific groups of people. In this way many geographers have been invading the field of the social psychologist to whom perception and the study of territorial space are familiar. The geographer is probably more at home when behavioural studies approach a scale which suggest communities, because there is a real problem of reductionism and involvement in another discipline.

Paradoxical though it may seem, at an aggregate level geographers still tackle behaviour from a positivist point of view, a sort of social physics in which the mathematically-based methodologies continue to flourish. The search is still for patterns and for generalizations. The idea that every individual has his own view of the world is frightening to contemplate—and perhaps misleading. There are few among us—poets and artists—who would wish to be so individualistic, though J. K. Wright (1947, 1966) stressed the incompleteness of a world that excludes this view. Most people are prepared to accept a taught view of the world, so that we are really exploring shared experiences and group values. Sometimes these are diffuse throughout a group because, for example, they share a common language; sometimes they are accepted from one person as a divine revelation, as the Mormons did from Smith, and later from Brigham Young. By this argument, social geographers must focus their studies on groups and on the value-systems of groups. It is a corollary, also, that the universal—the very aim of the scientific approach—does not exist. Generalizations can be no better than culture-specific.

There is no doubt about the important dimension that behavioural studies have added to social geography, drawing our attention away from simple rational models to explanations at individual or group level. As Wolpert (1965) reminded us, migration is more than a book-keeping exercise, for decisions not to migrate may be as important as decisions to migrate in trying to understand the complexities lying behind movements of population. This has become particularly important in studies of intra-urban movements, where the former assumptions about choice in decision-making have almost entirely been replaced by the notion of constraint and the prevention of moving: so much so that the world of gatekeepers and obstacles in which we seem to live appears not unlike the old environmental-deterministic world which we thought had disappeared for good!

It has also posed a major problem. There seem to be two levels at which explanation may be offered. One is at the macro-level where, as we have seen, positivist approaches seem promising, and the other is at the micro-level, where one is approaching the unique. How does one interpret the macro-level in the light of what is discovered at the micro-level, where human behaviour seems to deny generalizations? I have suggested elsewhere (Jones 1956) how significant indeterminacy is at the micro-level, and that it is no more than hidden statistically at the macro-level. If we wish to raise the level of explanation of the latter it can only be done by understanding the former. Probability has been put forward as one way of bridging the gap, and so has the idea of 'summation man' (Davies 1972, Introduction). The latter is not so far removed from an approach to social geography via the social group (Jones and Eyles 1977, Chapter 1), which deals with the idosyncratic in a variety of ways without ignoring it.

By the late 1960s there were many who argued that social geographers were losing themselves in the niceties of academic argument and ignoring society at the expense of solving the problems of geography. Some disciplines distinguish between 'pure' and 'applied' approaches. Whether social geography has advanced enough theoretically to claim a 'pure' side is very debatable, but there is no doubt that it does deal with the raw materials of issues which are eminently applicable, such as health and housing, ethnicity and segregation. Social geography seemed content with dealing with all these as if they were subsumed in academic problems of measurement, classification, and models. This of course was the essence of a scientific approach which implied objectivity and expected neutrality: social scientists were people who could stand back dispassionately and speculate on how society worked without necessarily committing themselves to any course of action. It is now agreed that a value-free social science is a myth, but the corollary to some was that the justification for social geography was its relevance to the problems of society. The movement towards this social awareness coincided, in the United States especially, with the popular identification of most students and some staff with sections of the society which were thought to be underprivileged—and 'relevance' became a major issue. *Social Justice and the City* (Harvey 1973) became a key text, 'well-being' and 'ill-being' and 'social indicators' the themes of books and articles (Knox 1975).

Once again, it is necessary to remind ourselves that although the volume of such work is now impressive and the commitment of its authors vehement, the concern is not novel in geographical thought. Geographers in the past were not unaware of the implications of their work; they often immersed themselves in practical issues and were as often motivated by a desire to improve society. We are reminded by Dunbar (1978) that in nineteenth-century France 'social' and 'socialist' were often used synonymously, and that 'Reclus could have been thinking of social geography in that sense, as implying socialistic geography'. Without committing themselves to the particular ideology implied in this, many British geographers had been anxious to be more involved in social change; but this was done mainly either by the participation of geographers on bodies which affected public policy, or by the preparation of factual material which provided a better basis for decision-making (e.g. Dudley Stamp's *Land Use Map of Great Britain*. 'Well-being' is an echo of Gilbert and Steel's (1945) call for a 'geography of happiness'). There has never been a dearth of geographers who have been very concerned with the state of society and did not see their activities as mere ivory tower exercises; but geography as a discipline was regarded by most as a marginal technical input to any proposed social changes.

In the 1970s social concern became a much more potent force as young geographers have immersed themselves in contemporary issues. To have a seminar on squatting is to discuss the issue with students who *are* squatters. There is still something very academic—indeed scientific—in the manipulation of social data, but the aim now is to produce social indicators to reveal spatial elements in social inequalities. One other element has helped to clarify the social implications of geographical work, and that is the tremendously renewed interest in the environment. The geographer is on home ground here, but the interest is focused on problems and issues under debate, particularly on pollution, conservation, management, diminishing resources, alternative uses. Problem orientation here means policy orientation.

Eyles (1977) sums up the three areas of impact which the relevance debate has had on social geography as follows:

(a) Intensifying the focus on social problems by discovering the extent and intensity of social ills and the quality of life, and their uneven distribution.
(b) Fostering the desire to understand and interpret spatial aspects of situations in terms of the actions, rules, and meanings of the participants.
(c) Asserting that the root causes of spatial inequalities are to be found in the structure of society.

This last point brings us to the final strand which must be briefly examined in order to understand the entire spectrum of work in social geography today. The political economy approach concentrates almost entirely on social processes and its methods are mainly Marxist or neo-Marxist (Pickvance 1976). The holistic element in this approach is not unfamiliar, and indeed may be an attraction to many geographers (Ley 1978). What many are not prepared to accept are the ideological implications which are that society—and, by implication, spatial inequalities—cannot be modified but must be replaced by a different social order. Past and present geography, in this view, is a capitalist geography. This is no more than stating that social geography, certainly in so far as it assumes a set of values, is culture specific, although the nomenclature of the argument is one we associate with power structure and political decision-making. Where issue could be taken with the political economy approach is that, in spite of the systems structure of Marxist philosophy, emphasis is always focused on mode of production as the key element in transforming structures, so that there is a strong deterministic base. The sets of relationships between structures are, moreover, not in harmony with one another, so there is a basic concept of conflict. The latter is very much in evidence in social geography in the last decade (Doherty 1969, Boal 1972), studies of segregation and of the distribution of housing groups, for example, tending to replace consensus models by conflict models: in other words the emphasis is now on constraints rather than on choice.

Possibly the most disconcerting attribute of the political economy approach is its increasing withdrawal from the patterns of distribution which are usually deemed the starting-point of geographical enquiry. In housing, for example, most managerial studies concentrated on the outcome of a process, taking the institutions as given (i.e. accepting the capitalist society). Now the problem is pushed back to examining what those institutions ought to be, and this leaves very little to geography. Herbert (1979) makes the point that the approach becomes a-disciplinary, and any spatial component constitutes an artificial constraint on a problem which is essentially one of social reproduction. Paradoxically, too, it is the most remote from day-to-day problems because it is the most theoretical, so that the meaning of the word relevant becomes strained.

The various strands which I have suggested as making up the corpus of social geographical studies—the sociological rural, the positivist urban, the phenomenological approach to space, the relevance debate, structuralist/Marxist studies—have been discussed in a convenient historical sequence, because the peaks of interest, and of output, they represent are sequential to some extent, even at times representing reactions against a previous set of assumptions. Quantification, and its theoretical scientific paradigm, because it was linked with the computer, came with a suddenness and novelty which seemed to merit the word 'revolution'. We went through an era of over-indulgence in statistical manipulation and emerged with ever more sophisticated patterns—but understanding seemed as elusive as ever. The revolution was pronounced over and done with almost before a second generation of students had mastered the techniques, and the *avant-garde* heralded the behavioural counter-revolution. Relevance came like a revelation at an AAAG meeting in 1970. A few years later *Social Justice and the City* seemed like tablets brought down from Sinai.

Even if they are sequential in that sense, these approaches should not be seen as stages, i.e. as a progression in understanding or a continuingly greater appreciation of 'truth'. The novelty of each movement certainly fired the imagination. Geography as a whole had been caught in a traditional mould and had become rather sterile: it was out of phase with exciting things happening in other social sciences: change was welcomed.

The new ideas, however, were not as revolutionary as they seemed to be, either in their novelty or in the suggestion that they were replacing old ideas which could now be relegated to the history of the subject. Too many paradigms tripped off the tongue too readily, and one could easily be caught unawares by a new bandwagon rolling by. The succeeding outbursts of interest were, each one, also part of long-established and recurring themes which are linked to fundamentally different views of the nature of knowledge. They centred on well-established modes of thought about the nature of man. Social geography can be interpreted from standpoints on a whole range of ideas extending from the deterministic and the positivist on the one hand to the humanist on the other. One standpoint is no more true than another, but it is understandable that people identify themselves with a particular view which seems to them the best interpretation of the problems they deem important.

This may well be the strength and the weakness of social geography at the present time, when it is still trying to establish an identity. There is a very good reason why this essay did not begin with a definition of social geography: it is because only most diffidently can we begin to express a certain agreement on topics and aims, and even more hesitantly on methodology. There are many who believe that even this modest step is trying to capture a chimera.

A discipline—if we are looking for this—may be thought of as a group of related processes regulated by its own paradigm and based on its own theoretical structure. So one of the necessary conditions of identification of a discipline is a commonly accepted theoretical framework. The preceding paragraphs have referred to the theoretical constructs which underlie various approaches, but it would be mistaken to suppose that social geography in the last two decades has been theoretically oriented. Its strength so far has been in its empirical work, and theoretical implications have been almost incidental. It is a truism to quote Buttimer (1968) in saying that 'social geography is not an academic discipline, but a field created and cultivated by a number of scholars'. Leaving aside theory for a moment, we can certainly distinguish elements common to an extensive range of topics, common concerns and common aims which encourage us to think of the work as something other than a rag-bag of disparate studies. What is disconcerting is the range of methodology. Paradigm has replaced paradigm so rapidly that it is impossible to discern a consensus of opinion which will support one view of social geography. No one would dispute that it focuses on spatial aspects of social processes. I would like to add that the unit of study must be the social group and that space must be seen phenomenologically. This leads to a tentative definition: the understanding of patterns which arise from the use social groups make of space as they see it and of the processes involved in making and changing such patterns (Jones 1975). Eyles (1974) focuses much more sharply on the current concern with social issues, and sees social geography as the analysis of the social patterns and processes arising from the distribution of and access to scarce resources. In some ways the latter definition is subsumed in the former.

Arriving at a definition by no means resolves the difficulties, and even contradictions, implicit in the range of methodologies discussed above and in their theoretical implications. On the face of it, social geographers are torn between several approaches, and are unlikely to agree that one is right and the others wrong. But is it necessary that they should agree in this way? Fundamentally it has been assumed that we have chosen to deal with man in a scientific way. That, I take it, is the meaning of being a 'social science'— and the problem is that in so many ways man transcends such a method. The scientific paradigm is part and parcel of modern Western thought, and it was a relief to many that these basic precepts were applicable in geography. The quantitative approach certainly gave geographers an academic respectability by adopting the same rules as other sciences (Chisholm 1975). The inadequacy of this approach in explaining so many problems in social geography does not necessarily invalidate it in all respects. The common-sense world of classical physics did not fall apart with the discovery of indeterminacy in the atom. But for social geographers working at 'atomic' level indeterminacy may overshadow all else; individual decisions and group motivations may provide a much richer understanding than the numerical generalizations which seem to give adequate explanation at the macro-level. Kirk's two worlds do not constitute a split in reality, but reveal two aspects of the whole, a view that is echoed by many scholars grappling with this problem—e.g. Pierre Teilharde de Chardin's (1959, Chapter 3) view of the external and internal world which correspond with the quantitative and the qualitative. It may well be that social geographers, in spite of emphasizing disaggregated society, have given too little consideration to the qualitative.

Gregory (1978) suggests that we might make Kirk's distinction, which is ontological, into an epistemological one, and thus arrives at a distinction between 'understanding' in the human sciences and 'explanation' in the natural; though I would disagree that this means two geographies, because they are facets of one reality. In the definition of social geography I gave above, I deliberately used the word 'understanding' and it corresponds to Gregory's use of the word. I am totally in sympathy with an approach which would be essentially hermeneutic—i.e. emphasizing interpretation and empathy. Taken to its logical conclusion such a view would be unwilling to countenance a theoretical structure at all. But Gregory makes it plain that this is not a retreat into the 'soft' geography of the past 'hermeneutics regards theories which are (partially) external to the life-world under investigation, as legitimate prejudices which furnish the necessary condition for any understanding.' Understanding depends on the 'reciprocation between two frames of reference rather than the replacement of one by the other'. This closely parallels Buttimer's (1974) demand that there should be an 'interplay of *inside* and *outside* views on situations'.

But this is merely delving deeper into one person's interpretation of social geography and to the philosophy behind one's view of man and the universe. What our topic has revealed over the last two decades is a series of interpretations which range from the 'hard' scientific to the existential: they are neither sequential nor do they cancel one another out. The strands in each mode of thought have long antecedents, and all will continue to enrich the subject in the future. What will bind the studies we call social geography together are their focus on the social group as a unit of study, and their common concern with the spatial implications of social processes.

# REFERENCES

BARROWS, H. H. 1923. Geography as human ecology. *Annals AAG.* 13.
BOAL, F. 1972. The urban residential out-community, a conflict interpretation. *Area,* 4.
BOWEN, E. G. 1959. Le Pays de Galles. *Trans. Inst. Brit. Geogr.* 26.
BUTTIMER, A. 1968. Social geography, in Sills, D. (ed.), *International Encyclopaedia of the Social Sciences,* 6. New York.
BUTTIMER, A. 1969. Social space in interdisciplinary perspective. *Geog. Rev.* 59.
BUTTIMER, A. 1978. Values in geography. *Ass. American Geogs. Commission on College Geography,* Resource Paper No. 24.
BUTTIMER, A. 1978. Charism and context: the challenge of *la géographie humaine,* Chapter 4 in Ley, D. and Samuels, M. (ed.), *Humanistic Geography.* Chicago.
DE CHARDIN, P. T. 1959. *The Phenomenon of Man.* London.
CHISHOLM, M. 1975. *Human Geography: Evolution or Revolution.* London.
DAVIES, W. K. D. 1972. *The Conceptual Revolution in Geography,* Part IV. London.
DICKINSON, R. E. 1957. *The West European City.* London.
DOHERTY, J. 1969. The distribution of immigrants in London. *Race Today,* 1.
DUNBAR, G. S. 1978. Some early occurrences of the term 'social geography'. *Scot. Geog. Mag.* 94.
EVANS, E. E. 1942. *Irish Heritage.* Dundalk.
EVANS, E. E. 1951. *Mourne Country.* Dundalk.
EYLES, J. D. 1974. Social theory and social geography. *Progress in Geography,* 6.
EYLES, J. D. 1977. After the relevance debate: the teaching of social geography. *Journal of Geog. in Higher Education,* 1.
FIREY, W. 1947. *Land Use in Central Boston.* Cambridge, Mass.
FITZGERALD, W. 1946. Geography and its components. *Geog. Journal,* 107.
GILBERT, E. W. and STEEL, R. W. 1945. Social geography and its place in colonial studies. *Geog. Journal,* 106.
GLASS, R. 1948. *The Social Background of a Plan: a Study of Middlesborough.* London.
GREGORY, D. 1978. *Ideology, Science and Human Geography.* London. pp. 59 ff.
HAJDU, J. G. 1968. Towards a definition of post-war German social geography. *Annals AAG* 58.
HARVEY, D. 1973. *Social Justice and the City.* London.
HERBERT, D. T. 1979. Introduction to Herbert, D. T. & Smith, D. M. (ed.), *Social Problems and the City.* Oxford.
HOKE, G. W. 1907. The study of social geography. *Geog. Journal,* 29.
JAMES, P. E. 1972. *All Possible Worlds.* Indianapolis.
JENKINS, D., JONES, E., JONES HUGHES, T., and OWEN, T. M. 1960. *Welsh Rural Communities.* Cardiff.
JONES, E. 1956. Cause and effect in human geography. *Annals AAG* 46.
JONES, E. 1960. *A Social Geography of Belfast.* Oxford.
JONES, E. 1975. Introduction, in Jones, E. (ed.), *Readings in Social Geography.* Oxford.
JONES, E. and EYLES, J. D. 1977. *An Introduction to Social Geography.* Oxford.
KIRK, W. 1951. Historical geography and the concept of the behavioural environment. *Indian Geog. Journal Silver Jubilee Vol.*
KIRK, W. 1963. Problems of geography. *Geography,* 48.
KNOX, P. 1975. *Social Well-being: a Spatial Perspective.* Oxford.

LEY, D. 1978.   'Social geography and social action'. Chapter 3 in Ley, D. and Sam-
muels, M. (ed.), *Humanistic Geography: Prospect and Problems*, 1978.
MOGEY, J. M. 1956.   *Family and Neighbourhood: Two Studies in Oxford*. Oxford.
MOSER, C. A. and SCOTT, W. 1961.   *British Towns: a Statistical Study of their Social
and Economic Differences*. Edinburgh and London.
PAHL, R. 1967.   Sociological models in geography. In Chorley, R. J. and Haggett,
P., *Models in Geography*. London.
PICKVANCE, C. G. (ed.) 1976.   *Urban Sociology: Critical Essays*. London.
REES, A. D. 1951.   *Life in a Welsh Countryside*. Cardiff.
ROBSON, T. R. 1975.   *Urban Social Areas*. London.
ROXBY, P. M. 1930.   The scope and aims of human geography. *Scot. Geog. Mag.*
46.
TIMMS, D. W. G. 1971.   *The Urban Mosaic*. Cambridge.
TUAN, Y. F. 1974.   *Topophilia*. Englewood Cliffs.
WATSON, J. W. 1957.   The sociological aspects of geography. Chapter 20 in Taylor,
G. (ed.), *Geography in the Twentieth Century* (3rd edn.). New York.
WILLIAMS, W. M. 1956.   *The Sociology of an English Village: Gosforth*. London.
WOLPERT, J. 1965. Behavioural aspects of the decision to migrate. *Papers Reg. Studies
Ass.* 15.
WRIGHT, J. K. 1947.   Terrae incognitae: the place of imagination in geography.
*Annals AAG* 37.
WRIGHT, J. K. 1966.   *Human Nature in Geography*. Cambridge, Mass.

# 12 The Geography of Leisure and Recreation

J. T. COPPOCK   *University of Edinburgh*

OVER the past fifteen years there has been a major expansion of interest among geographers in how people, especially in developed countries, spend their leisure time. In this chapter, the emergence of this interest, its changing character, and future prospects in this field are considered, primarily in a British context and with only passing reference to similar developments in other countries, notably those in North America, which have often anticipated those in Europe by a decade or more, but where the size of the continent and its wealth of recreational land have given leisure and recreation a somewhat different character. This appraisal will also focus primarily on the work of geographers writing about the United Kingdom, although much significant research has been undertaken in North America and such distinctions are not always readily made in a field that involves many disciplines (Patmore 1973b). The field is also unusual in its degree of dependence on the needs (and finance) of central and local government, which has greatly influenced the work undertaken.

Much of this chapter will concern outdoor recreation in the countryside, not because this is necessarily the most important aspect of leisure—though some 10 million people are estimated to visit the countryside of England and Wales on a summer Sunday (Countryside Commission, 1979, Personal Communication)—but because it has been received the greatest attention, not only from geographers but also from most researchers in this field. This emphasis is still marked, but the past decade has seen an appreciable shift towards a wider interpretation of recreation, as well as from description towards explanation, and a changing view of leisure, from simply residual time once the essential disciplines of work, sleep, and other basic needs have been met (Countryside Recreation Research Advisory Group 1970) to an opportunity for personal development and fulfilment, from a period of time to a process (Kaplan 1975). Geographers have not been in the forefront of these later developments and their future role may well depend on their recognition of these trends.

Leisure, whether regarded as a residuum or an opportunity, provides the context for both recreation, which is the positive use of leisure time for a wide variety of pursuits, and tourism, which is that part of leisure involving a stay of one or more nights away from home. By far the greater part of leisure time and much of that devoted to recreation, is spent in the home, but the emphasis here, as in most studies of recreation, will be on pursuits outside the home. Nor will a sharp distinction be made between tourists and day recreationists since, apart from the needs of the former for accommodation, it is less meaningful in the United Kingdom than in much larger countries.

### GROWING INTEREST IN THE GEOGRAPHY OF LEISURE

This growing interest in the geography of leisure is clearly shown by a perusal of geographical writing over the past fifty years. Before 1965 there were no textbooks or other general surveys in this field, few articles had appeared, and there were only scattered references, and then generally to tourism, in either systematic studies of human geography or in regional descriptions, though it is interesting to note Demangeon's comment of 1927 that 'British civilization wears an urban face even in its recreation' (1949, p. 380–1). In contrast, six books on various aspects of the geography of recreation and tourism (Patmore 1970; Lavery 1971; Cosgrove and Jackson 1972; Simmons 1975; Coppock and Duffield 1975; and Robinson 1976) have appeared since then, as well as increasing numbers of articles, and, more especially, mimeographed reports. Even this rapid growth in the volume of published work probably understates the rise in interest. The appearance of textbooks is explicit recognition of the increasing amount of teaching in the geography of recreation and tourism and the former is a popular topic for both undergraduate dissertations and postgraduate theses.

This growth of interest is not confined to geography. Similar trends have occurred in ecology, economics, landscape architecture, planning, psychology, and sociology, though the level of commitment varies. There has also been a tendency for leisure, recreation, and tourism to emerge as fields of study in their own right, to which many disciplines make valuable contributions. Not surprisingly, the forum of interest for many active researchers in the leisure field lies in newly formed, multidisciplinary bodies such as the Landscape Research Group, the Leisure Studies Association, and the Tourism Society.

The scale of this growth, from small beginnings, should not be exaggerated. Few British geographers would identify themselves as specialists in the geography of recreation or tourism, only a few articles have appeared in the principal geographical journals, and no established forum exists analogous to the Committee on Recreation, Tourism, and Sport of the Association of American Geographers, or the Working Group on Recreation and Tourism of the International Geographical Union. Nevertheless, it can be fairly claimed that academic geographers and those trained in geography and now practising as planners or in other professional roles have played a leading part in developments over the past fifteen years, a fact demonstrated by the selection of two geographers, Patmore and Rodgers, as specialist advisers to the Select Committee on Sport and Leisure (1973). This widespread interest, both among geographers and elsewhere, has been stimulated by a growing awareness of the significance, actual and potential, of what Dower (1965) has called the Fourth Wave, the massive increase in personal leisure and mobility that has occurred since the Second World War as a consequence of rising standards of living, increased discretionary spending, shorter working hours, longer holidays, and greater car ownership. Leisure is now estimated to account for over a fifth of the gross national product (Martin and Mason 1979) and tourism is now regarded as a major contributor to the balance of payments. Before the oil crisis of 1973 the Director of the Countryside

Commission expressed the view that outdoor recreation was growing at a compound rate of some ten to fifteen per cent per annum (Hookway 1973) and many other examples could be cited of similar growth rates in membership of recreational bodies and in participation in individual leisure pursuits. It is true that there has since been some check, though the causes and the actual course of events are complex, and one of the uncertainties about the future is whether such growth rates will return. Whether they do or not, there is widespread agreement that leisure will play an increasingly important part in the lives of those living in developed countries.

The growth of interest among geographers can broadly be divided into three phases. In the 1950s and early 1960s, public and professional interest largely centred on the countryside and only a few senior geographers were involved, and then primarily for their expertise in related fields. From the mid-1960s to the early 1970s increasing numbers of geographers, mainly with interests in the use of resources, undertook research, chiefly of a fact-finding kind in the field of rural recreation. Since the mid-1970s the range of approaches to countryside recreation has widened and the geographer's contribution diminished (at least in relative terms), and questions are being increasingly asked about motivation, behaviour, and prediction. These trends reflect not only a growing interest within the profession as the scale and significance of the leisure explosion came to be appreciated, but also the increasingly involvement of other professions, a shift in government policy as the need for a wider view was accepted, and, less certainly (since good time series are lacking), changes in leisure patterns.

## ANTECEDENTS

Before the Second World War the little research by geographers into leisure was mainly into two topics, one of which, the visual amenity of the countryside, has since become a major research interest and the other, the development of resort towns, has made little progress. Indeed, Gilbert's (1939) paper on the growth of resorts is perhaps seen more appropriately as a contribution to social historical geography, as can his later monograph on Brighton (1954). Cornish's interests in countryside recreation and the preservation of scenery were but one of many. He wrote extensively, including a pamphlet on National Parks (1930), a book on English scenery (1932), and articles on scenery (1934) and the preservation of cliffs (1935), and he was active both in the conservation movement and in promoting public enjoyment of the countryside (Goudie 1974). Somewhat surprisingly, in view of his recognition of recreation as one of the five basic needs that land should meet, Stamp (1948) did not include such a category in the first Land Utilisation Survey; it is interesting to note that, at about the same time, geographers in Michigan contributed at least three papers on the recreational use of land (Morrison 1964). Yet, with hindsight, it can hardly be said that the age of mass leisure in the countryside had arrived.

## ACCESS TO THE COUNTRYSIDE

Although there had long been a concern with the dual, and sometimes opposing, aims of protecting the countryside and promoting public enjoyment of it, the matter came to a head during the Second World War, culminating in 1949 in the enactment of the National Parks and Access to the Countryside Act. This established the first of the statutory agencies that were to influence the direction of research, the National Parks Commission, and although its emergence has been documented by geographers (Cherry 1975; Sheail 1975), those in the profession undertook little actual research. Two leading geographers, Darby and Steers, with expertise in the making of human and physical landscapes respectively, served on the Commission and Darby (1961, 1963) published accounts of its work; it is also interesting to note that Steers' (1944) appraisal of the conservation value of the coast subsequently formed a basis for the Countryside Commission's identification of Heritage Coasts (Patmore 1973b). Darby also served on the Water Resources Board and wrote a similar article on the recreational and amenity use of water (1967). Other discussions of recreation appeared almost incidentally. For example, Best assessed the extent of open space in his estimates of the composition of the urban area (Best and Coppock 1962) and Coppock, in the context of land-use studies in the Chilterns, attempted to find a logical basis for identifying an Area of Outstanding Natural Beauty (1959). Articles on tourism or resorts (e.g. House 1958) appeared from time to time, but there was no clearly identifiable interest in research into recreation and leisure.

INVENTORIES AND SURVEYS

During the 1950s and early 1960s living standards, levels of car ownership, and participation in leisure pursuits were growing steadily, but so was a concern with recreational pressures on the countryside, as the reports on the three Countryside in 1970 Conferences reveal. The recreationists who had been expected to visit National Parks on foot and bicycle came increasingly in cars and existing machinery was seen to be inadequate. Following a White Paper on *Leisure in the Countryside* (Cmnd. 2928), the National Parks Commission was replaced by a Countryside Commission in 1968 and Scotland, which had not appeared to need legislation in 1949, was given its own Countryside Commission in 1967. Tourism was also being increasingly recognized as potentially an important force in the poorer rural areas, notably by the newly formed Highlands and Islands Development Board (1966), and statutory tourist boards were created in 1969. The other leisure agencies, the Arts Council, formed in 1948, and the Sports Council, established in 1965, had little significance for recreation research at this stage, though the Sports Council was to become a major force in the 1970s.

The period of mass leisure had clearly arrived and increasing numbers of geographers in the 1960s were beginning to appreciate the importance of recreation and tourism (Rodgers 1969b; Mercer 1970). At the same time, the new agencies of central government, some local authorities, and, from 1965, the research councils provided potential sources of research funds. For those who looked across the Atlantic, there was the publication in 1962, after four years of work, of the report of the Outdoor Recreation Resources Review Commission, with its series of 27 special studies. Some of this interest was expressed in pure research, but increasingly, a large part was funded by public bodies that needed information in order to fulfil their duties or to formulate policies. The main themes of such commissioned work were the use of rural resources for recreation and the character of participation in such recreation: in short, supply and demand. In view of the long-established involvement of geographers in rural land use, it is not surprising that one strand of research should have been recreation as a use of land. Thomas (1970) made the first direct measurement of recreational land from aerial photographs and an attempt was made by Coppock in 1966 to assemble what was known about the recreational use of land and water in Great Britain. A more comprehensive review of outdoor recreation in England and Wales was undertaken between 1967 and 1969 by Patmore (1970), whose account remains the basic document for any consideration of recreation in the countryside. Much of the research and available data concern the public use of land, but a survey of central Scotland recorded the scale of recreational use of private land (Duffield and Owen 1970, 1971), an approach developed further in a sample of Highland estates by Millman (1971).

Three resources in particular attracted attention, forests, water (both coastal and inland), and scenery, in part at least because of the increasing involvement of statutory agencies in the leisure field. Attention was also focused on a number of problem areas, notably the National Parks (e.g. Blacksell 1971; Gittens, 1973). The Forestry Commission has long allowed public recreational use of its forests where this was compatible with other interests, but in the 1960s it began to play a more active role and commissioned Goodall and Whittow (1975) to examine the requirements of different kinds of recreation in a forest environment. There is also a long tradition of private recreational use of private woodland and, following a government assessment of the recreational value of both private and public woodlands, Duffield and Owen (1973) made a critical evaluation of the estimates. Geographical appraisals of the use of water for recreation have largely been the work of one man, Tanner, who has undertaken work for the Sports Council on recreational use of coasts (1967) and of coastal and inland waters (1973). For several years he was seconded as research officer to the Water Space Amenity Commission which subsequently published four reports he has produced (Tanner 1974).

Land and water came together in the third major topic, the assessment of scenery, which was seen as a significant resource for informal recreation, the principal recreational use of the countryside. Much of this interest, too, was stimulated by the statutory protection given to areas of beautiful scenery. In response to the Countryside (Scotland) Act, 1967, Linton (1968) made a single-handed appraisal of Scottish scenery, basing his assessment on his own deep knowledge and on map evidence. His methodology was adapted by Duffield and Owen (1970, 1971) in their studies of outdoor recreation in central Scotland, in which they used a square grid to facilitiate comparison with other data, a practice that subsequently made it easy to incorporate this assessment in a computer-based information system for recreational planning (PDMS 1979; Duffield and Coppock 1975).

The need for some systematic appraisal was clearly demonstrated by the very varied interpretations of the value of landscapes adopted by adjacent local authorities (Coppock 1968) and a wide variety of techniques of scenic assessment has since been devised (Penning-Rowsell 1973). Attempts have also been made to compare the assessments from different techniques (Crofts and Cooke 1973; Blacksell and Gilg 1975), and by 1975 sufficient interest had been aroused among geographers to warrant the holding of a symposium on scenic assessment (see, e.g. Appleton 1975a). It is interesting to note, in the light of all these developments, the Countryside Commission for Scotland's decision to base its assessment of Scottish scenery on judgements in the field (1978).

The second major component of recreational research in this period comprised surveys of users. As in many emerging fields, there is a lack of good data, and although not all surveys to establish the basic facts have been undertaken by geographers, they have been prominent in several important investigations. Rodgers (1967a, 1969a), in conjunction with the British Travel Association, played a leading role in the first national survey, which has provided a bench-mark for many subsequent regional and local surveys (Rodgers 1974). He and Patmore (1972) were also leading figures in the study of leisure in North West England, as were other geographers in the Scottish Tourism and Recreation Planning Study, Tourism and Recreation Research Unit, 1976a, and in the Study of Informal Recreation in South East England (Davidson and Sienkiecwicz 1975; Law and Perry 1971). Geographers have also been associated with many local surveys (e.g. Colenutt and Sidaway 1973).

Such surveys have often provided an input to recreational plans by local authorities, a trend that became more marked in the 1970s as the new local authorities became more aware of the significance of recreation (Palmer and Bradley 1971). Comprehensive appraisals of both demand and supply were undertaken in Lanarkshire and the Edinburgh area by Duffield and Owen (1970, 1971), and the recommendations in the resulting reports, described by the Countryside Commission for Scotland (1972) as models of what was required, provided a significant input to local plans. Because of the large contribution of those trained in geography to the planning profession, it is not easy to isolate geographical contributions, but a later study of the Chichester area (Tourism and Recreation Research Unit 1977d) was specifically designed as a guide to local authorities. Planning in a broader sense was not confined to local authorities (Coppock 1973) and in Scotland in particular a consortium of national agencies funded not only major surveys but also the development of a geographical information system which is being used for both planning and policy-making (Tourism and Recreation Research Unit 1974; PDMS 1979).

Although surveys and censuses dominated much of the work by geographers during this period, there was increasing interest in three related topics—the assessment of potential resources, the measurement of impact, and the analysis of recreational travel.

Much of the work on the physical impact of recreation has been concerned with vegetation and has been undertaken by ecologists (Speight 1973), but geographers have also made contributions (Goldsmith and Munton 1971; Goldsmith, Munton, and Warren 1970) and there was sufficient interest among active researchers to warrant the holding of a joint symposium with ecologists on the impact of recreation at the annual conference of the Institute of British Geographers in 1975. Attention has also been given to Scottish beaches (Mather and Ritchie 1978) and loch shores (Tivy, in press) in projects funded by the Countryside Commission for Scotland.

Interest has also been shown in the economic and social impacts of recreation, though studies of the former have primarily been undertaken by economists. A geographical influence is none the less evident in a study of the economic influence of tourism in Tayside (covering both day visitors and those staying overnight) in which local multipliers were constructed which could be applied to different types of rural communities (Tourism and Recreation Research Unit 1975; Coppock and Duffield 1975). Social impact is a relatively neglected field, but attempts have been made to ascertain the views of samples of residents about visitors to the countryside, whether these be tourists (Tourism and Recreation Research Unit 1977a) or owners of second homes (Coppock 1977), and more detailed studies have been undertaken in Skye (Brougham and Butler 1976) and Speyside (Mackinlay 1969, Getz 1977). A related, though rather different, impact has been that of migrant workers on provision for recreation for the local community; a study of workers in oil-related industry in the Highlands has highlighted how recreationally deprived some communities in sparsely populated areas are by comparison with those in towns and cities (Tourism and Recreation Research Unit 1976b; Coppock and Duffield 1979).

Closely associated with the concept of impact has been that of capacity, on which Tivy (1972) has undertaken a review of North American experience for the Countryside Commission for Scotland. Much of the work on this topic has been concerned with ecological capacity (Barkham 1973), but Burton (1974), in a pioneer study, has examined the concept of perceptual capacity and demonstrated both the wide range of perceptions among individuals and the unevenness of the distribution of recreationists within a site (a finding which is confirmed by many other studies). Where resources are limited or existing practices well established, such different perceptions may be a source of conflict, whether with other land users or with other recreationists, a topic that has been investigated mainly in the context of water-based recreation, for example, on the Broads (Owens 1978) and on Loch Lomond (Brown and Chapman 1974). Such studies of capacity have all been concerned with individual sites or localities. Regional capacities are much harder to define, except on Cracknell's (1971) criterion of road capacity, and presuppose some prior definition of purpose (Rodgers 1967).

The study of recreational use frequently overlaps that of potential, and the studies noted earlier by Goodall and Whittow and by Tanner include assessments of potential use; Johnstone and Tivy (in press) have also identified the potential of loch shores. In the main, such studies have applied empirical standards to known characteristics of land and water, an approach developed in Scotland (Duffield and Owen 1970, 1971) as a contribution to the identification of 'recreation environments', viz., areas with a similar capacity to support outdoor recreation on a continuing basis; these were based on criteria, agreed with experts, that could be recognized from topographic maps. A much more comprehensive evaluation of site requirements has been undertaken by Hockin, Goodall, and Whittow (1977), but nothing comparable to the nation-wide evaluation of capability undertaken for the Canada Land Inventory (1978) has yet been attempted in this country.

In view of the concern of geographers with spatial relationships it is not surprising that those interested in recreation should have paid particular attention to recreational travel. One of the characteristics of recreation, most markedly developed in outdoor recreation in the countryside, is the separation of residence (whether permanent or temporary) and recreation area; except for leisure pursuits undertaken at home, all recreational activity involves a journey to play. An increasing number of studies show that, for many journeys, the nature of the country that is traversed is important and those simply 'driving around' appear to have a certain minimum threshold distance in mind. In studies in central Scotland a secondary peak of journeys was recorded between 20 and 40 miles (32 and 64 km) (Duffield and Owen 1970, 1971; Duffield 1975), and Wall (1972) found similar evidence in his study of day trips from Hull. In Colenutt's survey of the Forest of Dean (1969, 1970) only a minority of visitors from major centres took the shortest route on either the outward or the return journeys (less than half of which were common). Such findings have obvious implications for the location of recreational facilities in the countryside.

Apart from such empirical observations, advances have also been made in techniques of estimating volumes of recreational traffic, both by observation and by modelling traffic flows. Owen and Duffield (1971) used traffic counters and linked cameras to provide estimates of the movement of touring caravans throughout Scotland, and Duffield (1976; Tourism and Recreation Research Unit 1977b) has developed a traffic-mix model that gives estimates of recreational traffic from counts of all traffic and limited field survey. These and other methods have been used to map the flows of such traffic (e.g. Duffield and Owen 1970, 1971) to identify the main tourist and recreational routes. Attention has also been directed at new or potential routeways, such as the railway lines abandoned during the rationalization of the railway system (Appleton 1970).

Although the research councils have funded research on many of these topics, by far the greater part of the work has been commissioned by public agencies. This source of funding has had two main consequences. Because of the difficulties faced by university teaching staff in completing such research to the tight timetables set by public agencies, and sometimes the scale of resources required, there has been a tendency for much of this research to be undertaken by units employing full-time researchers, both in the universities and elsewhere; although such teams are all multidisciplinary, the major direction in one has been geographical, a fact evident in the range of research undertaken and in the approaches adopted (Coppock 1975). Secondly, the topics studied are determined by the needs of the agencies, which are generally pragmatic and often short term. It is true that there is frequently a dialogue between sponsor and researcher before the final shape of the project is determined, so that it has been possible to modify projects in ways that bring academic benefits at no extra cost, and that the resulting data provide a good basis for further research in a field where reliable data have been scarce; but the needs have largely been factual and there has been little opportunity to develop more fundamental research, in particular about recreational behaviour (Rodgers 1972)).

## WIDENING HORIZONS AND THE SEARCH FOR EXPLANATION

This situation has been widely recognized and regretted by those working in the field and by some of the funding agencies. A joint Social Science Research Council/Sports Council working party on research needs (Patmore 1979) and a series of reviews of the state of the art commissioned subsequently, (e.g. Elson 1979) are now providing a much more comprehensive evaluation of what has been done and what research is needed. But although the Human Geography Committee of SSRC has taken the initiative within that council, the academic composition of these groups and of those undertaking the state of the art reviews is multidisciplinary. As the emphasis has swung from survey to explanation, so the contribution of other disciplines, notably sociology and social psychology, has increased in importance and the geographical focus of research has shifted from its almost exclusively rural orientation.

Although local authorities have long had powers to make provision for recreation in urban areas, through the construction of art galleries, museums, parks, swimming pools, and the like, research in these topics has been limited and confined largely to open space (Balmer 1971, 1973), though Thorpe's (1975) advocacy of the role of allotments as leisure gardens may also be included under this head. A growing appreciation by government, notably in the 1975 White Paper *Sport and Recreation* (Cmnd. 6200), that leisure extends beyond countryside recreation, the stimulus provided to local government by the restructuring of local government and the formation of comprehensive departments of recreation and leisure (Travis and Hudson 1978), and the increasingly important role of the Sports Council have stimulated an academic interest in leisure in these areas, and within geography, the skills of social geographers are being increasingly added to those of resource geographers (Long and Wimbush 1979).

Even within resource-based recreation there has been a shift of emphasis, reflecting in part public concern, although it has long been recognized that a rural/urban dichotomy is inappropriate in the study of recreation in highly urbanized communities (Rodgers 1969b). The view of recreational systems as a rural/urban continuum is explicitly recognized by the Countryside Commission for Scotland (1974) in its proposals for a park system, and a growing appreciation both of the implications of the energy crisis and the needs of that substantial minority of the population that lacks access to a private car have combined to focus attention on the urban fringe.

The importance of accessibility was recognized by Stamp in 1948 and Cracknell (1967), writing in the context of an expected massive increase in population, examined the significance of road capacity in defining the recreational zone around a city. Bowen (1974) and Davidson (1976) have both provided general overviews of recreation in the urban fringe, and this work has been complemented by more detailed studies of land for informal recreation in London's Green Belt (Ferguson and Munton 1978) and of parks on the urban fringe of Glasgow (Tourism and Recreation Research Unit 1978c; forthcoming)).

The widespread appreciation of the need for a conceptual framework has already been noted in respect of recreational trips (see also Elson 1976), and Appleton (1975a) and Lowenthal (1978) have similarly criticized the lack of any sound basis for the evaluation of scenery, a deficiency which the former has attempted to remedy (1975b). Models have also been devised for recreational planning (Perry 1973) and for informal recreational trips, and attempts have been made to apply locational optimizing models to the location of urban recreational facilities (Cargill and Hodgart 1978; Hodgart 1978). Statistical models have also been used to identify substitutability between recreational pursuits (Tourism and Recreation Research Unit 1977c). Many of these models are mechanistic, and there is widespread recognition of the need for models that take account of recreational behaviour.

Geographers have made some contributions to this widening horizon (and it should be stressed that many of the approaches discussed previously in this chapter could be matched by similar studies by those in other disciplines); but, perhaps through momentum, the main emphasis in geographical research continues to be in rural areas. In Patmore's (1978) view there is little that can be distinctively recognized as recreational geography, and he regards the field as neither inter- nor multi-disciplinary but extra-disciplinary, though he recognizes that there is a geography of recreation and a distinctive geographical contribution (1973a). Similarly, Rodgers (1973) saw a clear role for geographers in analysing spatial dimensions of demand.

Urban recreation is not the only aspect of leisure to be neglected by British geographers. It is true that internal movements of tourists have generally been included in surveys of recreational travel and that aspects of tourism have been investigated from time to time (e.g. Price 1967; Patmore 1968), but little attention has been paid to the analysis of international tourist flows. Geographical aspects of sport have also largely been ignored by academic geographers, except in so far as they concern publicly provided facilities in the countryside. Apart from the impact of informal recreation on private land in the countryside, private provision for leisure has similarly been neglected, although this is also true of a large part of research in the leisure field, irrespective of discipline.

PAST AND FUTURE

The emphasis on resource-based recreation in the countryside in the expansion of geographical interest in leisure is understandable; not only was this the principal concern of public bodies but such recreation, needing access to large areas, often having specific site requirements and involving long journeys to play, also required traditional geographical skills. The increasing need in leisure research to understand motivation and satisfaction and, from the viewpoint of public policy, to see leisure as a process which can be used to enhance the quality of life, requires different skills and has led to a much more active involvement by researchers in other disciplines. None the less, skills and concepts that have been developed in other aspects of geography during the past twenty years are equally appropriate to this wider view of leisure and to an analysis on its spatial manifestations in urban areas, not least in respect of the perception of opportunities (Mercer 1971).

The decision by the Social Science Research Council and the Sports Council to invest £80 000 a year for five years in support of research into leisure provides an opportunity to remedy some of the deficiencies (Patmore 1979). Priorities will be influenced both by the working party's report and by the state of the art reviews, of which more than twenty have been commissioned. While there will be a continuing need for information, not least to provide good time series, this is a task for public agencies rather than for individual researchers. For the latter, a dearth of information has been replaced by a glut (Rodgers 1973), even if its quality is uneven, and the future is likely to see further analysis of existing data and much more limited and selective surveys intended to remedy particular deficiencies and to test hypotheses.

There can be little doubt that the main thrust in future research will be conceptual and theoretical; for without an adequate conceptual framework little progress will be made beyond an ever increasing number of case studies.

It also seems likely that leisure in an urban setting will receive much more attention than in the past. Possibly, as Patmore (1979) has suggested, the massive National Urban Recreation Study (1978) in the United States, with its thirteen technical reports and its analyses of seventeen sample cities, will have the same kind of influence on leisure research in the next decade that its predecessor, the report of the Outdoor Recreation Resources Review Commission (1962), had sixteen years earlier.

What seems certain is that leisure time will play an increasingly important place in people's lives. Already the prospect of the micro-processor revolution has led trade unions leaders to press for shorter working weeks, longer holidays, and earlier retirement as ways of spreading a diminishing volume of work (Jenkins and Sherman 1979). At the same time, the increasing cost and possible shortage of petroleum may well affect the recreational patterns that have emerged in a period of cheap energy; alternatively, the decline of employment in manufacturing industry and the increasing importance and ease of transmitting information may increasingly permit people to live where they want to live rather than near their place of work (Berry 1970). Although the emphasis in this review has necessarily been placed on research, geography can also contribute both to the training of professional staff in the leisure field and to the better appreciation of leisure opportunities by the public at large. More than thirty years ago Darby (1946) asserted that his training should enable the geographer to take his country walks—and maybe his town walks—with understanding, and such an awareness can enhance the enjoyment of both town and country. In both teaching and research geographers thus have an opportunity, along with other disciplines, to contribute to the better understanding of leisure processes that is the foundation both of greater enjoyment and of sound policies. 'Only when spatial perspectives are as finely tuned as physical and financial will recreational provision make the contribution it could, and should, to the quality of human life' (Patmore 1975, p. 35).

# REFERENCES

APPLETON, J. H. 1970. *Disused Railways in the Countryside of England and Wales.* Report to the Countryside Commission, London.

APPLETON, J. H. 1975a. Landscape evaluation: the theoretical vacuum. *Trans. Inst. Br. Geographers*, 66: 120–3.

APPLETON, J. H. 1975b. *The Experience of Landscape.* Chichester.

BALMER, K. 1971. Urban open space and outdoor recreation, in Lavery, P. (ed.), *Recreational Geography.* Newton Abbot.

BALMER, K. n.d. *Use of Urban Space in Liverpool.* Liverpool.

BARKHAM, J. P. 1973. Recreational carrying capacity: a problem of perception. *Area*, 8: 218–22.

BERRY, B. J. L. 1970. The geography of the United States in the year 2000. *Trans. Inst. Br. Geographers*, 51: 21–53.

BEST, R. H. and COPPOCK, J. T. 1962. *The Changing Use of Land in Britain.* London.

BLACKSELL, M. 1971. Recreation and land use—a study in the Dartmoor National Park, in Gregory, K. J., and Ravenhill, W. (eds.), *Exeter Essays in Geography.*

BLACKSELL, M. and GILG, A. W. 1975. Landscape evaluation in practice—the case of south-east Devon. *Trans. Inst. Br. Geographers*, 66: 135–40.

BOWEN, M. J. 1974. Outdoor recreation around large cities, in Johnson J. H. (ed.), *Suburban Growth: Geographical Processes at the Edge of the Western City.* London, 225–48.

BROUGHAM, J. E. and BUTLER, R. W. 1976. *The Social and Economic Impact of Tourism. A Case Study of Sleat, Isle of Skye.* Scottish Tourist Board, Edinburgh.

BROWN, R. and CHAPMAN, V. 1974. *Loch Lomond Recreation Report.* Countryside Commission for Scotland, Occ. Pap. No 5, Battleby.

BURTON, R. J. C. 1974. *The Recreational Carrying Capacity of the Countryside.* Keele Univ. Library. Occ. Pub. No. 11, Keele.

CANADA LAND INVENTORY 1978. *Land Capability for Recreation Summary Report.* Rep. No. 14, Environment Canada, Ottawa.

CARGILL, S. and HODGART, R. L. 1977. *A Strategy for the Provision of Swimming Pools in the Lothian Region.* Dept. of Geography, Univ. Edinburgh.

CHERRY, G. C. 1975. *National Parks and Recreation in the Countryside.* Environmental Planning, vol. ii. HMSO, London.

COLENUTT, R. J. 1969. Modelling traditional patterns of day visitors to the countryside. *Area*, 2: 43–7.

COLENUTT, R. J. 1970. An Investigation into the Factors Affecting the Pattern of Trip Generation and Route of Choice of Day visitors to the Countryside. Unpublished Ph.D. thesis. Univ. Bristol.

COLENUTT, R. J. and SIDAWAY, R. M. 1973. *Forest of Dean Day Visitor Survey.* For. Comm. Bull. No. 46, HMSO, London.

COPPOCK, J. T. 1959. The Chilterns as an Area of Outstanding Natural Beauty. *J. Tn. Plann. Inst.*, 45: 137–41.

COPPOCK, J. T. 1966. The recreational use of land and water in rural Britain. *Tijdschr. econ. soc. Geogr.*, 57: 81–96.

COPPOCK, J. T. 1968. The Countryside (Scotland) Act and the geographer. *Scott. Geog. Mag.* 84: 201–11.

COPPOCK, J. T. 1973. Planning for Outdoor Recreation in the Scottish Countryside, in Select Committee of the House of Lords on Sport and Leisure, *Second Report.* House of Lords Paper 193–III, HMSO, 541–6.

COPPOCK, J. T. 1974. Leisure in Scotland: a synoptic view, in Appleton, I. (ed.), *Leisure Research and Policy*. Edinburgh, 233–43.

COPPOCK, J. T. 1975. Research and policy in outdoor recreation: the contribution of the Tourism and Recreation Research Unit, in Coppock, J. T., and Sewell, W. R. D. (ed.), *Spatial Dimensions of Public Policy*. Oxford, 104–23.

COPPOCK, J. T. (ed.) 1977. *Second Homes: Curse or Blessing*. London.

COPPOCK, J. T. and DUFFIELD, B. S. 1975a. *Outdoor Recreation: a Spatial Analysis*. London.

COPPOCK, J. T. and DUFFIELD, B. S. 1975b. The economic impact of tourism—a case study, in *Tourism as a Factor in National and Regional Development*, Occ. Paper 4, Dept of Geography, Trent Univ., Peterborough.

COPPOCK, J, T, and DUFFIELD, B. S. 1979. Provision for recreation in areas affected by oil-related industries in the Highlands and Islands of Scotland, in Sinnhuber, K., and Julg, F. (ed.), *Studies in the Geography of Tourism and Recreation*, vol. 2, Wiener Geographische Schriften, 53–4: 87–100.

CORNISH, V. 1930. *National Parks and the Heritage of Scenery*. London.

CORNISH, V. 1932. *The Scenery of England: A Study in Harmonious Groupings in Town and Country*. London.

CORNISH, V. 1934. The scenic amenity of Great Britain. *Geography*, 19: 192–202.

CORNISH, V. 1935. The clifflands of England and the preservation of their amenities, *Geography*, 3: 86, 503–11.

COSGROVE, I. and JACKSON, R. 1972. *The Geography of Recreation and Leisure*. London.

Countryside Commission for Scotland 1972. *Memorandum to the Select Committee on Scottish Affairs*. Minutes of Evidence, House of Commons Paper 511, vol. 3, HMSO, London.

Countryside Commission for Scotland 1974. *A Park System for Scotland*. Battleby.

Countryside Commission for Scotland 1978. *Scotland's Scenic Heritage*. Battleby.

Countryside Recreation Research Advisory Group 1970. *Countryside Recreation Glossary*. Countryside Commission, London.

CRACKNELL, B. 1967. Accessibility to the countryside. *Reg. Stud.* 1: 147–61.

CROFTS, R. S. and COOK, R. U. 1974. *Landscape Evaluation: a Comparison of Techniques*. Dept. of Geography, Univ. College, London, Occ. Paper No. 25.

DARBY, H. C. 1946. *The Theory and Practice of Geography*. University of Liverpool.

DARBY, H. C. 1961. National Parks in England and Wales, in Jarrett, H. (ed.), *Comparisons in Resource Management*. Washington.

DARBY, H. C. 1963. Britain's National Parks. *Adv. Sci.*, 20: 307–18.

DARBY, H. C. 1967. The recreational and amenity use of water. *J. Inst. Water Eng.*, 21: 225–31.

DAVIDSON, J. 1976. The urban fringe. *Recreation News*, 1: 2–7.

DAVIDSON, J. M. and SIENKIEWICZ, J. 1975. Study of informal recreation in South-East England, in Searle, G. A. C. (ed.), *Recreational Economics and Analysis*. London, 151–9.

DEMANGEON, A. 1949. *The British Isles*. London (translated and revised by E. D. Laborde).

DOWER, M. 1965. The fourth wave. *Arch. J.*, 141: 123–90.

DUFFIELD, B. S. 1975. The nature of recreational travel space, in Searle, G. A. C. (ed.), *Recreational Economics and Analysis*. London, 15–35.

DUFFIELD, B. S. 1976. *A Method of Estimating Recreational Traffic Flows*. Transport and Road Research Laboratory, Supplementary Rep. 247.

DUFFIELD, B. S. and COPPOCK, J. T. 1975. The delineation of recreational landscapes: the role of a computer-based information system. *Trans. Inst. Br. Geographers*, 66: 141–8.

DUFFIELD, B. S. and OWEN, M. L. 1970. *Leisure + Countryside = A Geographical Appraisal of Countryside Recreation in Lanarkshire.* Tourism and Recreation Research Unit, Univ. Edinburgh.

DUFFIELD, B. S. and OWEN, M. L. 1971. *Leisure + Countryside = A Geographical Appraisal of Countryside Recreation in the Edinburgh Area.* Tourism and Recreation Unit, Univ. Edinburgh.

DUFFIELD, B. S. and OWEN, M. L. 1973. Forestry policy—recreation and amenity considerations. *Scott. For.,* 27: 114–34.

ELSON, M. J. 1976. Activity spaces and recreational spatial behaviour. *Tn. Plan., 5 Rev.,* 47: 241–55.

ELSON, M. J. 1979. *Countryside Trip Making.* SSRC/Sports Council Joint Panel on Leisure and Recreation Research, London.

FERGUSON, M. J. and MUNTON, R. J. C. 1978. *Informal Recreation in the Urban Fringe: the Provision and Management of Sites in London's Green Belt.* Working Paper No 2, Dept. of Geography, Univ. College London.

GETZ, D. 1977. The impact of tourism on host populations. A research approach, in Duffield, B.S. (ed.), *Tourism, A Tool for Regional Development.* Leisure Studies Association Conference, Edinburgh, 9.1–9.13.

GILBERT, E. W. 1939. The growth of inland and seaside health resorts in England. *Scott Geogr. Mag.,* 55: 16–35.

GILBERT, E. W. 1954. *Brighton. Old Ocean's Bauble.* London.

GITTENS, J. W. 1973. Conservation and capacity. A case-study of the Snowdonia National Park. *Geogrl. J.,* 139: 482–6.

GOLDSMITH, F. B. and MUNTON, R. J. C. 1971. The ecological effect of recreation in Lavery, P. (ed.), *Recreational Geography.* Newton Abbot.

GOLDSMITH, F. B., MUNTON, R. J. C., and WARREN, A. W. 1970. The impact of recreation on the ecology and amnity of semi-natural areas. Methods of investigation used in the Isles of Scilly. *Biol. J. Linn. Soc.,* 2: 287–306.

GOODALL, B. and WHITTOW, J. B. 1975. *Recreation Requirements and Forest Opportunities.* Dept. of Geography, Univ. Reading, Geogr. Pap. No. 37.

GOUDIE, A. 1972. Vaughan Cornish: Geographer, with a bibliography of his published works. *Trans. Inst. Br. Geographers,* 55: 1–16.

HIGHLANDS AND ISLANDS DEVELOPMENT BOARD 1966. Annual Report, Inverness.

HOCKIN, R., GOODALL, B., and WHITTOW, J. 1977. *The Site Requirements and Planning of Outdoor Recreation Activities,* Univ. Reading, Geog. Pap. No. 54.

HODGART, R. L. 1978. Locating public services. *Progress in Human Geography,* 2: 17–48.

HOOKWAY, R. 1973. Q1571, Select Committee of the House of Lords on Sport and Leisure, *Minutes of Evidence.* House of Lords Paper 193, HMSO, London.

HOUSE, J. W. 1958. Whitby as a resort, in Daysh, G. H. J. (ed.), *A Survey of Whitby.* Windsor.

JENKINS, C. and SHERMAN, B. 1979. *The Collapse of Work.* London.

JOHNSTONE, M. and TIVY, J. (in press). Assessment of the physical capability of land for rural recreation, in Thomas, M. F., and Coppock, J. T. (eds.), *Land Assessment in Scotland.* Aberdeen.

KAPLAN, M. 1978. *Leisure: Theory and Policy.* London.

LAVERY, P. (ed.) 1971. *Recreational Geography.* Newton Abbot.

LAW, S. and PERRY, N. H. 1971. Countryside recreation for Londoners: a preliminary research approach. *GLC Intelligence Bull.,* 14: 11–16.

LINTON, D. L. 1968. The assessment of scenery as a natural resource. *Scott. Geogr. Mag.,* 84: 219–38.

LONG, J. A. and WIMBUSH 1979. *Leisure and the over-50s:* SSRC. Sports Council Joint Panel on Leisure and Recreation Research, London.

LOWENTHAL, D. 1978. Finding valued landscapes. *Progress in Human Geography,* 2: 373–418.

MACKINLAY, D. A. 1969. *Upper Speyside.* Dept of Geography, Univ. Edinburgh.

MARTIN, W. H. and MASON, S. 1979. *Broad Patterns of Leisure Expenditure.* SSRC/Sports Council Joint Panel on Leisure and Recreation Research, London.

MATHER, A. S. and RITCHIE, W. 1978. *The Beaches of the Highlands and Islands of Scotland.* Countryside Commission for Scotland, Battleby.

MERCER, D. C. 1970. The geography of leisure—a contemporary growth point. *Geography,* 55: 261–73.

MERCER, D. C. 1971. The role of perception in the recreation experience. *J. Leisure Research,* 3: 261–70.

MILLMAN, R. 1971. *Outdoor Recreation in the Highland Countryside.* Mimeographed, Aberdeen.

MORRISON, P. C. 1965. Geographers' mirror of Michigan: a bibliography of professional writings. *Proc. Michigan Acad. Sci. Arts, Letters,* 50: 493–518.

National Urban Recreational Study 1978. *Technical Reports.* Vols. 1–13, Dept of the Interior, Washington.

Outdoor Recreation Resources Review Commission 1962. *Study Reports,* 1–27, Government Printer, Washington.

OWEN, M. L. and DUFFIELD, B. S. 1971. *The Touring Caravan in Scotland.* Scottish Tourist Board, Edinburgh.

OWENS, P. L. 1978. Conflict between Norfolk Broads coarse anglers and boat users: a managerial issue, in Moseley, M. J. (ed.), *Social Issues in Rural Norfolk.* Centre of East Anglia Studies, Norwich, 123–43.

PALMER, J. and BRADLEY, J. 1971. Planning for outdoor recreation, in Lavery, P., *Recreational Geography.* Newton Abbot, 270–96.

PATMORE, J. A. 1968. Spa towns of England and Wales, in Beckinsale, R. P., and Houston, J. M. (ed.), *Urbanisation and its Problems.* Oxford, 47–69.

PATMORE, J. A. 1970. *Land and Leisure.* Newton Abbot.

PATMORE, J. A. 1973a. Patterns of supply. *Geogrl. J.,* 139: 473–82.

PATMORE, J. A., 1973b. Recreation, in Dawson, J. A., and Doornkamp, J. C. *Evaluating the Human Environment.* London, 224–48.

PATMORE, J. A. 1975. *People, Place and Leisure.* Univ. Hull.

PATMORE, J. A. 1977. Recreation and leisure. *Progress in Human Geography,* 1: 111–17.

PATMORE, J. A. 1978. Recreation and leisure. *Progress in Human Geography,* 2: 141–7.

PATMORE, J. A. 1979. Recreation and leisure. *Progress in Human Geography,* 3: 126–32.

PATMORE, J. A. and RODGERS, H. B. (eds.) 1972. *Leisure in the North West.* North West Sports Council, Salford.

PDMS 1979. *Service Description.* Univ. Edinburgh.

PENNING-ROWSELL, E. C. 1973. *Alternative Approaches to Landscape Appraisal and Evaluation.* Middx. Polytechnic Planning Research Group, Rept. No. 11, Enfield.

PERRY, N. H. 1973. Models in recreation planning. *Recreation News Supplement,* 8: 2–10.

PRICE, W. T. R. 1967. The location and growth of holiday caravan camps in Wales 1956–65. *Trans. Inst. Br. Geographers,* 42: 127–52.

ROBINSON, H. 1976. *A Geography of Tourism.* London.

RODGERS, H. B. 1967. *British Pilot National Recreation Survey.* Rept. No. 1, British Travel Association—Univ. Keele, London.

RODGERS, H. B. 1969a    *British Pilot National Recreation Survey*, Rept. No. 2, British Travel Association—Univ. Keele, London.

RODGERS, H. B. 1969b.    Leisure and recreation. *Urban Stud.*, 6: 368–84.

RODGERS, H. B. 1972.    Problems and progress in recreation research. A review of some recent work. *Urban Stud.*, 9: 223–8.

RODGERS, H. B. 1973.    The demand for recreation. *Geogrl. J.*, 139: 467–73.

RODGERS, H. B. 1974.    Regional recreational contrasts, in Appleton, I. (ed.), *Leisure Research and Policy*. Edinburgh. 109–18.

SELECT COMMITTEE OF THE HOUSE OF LORDS ON SPORT AND LEISURE 1973.    *Second Report*. House of Lords Paper 193, HMSO, London.

SHEIAL, J. 1975. The concept of National Parks in Great Britain 1900–1950. *Trans. Inst. Br. Geographers*, 52: 41–56.

SIMMONS, I. G. 1975.    *Rural Recreation in the Industrial World*. London.

SPEIGHT, M. C. D. 1973.    *Outdoor Recreation and its Ecological Effects: a Bibliography and Review*. Discussion Paper on Conservation No. 4, Univ. College, London.

STAMP, L. D. 1948.    *The Land of Britain. Its Use and Misuse*. London.

STEERS, J. A. 1944. Coastal preservation and planning. *Geogrl. J.*, 104: 7–27.

TANNER, M. F. 1967.    *Recreation on the Coast*. Sports Council, London.

TANNER, M. F. 1973a.    *Water Resources and Recreation*. Sports Council Water Recreation Series No. 3, London.

TANNER, M. F. 1973b.    The recreational use of inland waters. *Geogrl. J.*, 139: 486–91.

TANNER, M. F. 1974a.    *The Potential of Towpaths as Waterside Footpaths*. Water Space Amenity Commission Research Rep. No. 1, London.

TANNER, M. F. 1974b.    *Water Recreation in Country Parks*. Water Space Amenity Commission, Research Rep. No. 2, London.

TANNER, M. F. 1974c.    *Recreational Use of Water Supply Reservoirs in England and Wales*. Water Space Amenity Commission, Research Rep. No. 3, London.

TANNER, M. F. 1974d.    *Permit Sailing on Enclosed Water*. Water Space Amenity Commission, Research Rep. No. 4, London.

THOMAS, D. 1970.    *London's Green Belt*. London.

THORPE, H. 1975.    The homely allotment: from rural dole to urban amenity: a neglected aspect of urban land use. *Geography, 60:* 169–83.

TIVY, J. 1972.    *The Concept and Determination of Carrying Capacity of Recreational Land in the USA*. Countryside Commission for Scotland, Occ. Papers No. 3, Perth.

TIVY, J. (in press).    *The Effects of Recreation on Freshwater Lochs Throughout Scotland*. Countryside Commission for Scotland, Battleby.

TOURISM AND RECREATION RESEARCH UNIT (TRRU) 1974. *TRIP Series No. 1 System Description*, Rept. No. 11, Edinburgh.

TOURISM AND RECREATION RESEARCH UNIT 1975.    *The Economic Impact of Tourism: a Case Study in Greater Tayside*. Rept. No. 13, Edinburgh.

TOURISM AND RECREATION RESEARCH UNIT 1976a.    *STARS Series No. 2: Summary Report*. Rept. No. 18, Edinburgh.

TOURISM AND RECREATION RESEARCH UNIT 1976b.    *A Research Study into Provision for Recreation in the Highlands and Islands*. Oil Leisure and Society, Vol. A. Area Studies and Recommendations, Vol. B, Rept. No. 22, Edinburgh.

TOURISM AND RECREATION RESEARCH UNIT 1977a.    *Tourism in the Highlands and Islands*, Rept. No. 27, Edinburgh.

TOURISM AND RECREATION RESEARCH UNIT 1977b.    *Traffic Mix Model: a Methodology for Estimating Traffic Flows*, Rept. No. 30, Edinburgh.

TOURISM AND RECREATION RESEARCH UNIT 1977c.    *Recreation Activity Substitution in Scotland*. Rept. No. 32, Edinburgh.

TOURISM AND RECREATION RESEARCH UNIT 1977d. *A Research Study into Tourism and Recreation in the Chichester Area: a Basis for Planning*. Rept. No. 35, Edinburgh.

TOURISM AND RECREATION RESEARCH UNIT 1977e. *Strathclyde Park*. Rept. No. 39, Edinburgh.

TOURISM AND RECREATION RESEARCH UNIT (forthcoming). *Glasgow Parks*. Edinburgh.

TRAVIS, A. S. and HUDSON, S. 1978. The influence of local authority structures in Rees, B., and Parker, S. (eds.), *Community Leisure and Culture: Arts and Sports Provision*. Leisure Studies Association, London.

WALL, G. 1972. Socio-economic variations in pleasure trip patterns: the case of Hull car owners. *Trans. Inst. Br. Geographers*, 57: 45–57.

# 13 Medical Geography

G. M. HOWE *University of Strathclyde*

IN the UK the major but not the exclusive concern of medical geography is the analysis of spatial variations in human health (though, more often, it is the *lack* of it!), and of the environmental conditions which are, or may be the causes of them. Health as such is not a precise or simple concept. The states of health of individuals range from the ideal through different degrees of illness and disability to the brink of death. Judgements often have to be made about levels of health. What is acceptable varies from one individual to another, from one physician to another, and from one part of the country to another.

The World Health Organisation defines health as 'a state of complete physical, mental and social well-being and not merely the absence of disease or infirmity (WHO 1965). This represents a state of adjustment or harmony between man (his living tissues, cells, and components of cells) and his environment, one of ecological equilibrium. Ill health or disease, on the other hand, is maladjustment, *dis*harmony, or ecological disequilibrium. Health is an expression of the state of conflict between man and his environmental challenges as he endeavours to cope with them in order to survive. The environment is thus a reservoir of forces, physical, chemical, biological, and socio-cultural, which support or threaten and which have, among other powers, mutagenic properties over the genotype of all living things. The forces or factors which may place man's health in jeopardy are the *stimuli*, in contrast to the behaviour of man when exposed to or affected by, them, which is the *response*. The analysis of responses and stimuli, two opposite terms in the equation of life, provides a theory of environmental maladjustment and of medical geography.

The traditional and generally accepted view in the UK is that the level of health of both the individual and the community is equated with the quality of medicine and with the health care system generally. Close examination of the causes and underlying factors (stimuli) of sickness and death (responses) reveals that, on the contrary, environment, human biology and life style are as important as the quality of the health care system. Such a perspective on health, health status, and the health field generally, is really a re-emergence into contemporary thought of the time-honoured Hygeian tradition of the Greeks. Its revival, first in the UK (McKeown 1975; Draper, Best, and Dennis 1977) and later in Canada (Lalonde 1975), affords equal emphasis both to the factors which prejudice health, and to those which attempt to reconstitute health once it has been lost. As such it represents a radical revaluation in view of the clear pre-eminence health care organisation and delivery (e.g. the National Health Service) has had in other concepts of the health field. The revived Hygieian concept as presented by Lalonde (1975) is intellectually acceptable and sufficiently simple to allow for a quick

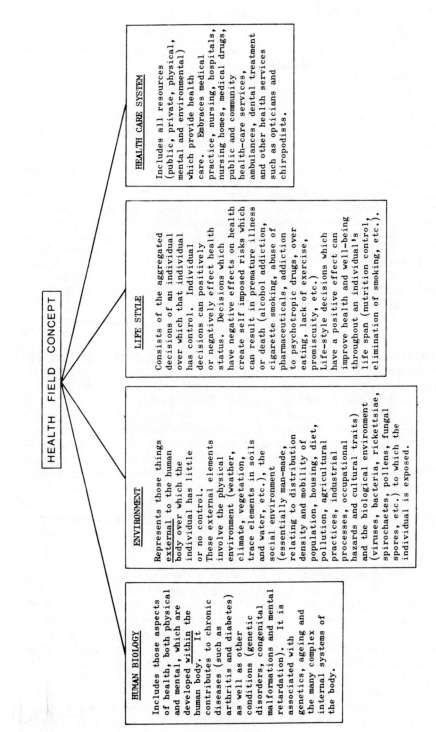

**HEALTH FIELD CONCEPT**

**HUMAN BIOLOGY**

Includes those aspects of health, both physical and mental, which are developed within the human body. It contributes to chronic diseases (such as arthritis and diabetes) as well as other conditions (genetic disorders, congenital malformations and mental retardation). It is associated with genetics, ageing and the many complex internal systems of the body.

**ENVIRONMENT**

Represents those things external to the human body over which the individual has little or no control. These external elements involve the physical environment (weather, climate, vegetation, trace elements in soils and water, etc.), the social environment (essentially man-made, relating to distribution density and mobility of population, housing, diet, pollution, agricultural practices, industrial processes, occupational hazards and cultural traits) and the biological environment (viruses, bacteria, rickettsiae, spirochaetes, pollens, fungal spores, etc.) to which the individual is exposed.

**LIFE STYLE**

Consists of the aggregated decisions of an individual over which that individual has control. Individual decisions can positively or negatively effect health status. Decisions which have negative effects on health create self imposed risks which can result in premature illness or death (alcohol addiction, cigarette smoking, abuse of pharmaceuticals, addiction to psychotropic drugs, over eating, lack of exercise, promiscuity, etc.) Life-style decisions which have a positive effect can improve health and well-being throughout an individual's life span (nutrition control, elimination of smoking, etc.).

**HEALTH CARE SYSTEM**

Includes all resources (public, private, physical, mental and environmental) which provide health care. Embraces medical practice, nursing, hospitals, nursing homes, medical drugs, public and community health-care services, ambulances, dental treatment and other health services such as opticians and chiropodists.

FIG. 13.1. Health field concept (adapted from Lalonde 1979).

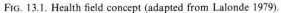

location within its framework of almost any idea, problem, or activity related to health/disease (Fig. 13.1). At the same time credence and status is given to medical–geographical inquiries.

Much of the subject matter of medical geography is as old as Hippocrates and his *De Aere, Aquis et Locis* (On Airs, Waters and Places).

> Whoever wishes to investigate medicine properly, should proceed thus: in the first place to consider the seasons of the year, and what effects each of them produces.... Then the winds, the hot and the cold, especially such as are common to all countries, and then such as are peculiar to each locality. We must also consider the qualities. In the same manner, when one comes into a city to which he is a stranger, he ought to consider its situation, how it lies as to the winds and the rising of the sun; for its influence is not the same whether it lies to the north or the south, to the rising or to the setting sun. These things one ought to consider most attentively, and concerning the waters which the inhabitants use, whether they be marshy and soft, or hard, and running from elevated and rocky situations, and then if saltish and unfit for cooking; and the ground, whether it be naked and deficient in water, or wooded and well watered, and whether it lies in a hollow, confined situation, or is elevated and cold; and the mode in which the inhabitants live, and what are their pursuits, whether they are fond of drinking and eating to excess, and given to indolence, or are fond of exercise and labour, and not given to excess in eating and drinking.

> The Genuine Works of Hippocrates
> Francis Adams, 1849, vol. i, p. 190

The Hippocratic concept of disease persisted in aetiology (causation) and treatment until the time of Pasteur and Koch and the introduction of the germ theory towards the end of the nineteenth century. With the advent of bacteriology the mystery of infectious disease seemed to have been swept away. The parasite became the new focus of attention and the importance of the environment forgotten. Bacteriology provided the new paradigm by which more and more disease problems might be solved. The internal *milieu* of the body and its balance (homeostasis) became the centre of interest and from it developed an 'introvert' attitude to disease. Now, however, since the most important principles of the co-ordinated functions of the various organs of the human body have been clarified, and additional knowledge of molecular biology accumulated, a more 'extrovert' attitude is being adopted. Attention is being focused increasingly on the health hazards or provocative factors (stimuli) of the environments in which man and communities live.

It is becoming increasingly evident that the actual state of the human organism, whether healthy or showing pathological traits, depends on the equilibrium of the internal *milieu* of the human body, external stimuli in the environment, and on the complex inter-relationships of the two. Each and every member of the ecosystem—and this must inevitably include man—is inherently bound to the existing physical, biological, and man-made environment with which they form an organic unit. Health problems are thus environmental problems and as such are amenable to techniques of spatial analysis used by geographers.

The contemporary thrust of medical–geographical studies in the UK is the demonstration and analysis of spatial patterns of morbidity (illness) and mortality (death) relative to local environments and life-style characteristics. It is hoped that, in collaboration with medical scientists, a contribution can be made to the understanding, possibly the explanation, and, it is to be hoped the prevention of the diseases or causes of ill health for which the aetiology is still unknown. In the UK the severely disabling and killing communicable diseases which were endemic during pre-industrial and industrial times have been largely eradicated. They have been controlled by improved water supplies, sewage disposal, better housing, hygiene, health legislation, education, and therapeutic advances. Instead, as in all technologically advanced and economically developed countries, the UK is scourged by diseases of the cardiovascular system, cancer, chronic bronchitis, accidents, and mental illnesses. These diseases have in common the fact that their prevention is still very much in its infancy.

Unlike the situation in most other countries, the comprehensive UK National Health Service ensures that medical care and attention is available to every citizen with a comparatively high degree of equality. It also means that medical data are available on a nation-wide basis. For purposes of analysing spatial variations of ill health morbidity data are preferable to mortality data as a community health measure. But morbidity means what it is defined to mean: whether a subjective malaise, measurable evidence of a disorder, pathology diagnosed by a doctor, or certified incapacity to work. Related, particularly in the chronic, non-communicable diseases, are the problems of when 'morbidity' begins, how long it lasts, how to define its severity, and the incapacity it causes. General practitioner consultation rates and hospital data are used as a basis for estimating community morbidity levels. Such data are far from adequate since the level of disease in a population may not be reflected in the number of people seeking treatment. What may be regarded by some people in some areas or under some circumstances as sufficient to justify absence from work, a visit to the doctor, or admission to hospital may be regarded quite differently in another *milieu*. Mortality, on the other hand, is unconditional and mortality data are readily available. They provide information on age, sex, marital status, place of death, and the primary and underlying causes of death. But they are a direct measure of incidence for only those diseases which have a high fatality rate. For other diseases such as 'rheumatoid arthritis' or peptic ulcer, for which the fatality rate is extremely low, they are, of necessity, almost useless. As regards 'cause of death' a great deal has been done to standardize what can be put by the certifying doctor on the death certificate. The fact remains, however, that different doctors, when faced with the same history, signs, and symptoms will, on occasion, fill in the death certificate differently. In relatively few cases does this reflect ignorance; in most instances it is due to real difficulties of interpretation and the extent of such 'errors' must not be exaggerated. In the younger age groups it is relatively small: it is really only in old age, when several different physiological systems are collapsing at the same time and death is the result of a multiplicity of aetiological factors, that the error is appreciable. For this reason it is usual, for purposes of spatial analysis, to restrict the use of mortality data to specific age ranges or groups, e.g. 15–64 years, depending on the disease being studied.

The illness and death of an individual may have more than one cause. Indeed they are more likely to be the result of the operation of a multiplicity of factors or prior circumstances, including those known as chance  Even so diagnosticians usually identify a primary cause, which is subsequently classified at offices of the Registrars General in London, Edinburgh, and Belfast according to the *WHO International Statistical Classification of Diseases, Injuries and Causes of Death* (the Ninth Revision of 1975 was published by WHO in 1977). The 'place of death' or 'usual place of residence' given on death certificates is obviously important for purposes of spatial analysis. It is an indication of the place of exposure to environmental stimuli or stress but is often difficult to interpret. Difficulties arise with a highly mobile population or when people are in hospital and institutions (which may be distant from the locality of their usual residence). The general solution in such cases is to 'transfer' these deaths to the 'usual place of residence'; this is not always possible and regulations concerning these transfers have varied from time to time. A further limitation follows the collecting of basic medical diagnostic information relating to ill health or death on the basis of local government or other administrative units rather than on the basis of map grids or postal codes. It seems highly unlikely that actual disease experience will conform to, or be influenced by, such arbitrary constraints.

Despite such problems useful contributions have been made in demonstrating and analysing spatial variations in mortality experience in the UK. Emphasis to date has been concentrated on the medical geography of chronic, non-infectious diseases such as the cancers, coronary artery disease, cerebrovascular disease, and bronchitis (Howe 1959, 1961, 1963, 1968, 1969, 1970, 1971, 1972, 1973, 1974, 1976, 1977, 1980; Murray 1967; Coates and Rawstron 1971). Pointers to anomolous localities have been made and environmental associations which may have aetiological connections hypothesized.

The value of cartographic presentations of medical data for purposes of description, analysis, and aetiological enquiry has long been recognized. Petermann (1852) well explains the advantages of maps for medical purposes.

The object, therefore, in constructing Cholera Maps is to obtain a view of the Geographical extent of the ravages of this disease, and to discover the local conditions that might influence its progress and its degree of fatality. For such a purpose, Geographical delineation is of the utmost value, and even indispensable; for while the symbols of the masses of statistical data in figures, however clearly they might be arranged in Systematic Tables, present but a uniform appearance, the same data, embodied in a Map, will convey at once, the relative bearing and proportion of the single data together with their position, extent, and distance, and thus, a Map will make visible to the eye the development and nature of any phenomenon in regard to its geographical distribution.

But there are problems; since maps may be read, used, and acted upon by other professions, it is of the utmost importance that their limitations and total dependence on the quality of the data employed be realized. Otherwise an impression of totally spurious reliability may be conveyed. In the first instance there is no single epidemiological index which completely characterizes the impact of a disease in a community. The attack or incidence rate measures the rate of appearance of the disease in a defined population; prevalence is the measure of the number in a defined population suffering from a given disease at a given time; the mortality rate measures the rate at which people in a defined community die; case fatality measures the rate at which people with a particular disease die. Different indices are used by medical geographers to investigate different problems.

The most satisfactory single-figure comparison of the mortality in each local area with that of the UK as a whole is the Standardised Mortality Ratio (SMR) calculated for each sex separately. The SMR makes allowance for variations in age structure of local populations compared with the national population which is regarded as having a standard structure of age groups. Local SMRs are expressed in terms of a standard national or UK rate taken as 100. A range of values of local SMR's around 100 is to be expected, some doubtless arising through chance. For this reason statistical tests have to be applied e.g. Poisson probability test, to ensure that any spatial patterns revealed by mapped SMR's are not simply the result of the operation of chance.

The different ways in which SMR's are represented on maps suggest different spatial patterns and environmental associations. Choropleth maps of SMRs, using a geographic base, suffice reasonably well for descriptive or illustrative purposes but they present a visual impression of even population spread, giving prominence to SMRs of extensive, sparsely, and unevenly populated areas and insufficient weighting to localized concentrations of population associated with cities and large towns. An incorrect visual impression of regional intensities of mortality experience is thereby created. One means of overcoming this weakness has been the use of a demographic map on which the area assigned to each locality is proportional to its population size (Forster 1966; Howe 1963, 1969, 1970, 1971). The disadvantage of such demographic maps is their distortion of geographical reality by loss of shape and/or contiguity. The loss of contiguity of neighbouring areas is a particular weakness when a disease condition involves diffusion.

Several cartographic methods, including computer graphics, have been employed to demonstrate real patterns of mortality resulting from the variety of conditions—environmental and human—at the local and national scales (Howe 1970(d), 1971(c), 1980). Maps or cartograms explain nothing but they all pose the question 'Why?' This inevitably leads to a search for explanations of the spatial patterns revealed, involving an examination of a wide range of possible causal factors, singly or in multiple association and requiring a preparedness to disregard tentative correlations and preconceived hypotheses. Not until an examination of all suspected environmental stimuli fails to provide a satisfactory solution can research begin, involving the search for and detection of hitherto unsuspected predisposing or causal factors and the testing of hypotheses. It is in this direction and by the use of all appropriate statistical techniques, that geographical or spatial analysis of mortality data may possibly make a real contribution to knowledge. By demonstrating the influence of environmental stimuli in the aetiology of the chronic diseases which currently scourge British society, prevention then becomes possible.

The study of variations in the distribution of mental disorders and mental subnormality is a relatively new phenomenon in medical geography in the UK. This is somewhat surprising since as early as 1962–3 nearly half of the hospital beds in England and Wales were occupied by the mentally ill or retarded. At the same time 32 million working days were lost due to mental illnesses and the cost in terms of National Insurance funds alone amounted to £21 million (Office of Health Economics 1965). Research to date has been hindered by lack of adequate definitions of mental health and mental illness and the difficulty of categorizing rigidly the problems that fall within the purview of geography and mental health. Nevertheless geographers have made a number of useful environmental and ecological studies (Bain 1971, 1974, 1976; Giggs 1977, 1979, 1980; Dean 1979).

Geographical studies of disease diffusion apply to communicable rather than to chronic ailments. Historical reconstructions have been made of the transmission across space of plague ('the Great Pestilence', 'Black Death') in fourteenth-century Britain, and of the epidemics of cholera in the nineteenth century (Howe 1972(a)). Knowledge of the historical progression of such diseases sheds light on the mechanisms contributing to disease diffusion at the present time. Waves of diffusion were exemplified during the 1918–19 pandemic of influenza (Min. of Health 1920). Hunter and Young's (1971) examination of the diffusion of influenza in England and Wales in 1957 revealed a relationship with population potential. Haggett (1972) has demonstrated how a contagious process can be represented on a planar graph (utilizing some principles of network theory) and shown how the study of spatial diffusion can be used in understanding the spread of disease.

Outside the UK, British geographers have made valuable contributions to the study of medical geography. Prothero (1961, 1965, 1972, 1977), well versed in the migratory movements of people in northern Nigeria and elsewhere, was able to advise on possible appropriate courses of action when a malaria control project met unexpected difficulties related to the seasonal migration of thousands of people from controlled to uncontrolled and malarious areas. Coles (1972) contributed greatly towards basic malaria survey work in the former Trucial States in the Arabian peninsula, while Fox (1972) has provided an in-depth study of morbidity and mortality in Mexico City.

Learmonth's long association with medical geography is represented by now classic studies of geographical factors contributing to the ecology of malaria in India and Pakistan, various models of a hypothetical malarial continent, as well as more recent contributions pertaining to differences in health in different parts of the world, and more particularly the geography of hunger (Learmonth 1957, 1958, 1961, 1965, 1969, 1972, 1977, 1978, 1979). Howe, has provided a survey of the multiplicity of geographical factors involved in disease in Britain throughout the ages (Howe 1972(a), 1976) and, in collaboration with J. Loraine and other medical authorities, discussed, at length, how certain environmental factors may operate in combination to produce well-known disease states (1973, 1980). A comprehensive study, edited by Howe (1977) deals with the main diseases of mankind on a global basis. It brought together a group of leading experts whose approach was varied and multi-disciplinary but who shared a particular interest in the problems of man's interaction with his environment, and in the global variability and spatial patterns of disease.

If there is indeed a single British approach to medical geography, its main strengths are in the areas of disease mapping and the regional synthesis of health problems. Spatial studies of the administration and provision of treatment facilities in Britain—the geography of health care—are still very much in their infancy. This is possibly due to the general availability of a National Health Service, a situation similar to that of Sweden, France, and, to some extent, Canada. It is essentially a welfare-state health-care delivery system whereby the provision of health services is part of the whole governmental provision of social services.

A complete reorganization of the NHS took place in 1973 and 1974 but few have greeted it as an unqualified success. Some parts of the UK are still better served than others in terms of expenditure in health service resources and numbers of medical staff. A few geographers, impatient with the gross regional disparities in resource provision, are now turning their attention to what they consider to be more relevant medical geography. Geographical concepts and techniques are being applied to the study of the spatial aspects of health service provision (Woods 1978; Smith 1979). In this respect they lean towards their North American colleagues who are publishing an increasing number of works concerned with the geography of health care.

And what are the future prospects for medical geography in the UK? In so far as geographers, by their training, are familiar with a wide range of environmental factors and their associated processes it seems not unreasonable to assume that geographers might add a new dimension to medical science by dealing with the geography and environmental associations of the disabling and killing diseases. Courses in medical geography in university teaching are, therefore, to be encouraged (Hellen 1979). Röntgen, who discovered X-rays, was a physicist; Pasteur, the principal architect of the germ theory, was a chemist; Mendel, who laid the foundations of genetic theory, was a biologist and monk; Chadwick, 'Father of the Sanitary Idea', who waged war against filth, bad housing, and inferior sanitation, was a civil servant. They and others are instances of people from outside medicine who have contributed as much as any doctor to the understanding of disease and how to deal with it.

Without a knowledge of causes the best means of prevention will never be known. Environmental or ecological medical geographers have the potential to highlight those factors of the environment which may prove to be of aetiological significance, or those aspects of the ecological web which provide the key to breaking a particular disease cycle. They may thus play an undoubted part in social, community, and preventive medicine. Indeed it is already conceded (Royal Commission 1979) that preventive measures are by no means the exclusive responsibility of the National Health Service.

Given appropriate geographically–based information, health-care geographers can provide measures of distribution of health-care resources and facilities in relation to availability, accessibility, and need, and assist in the determination of optional locations of new treatment facilities (Forster 1975; Rigby 1978).

Improvements in environmental quality and the general quality of life may well constitute the ultimate aim of medical geographical studies. This could well involve a change in philosophy from one of economic growth to one of population and environmental stability. Medical geographers are imbued with a genuine concern for the components of health implicit in the WHO definition of 'positive health', a concern for human welfare, and for the general well-being of society at large.

# REFERENCES

BAIN, S. 1971. The geographical distribution of psychiatric disorders in the North East Region of Scotland. *Int. Journ. Med. Geog.* (Hungary), **2.**

BAIN, S. 1974. A geographers approach in the epidemiology of psychiatric disorder. *J. Biosoc. Sci.* 6 (2), 195.

BAIN, S. 1976. Psychiatric disorder and density of population, in *Methoden und Modelle der Geomedizinischen Forschung.* Wiesbaden.

COATES, B. E. and RAWSTRON, E. M. 1971. *Regional variations in Britain: studies in social and economic geography.* London.

COLES, A. 1972. Malaria in the Trucial States. *Inst. Brit. Geog.,* Med. Geog. Study Group Symposium. Univ. of Strathclyde, Glasgow.

DEAN, K. G. 1979. The geographical study of psychiatric illness: the case of depressive illness in Plymouth. *Area,* **11**(2): 167.

DRAPER, P., BEST, G., DENNIS, J. 1977. Health and Wealth. *Roy. Soc. Health J.*

FORSTER, F. 1966. Use of a demographic base map for the presentation of areal data. *Brit. Journ. Prev. and Soc. Med.* 20: 156.

FORSTER, F. 1975. The concept of an atlas of health and social services. *Health Bulletin,* 33 (2).

FOX, D. 1972. Patterns of morbidity and mortality in Mexico City. *Geog. Rev.,* 62(2) 151.

GIGGS, J. A. 1973(a). The distribution of schizophrenics in Nottingham. *Trans. Inst. Brit. Geogrs.* 59: 55.

GIGGS, J. A. 1973(b). High rates of schizophrenia among immigrants in Nottingham. *Nursing Times* 1210.

GIGGS, J. A. 1977. Mental disorders and mental subnormality, in Howe, G. M. (ed.), *A World Geography of Human Diseases.* London.

GIGGS, J. A. 1979. Health problems in cities, in Herbert, D. T. and Smith, D. M. (eds.), *Social Problems in the City: A Geographical Approach.* London.

GIGGS, J. A. 1980. Mental health and the environment, in Howe, G. M. and Loraine, J. (eds.), *Environmental Medicine.* London, second edition.

GIGGS, J. A., and MATHER, P. M. 1976. The geography of mental disorders and mental health services in England and Wales, in Diesendorf M. and Furnass B. (eds). *The Impact of Environment and Lifestyle on Human Health.* Proceedings of a Symposium to mark United Nations Habitat Year, ANU, Canberra, 134.

GIGGS, J. A., BOURKE, J. B., and EBDON, D. S. 1979. Variations in the incidence and the spatial distribution of patients with primary acute pancreatitis in Nottingham 1969–76. *Gut* 20: 366.

GILBERT, E. W. 1958. Pioneer maps of health and disease in England. *Geog. J.* 124; 172.

HAGGETT, P. 1972. Contagious processes in a planar graph: an epidemiological application, in McGlashan, N. D. (ed.), *Medical Geography: techniques and field studies.* London.

HELLEN, J. A. 1979. Medical geography and its place in University teaching, in Jusatz, H. J. (ed.), *Geomedizin in Forschung und Lehre,* Wiesbaden.

HOWE, G. M. 1959. The geographical distribution of disease with special reference to cancer of the lung and stomach in Wales. *J. Prev. Soc. Med.* 13: 204.

HOWE, G. M. 1961. The geographical variations of disease mortality in England and Wales in the mid-20th century. *Adv. Sci.* 17: 415.

Howe, G. M. 1963(a).   The geography of death in England and Wales, 1960. *Lancet* 1: 818.

Howe, G. M. 1963(b).   *National Atlas of Disease Mortality in the United Kingdom*, London. Second edition, 1970.

Howe, G. M. 1968.   The geography of death. *New Sci.* 40: 612.

Howe, G. M. 1969.   Putting disease on a map. *Nature* 223: 891.

Howe, G. M. (1970(a).   Some aspects of social malaise in Scotland. *Health Bull.* 28: 1.

Howe, G. M. 1970(b).   Geography looks at death. *Spectrum*, 71: 5.

Howe, G. M. 1970(c)   Disease patterns and trace elements. *Spectrum* 77: 2.

Howe, G. M. 1970(d).   Some recent developments in disease mapping. *Roy. Soc. Health*, 90: 16.

Howe, G. M. 1971(a).   The geography of life in Britain in the mid-twentieth century. *Update* 3: 429.

Howe, G. M. 1971(b).   The geography of lung-bronchus and stomach cancer in the United Kingdom. *Scot. Geog. Mag.* 87: 202.

Howe, G. M. 1971(c).   The mapping of disease in history, in Clarke, E. (ed), *Modern Methods in the History of Medicine*. London.

Howe, G. M. 1972(a).   *Man, Environment and Disease: a medical geography through the ages*. Newton Abbot. (Harmondsworth, 1976).

Howe, G. M. 1972(b).   Some aspects of social malaise in South Wales. *Int. J. Env. Studies* 4: 9.

Howe, G. M. 1972(c).   Social malaise and environmental quality in Glasgow and West Central Scotland. *Int. Geog. (IGU)*, Vol. 1, Toronto.

Howe, G. M. 1972(d).   London and Glasgow: a comparative study of mortality patterns. *Int. Geog. (IGU)*, Vol. 2, Toronto.

Howe, G. M. 1972(e).   Plotting the scale of disease. *Geog. Mag.* 44: 304.

Howe, G. M. 1973.   The geography of life and death. *J. Biosoc. Sci.* 5: 285.

Howe, G. M. 1974   Disease and the environment in Britain. *J. Roy. Coll. Phys.* 8: 127.

Howe, G. M. 1976(a).   The geography of disease, in Carter, C. O. and Peel, J. (eds.), *Equalities and Inequalities in Health*. London.

Howe, G. M. 1976(b).   Aspects of medical geography in Great Britain, in Harrison, G. A., and Gibson, G. B. (eds.), *Man in urban environments*. Oxford.

Howe, G. M. (ed.) 1977.   *A World Geography of Human Diseases*. London.

Howe, G. M. 1979(a).   Statistical and cartographic methods of analyzing spatial patterns of mortality. *Geoforum* 10(3), 311.

Howe, G. M. 1979(b).   Mortality from selected malignant neoplasms in the British Isles: the spatial perspective, *Geog. J.* **145**(3), 401.

Howe, G. M. and Loraine, J. (eds.) 1973.   *Environmental Medicine*. London, second ed. 1980.

Hunter, J. and Young, J. 1971.   Diffusion of influenza in England and Wales. *Ann. Assoc. Amer. Geogrs.* 6: 637.

Lalonde, M. 1975.   *A new perspective on the health of Canadians*. Ottawa.

Learmonth, A. T. A. 1957. Some contrasts in the regional geography of malaria in India and Pakistan. *Trans. Inst. Brit. Geogrs.* 23: 37.

Learmonth, A. T. A. 1958.   Medical geography in Indo-Pakistan: a study of 20 years' data for the former British India. *Indian Geog. J.* 33: 1.

Learmonth, A. T. A. 1961.   Medical geography in India and Pakistan. *Geogrl. J.* 127: 10.

Learmonth, A. T. A. 1972.   Atlases in medical geography, in McGlashan (ed.), *Medical Geography: techniques and field studies*. London.

Learmonth, A. T. A. 1977.   Malaria, in Howe, G. M. (ed.), *A World Geography of Human Diseases*. London.

LEARMONTH, A. T. A. and NICHOLS, G. C. 1965. *Maps of some Standardised Mortality Ratios for Australia, 1959–63.* Canberra.

LEARMONTH, A. T. A. and GRAU, R. 1969. *Maps of some Standardised Mortality Ratios for Australia, 1965–66 compared with 1959–63.* Canberra.

LEARMONTH, A. T. A. and AKHTAR, R. 1979. India's malarial resurgence 1965–78. *Geography* 64(3): 221.

MCKEOWN, T. 1965. *Medicine in modern society: medical planning based on evaluation of medical achievement.* London.

MCKEOWN, T. 1976. *The role of medicine,* London.

MCKEOWN, T. 1979. *The role of medicine: dream mirages or nemesis,* Oxford.

MINISTRY OF HEALTH 1920. Report on the pandemic of influenza, 1918–19. *Report on Public Health and Medical Subjects,* No. 4, London.

MURRAY, M. A. 1967. Geography of death in the United States and the United Kingdom. *Annal. Ass. Amer. Geogrs.* 58: 301.

OFFICE OF HEALTH ECONOMICS 1965. *The cost of mental care.* London.

PETERMANN, A. H. 1852. *Cholera map of the British Isles, showing the districts attacked in 1831, 1832 and 1833.* London.

PROTHERO, R. M. 1961 Population movements and problems of malaria eradication in Africa. *Bull. World Hlth. Org.* 24: 405.

PROTHERO, R. M. 1965. *Migrants and malaria.* London.

PROTHERO, R. M. 1977. Disease and mobility: a neglected factor in epidemiology. *Int. Journ. Epidem.* 6(3): 259.

RIGBY, J. P. 1978. *Access to hospitals: a literature review.* Transport and Road Research Laboratory Report 853. Crowthorne, England.

ROYAL COMMISSION 1979. *Report on the National Health Service.* London.

WORLD HEALTH ORGANISATION (WHO), 1965. *Basic Documents,* sixteenth edition, 1.

# 14 Regional Geography

W. R. MEAD *University College, London*

A MATTER OF SEMANTICS

LINGUISTICS and semantics are increasingly the concern of the geographer. The language of geography has changed greatly during the last generation. The vocabulary has increased, some words have been revalued, some have been devalued, some have acquired new meanings. Words and concepts from other disciplines have been freely introduced to the subject and, as in both the natural sciences and social sciences, wordsmiths have also been busy. As a result, the subject has acquired a private language, in some respects a metalanguage. Reciprocally, some of its vocabulary has been adopted for popular use and given a new public meaning. All these developments, rather than clarifying the language of geography have resulted in growing ambiguity. Not surprisingly, as epistemological discussion has sharpened the concern over ambiguity has deepened.

In the process, a number of facts have become clear. Firstly, there is a widespread concern about terminology and terms of reference which, in turn, has called into question a number of old-established geographical concepts. Among them, is the concept of the region. Secondly, the very words 'region' and 'regional' are misunderstood: so, too, regional geography as it has been traditionally practised and taught. Thirdly, the regional approach has acquired a completely new interpretation through the swift rise of regional science. Fourthly, the regional concept is experiencing a reappraisal with the emergence of humanistic geography in general and of behavioural studies in particular.

The popularization of the vocabulary of geography has inevitably devalued it. The quintessentially geographical word 'environment' is now employed so widely and with so many different meanings that many geographers are reluctant to use it. A similar fate has befallen the word 'region'. Publishers have increased its loose and casual employment by grouping an ever widening series of titles under the umbrella of regional geography (e.g. Harrison Church 1973). When radio, television, and the press took up the word, every area became a 'region'. The demise was hastened by the rise of quantification, some practitioners of which took over the adjective 'regional' and married it to the noun 'science'.

This short review of the situation looks first at the debate that surrounds the concept of the region—a debate which is independent of regional geography *per se*, but which has relevance for geography *in toto*. Secondly, it urges the continuing need to project the substance that was once so widely taught as regional geography—a geography which drifted away from the geography of regions. Thirdly, it dwells briefly on the academic pursuit of area studies. Finally, it pleads for an *entente cordiale* between those who with equal legitimacy but in their widely differing ways employ the word 'region' and its derivatives to describe the fields of their work.

## FROM THE TAXOMETRIC TO THE POETIC

The shift from the traditional regional approach to the more formal enquiry into regional territorial characteristics received a powerful thrust a generation ago through Walter Isard (1960, 1975) with the establishment of the Regional Science Association and its associated *Journal of Regional Science*. In seeking to bring greater precision and authority to spatial investigation the Association issued a challenge and provided a considerable stimulus, though at the same time it posed another threat to the unity of geography. The establishment of the International Geographical Union's Commission on methods of economic regionalization (cf. Hamilton 1969), the appearance of the *London Papers in Regional Science*, and the publication of *Regional Studies* (the journal of the *Regional Studies Association*), have been expressions of the gathering appeal of this approach to spatial investigation. In the process, a clear-cut distinction has emerged between 'arithmomorphic' regions and 'non-arithmomorphic' regions (Geogrescu-Roegen 1972). The former are theoretical constructs, with geometrical and legalistic connotations. The latter, empirically derived from observation, context-dependent, and indeterminate, are subject to all the criticisms of value judgements. Regional science looks to the mathematical representation and analysis of economic and spatial relationships. Central to its methods is the whole range of taxometrics that have become basic to the identification and understanding of regional structures, and fundamental to the analytical regionalization that lies behind planning programmes.

As a result of the rise of regional science, there are some for whom the difference between the arithmomorphic region and the non-arithmomorphic region is so great that they constitute 'two mutually incompatible ontologies of region' (Grigg 1967). For them, the traditional region is dismissed by the argument of A. N. Whitehead that 'wherever there is ambiguity as to the contrast of boundedness between inside and outside, there is no proper region'. In Grigg's words, 'if a region cannot be delineated accurately can it really exist?' And, since 'there are more intermediate areas than there are "regions" ', it is theoretically arguable that 'complete partitioning of space implies mis-classification in areas where features are not uniquely separable' (Grigg 1967).

Be that as it may, several points are clear. Firstly, regional feeling among groups of people is as widespread as ever, and the regional approach to problems was never more relevant. The functional region and the developmental region may have replaced the traditional region as frameworks for enquiry, but they too proclaim that, in one way or another, the regional concept is still very much alive (Dickenson 1978). The emotion that prevails over reason and the myth that takes precedence over reality are as significant today as ever they were when Georges Duhamel (1926) and Estyn Evans (1973) evoked the habitat and heritage of the Île de France and Ireland respectively. But the attachment to place and the community feelings that they touched upon are now more articulately expressed. Whether the feelings are rooted in ethnographic facts or whether they spring from the centrifugal forces that characterize areas remote from the core of a country (Lafonte 1967), they tend to be stronger among the older than among the younger members of the community. It is increasingly recognized that a familiar environment, for all its economic insecurity and lack of opportunity, may mean more to a community of people than greater wealth in another setting. The widespread contemporary feeling of rootlessness is inseparable from population mobility and loss of local or regional identity (Olsson 1979). While the humanistic geographers detect attempts to recapture the lost identities (Buttimer 1978), the planners have turned their attention increasingly to the areas of stagnation and population outflow. In the welfare states of western Europe, such areas have become development regions.

Secondly, and at a more theoretical level, regional studies of all kinds are carried along on the changing tide of thought. It is impossible for the regionalist to be described as static—even isolationist. He accepts the fact that there exists a graduated continuum of facts and features. No less than the arithmomorphic, the non-arithmomorphic regionalist recognizes that the classifications basic to his considerations must change as knowledge changes. The increasing speed of technical change, with all that it implies, often suggests a return to the state of flux that existed before the development of modern scientific thought. The data matrix of Brian Berry (Berry 1964) into which regional geography comfortably fits is not enough, the need is for some kind of multidimensional co-ordinate system. Meanwhile, the arithmomorphic regionalists themselves face a situation in which, as Peter Gould notes, models designed for posterity may last for but a few weeks (Olsson and Gale 1979).

Change is generated by spiritual forces as well as by technical. Poets, philosophers, and artists continue to play a role in creating a regional feeling. Historical experience suggests that what begins as poetry often finishes as politics (Weaver 1978). Independently of historical considerations, there is at least one contemporary school of thought which considers that the substance of taxometrics might be justifiably expressed in poetry. Might we not 'throw away our equations and speak as poets?' asks Gunnar Olsson (Olsson and Gale 1979). Exchange our seven lamps of reason, perhaps, for seven ambiguities of romance? *Plus ça change* ... much the same was being argued among the philosophical fraternity in Wittgenstein's Vienna before the fall (Janik and Toulmin 1973).

### OTHER WHERE IT IS

Pedagogical discussion runs side by side with philosophical debate. The subject matter of much of the old regional geography needs to be projected at certain educational levels. Perhaps it should never have been called regional geography—simply the geography of other lands. While it is true that in the longer term a minority may find greater intellectual appeal in the truth and beauty of equations, in the shorter term a majority is more likely to find its truth and beauty directly or vicariously in 'the splendour of earth'. It was for such a majority that the old-fashioned regional geography met a need and which, in its various transformations, it must continue to supply. Concern for other lands has sometimes been dismissed as 'otherwhereitis'. But it would be a bold critic who denied its significance for students. Adventurous human contact with the strange, the unfamiliar, the exotic, and the dramatic remain central to the appeal of geography in the schoolroom—and it is in the schoolroom that most geographers are born. There is no denying that the romance of other lands has played a part in the early upbringing of many academic geographers (Buttimer 1976), though it has scarcely been the romance incorporated in regional textbooks. The difficulty is to produce school texts or to construct university syllabuses which are both appealing and able to strike an appropriate intellectual level. In this context, all regional studies must make certain assumptions about an adequate body of facts— and the assumptions are increasingly wrong. It may be agreed that the understanding of relationships is more important than the amassing of facts, but one result of shuffling facts off to 'the servants' quarters' (Butterfield 1951) is that large numbers of students lack a basic knowledge of European, let alone world geography. The *Onomasticon* of the continental cadet school is scarcely a romantic solution to the problem; but more would be able to smile the smile of derision about Neville Chamberlain's knowledge of Czechoslovakia were they familiar with such a volume. It is an ironical denial of 'otherwhereitis' that the student body of a certain distinguished European university protested successfully against the requirement that they learn in their first year the location of a thousand listed place-names. Meanwhile, regional geographers must be increasingly perplexed as to what must be included (or dare be excluded) from their eclectic texts (Lukerman 1965). Space admits no more than reference to selected examples of recently published work on Europe and the developing world.

Academic geography was born in western Europe. Everyone travels to Europe and a knowledge of its geography is taken for granted. As a result, the geography of the continent—perhaps 'a culture rather than a continent' (Jordan 1973)—is often less than well known. The dinosauric age of European textbooks—country by country or region by region—has virtually passed. *Mitteleuropa* (Mutton 1970), that historical apotheosis of European regions, has become a ghost of its former self. The continent is politically and economically divided between East and West, though the compass points are not clearly defined. While the European Free Trade Area, a strangely dispersed group of states, never yielded its own geographical text, the East European bloc has produced its own particular responses (Mellor 1975; Turnock 1978). At the same time, the European Economic Community has claimed growing attention as a functional region (Commission of the European Communities 1973; Lee and Ogden 1976; Minshull 1978). A seal was also set on the geographical recognition of the EEC when its 'father', Jean Monnet was given the unusual distinction of honorary membership of the Royal Geographical Society in 1972. The challenge of grasping *Europa als Ganzes* has been taken up twice and in a unique manner by W. William-Olsson (also a Victoria medallist of the Society). His *Economic Map of Europe* (Stockholm, 1954) was the first attempt to produce a fully comparative statement about the continent. It is a measure of the diminished availability of statistical information for Europe as a whole that the second attempt (1974) proved harder than the first and excluded the USSR. At the same time, British geographers have enjoyed a number of bilateral seminars with European colleagues at which regional problems have provided the focus for discussion (e.g. Compton 1976). The old-established Anglo–Polish seminar has been complemented by joint seminars with geographers from Romania, Hungary, West Germany, and the USSR; while Anglo-French meetings have yielded their own distinctive text (Clarke 1976). Problem regions and functional regions frequently cut across international boundaries and many are dealt with in the series of short monographs bearing the title *Problem Regions of Europe* (Scargill 1973).

The challenge of traditionally accepted regions is still taken up in spite of the inhibitions that must occasionally inflict authors. J. M. Houston's unusually ambitious attempt to embrace the geography of *The Western Mediterranean* (1964) stands out. Although it lacks the compelling historical theme that makes Ferdnand Braudel's memorable volume a classic (Braudel 1972–3), Braudel might still learn something about the human geography of the area from Houston. For the north, Brian John has transformed Andrew O'Dell's *The Scandinavian World* (1980). Otherwise, it is countries rather than broader regions which claim attention and, when the latter appear, the single author tends to yield to the edited volume (e.g. Carter 1977). Countrywise, the coverage in uneven. France continues to attract a strong team (e.g. Clout 1972; Thompson 1970); Germany has been relatively neglected (but n.b. Blacksell 1978); the Low Countries have claimed historical rather than contemporary attention (Lambert 1971); Romania (Turnock 1974) is well served, and Finland is not forgotten (Jones 1977).

Unevenness also characterizes the concentration of interest in the 'third world'. Latin America appears to exert the strongest appeal. The emphasis is neither independent of the lively concern for Latin America among the geographers of North America nor of the political philosophies that have been given such colourful expression in it. Preston James and C. T. Smith (1971, 1972), both of whom have contrived to couple respectable scholarship with a gargantuan capacity for swallowing Latin America whole, set early examples. Chimborazo and Cotopaxi may have 'stolen souls away' in their time, while the Aztec and Inca civilizations have had a continuous appeal since their nineteenth-century popularization by W. H. Prescott. Today's geographers, however, either concentrate their attention on a single country (e.g. Dickenson, J. E. 1978) or look to topics such as the shanty town (Ward 1978), development studies (Gilbert 1974; Odell and Preston 1978), or the fate of the indigenous peoples (Hemming 1978). But, whatever the specific research, it tends to be rooted in field experience. Can the future yield a geographical complement to *Tristes tropiques* (Lévi-Strauss 1973)? Meanwhile it may be noted that geographers, similar to other social scientists who work in the area, tend to classify themselves as Latin Americanists rather than geographers pure and simple. In some respects this is a natural response to the rise of development studies, with the institutes and departments that foster them combining a variety of disciplines in a generalist approach.

A similar situation prevails among the Africanists, where more recent regional publication in English about the continent is often by African scholars who have had their early training in British and American universities.

Much of the interest that has been generated in the developing world springs out of fortuitous personal experience. It is epitomized in the southeast Asian work of Charles Fisher, in the experiences that set it in motion (Fisher 1979), and in the resulting specialist component in Asian geography at the School of African and Oriental Studies. The old-established research interest of W. B. Fisher in the Middle East, leading to the major contributions of the Institute of Middle Eastern Studies at the University of Durham (Fisher, ed. 1968), is paralleled by the initiatives of the Institute of Southeast Asian Studies at Cambridge, the director of which is no mean apologist for the regional approach (Farmer 1973). Yet, while the Malay peninsula has inspired a classical monograph (Wheatley 1966) and an unusual essay in nostalgia (Kirk 1978), the doors of China are being opened to a linguistically ill-equipped and somewhat reluctant British geographical fraternity. For, if there is one country more than any other which offers a challenge to the geographer—whatever kind of a regionalist he may be—it is China. The replacement of old Cathay by new China as a regional specialism presents peculiar challenges, but the potential that it offers to an enthusiastic and resilient research worker are infinitely promising.

## THE PURSUIT OF AREA STUDIES

There will always be a place for a cadre of geographers who are prepared to adopt other lands, to share other cultures, to acquire a specialist understanding of them, to make a contribution to the store of knowledge about them, and, in the process, to become the revised version of the regional geographer. For such people, a tolerable fluency in the appropriate languages is more relevant than technical skills (which others can supply), while a broad interdisciplinary understanding is likely to be as useful as a specialized knowledge of any one systematic branch of their own subject. Ideally, a systematic specialization should be married with an area commitment but if it is physically impossible to accommodate both, the former will probably have to be subordinated to the latter. The systematist is likely to be tempted to range widely in pursuit of his examples. The world will be his oyster. The area specialist, though not exactly cabined, cribbed, and confined, will usually have to deny himself the pleasures of other places. Such people are born as often as they are made. Intuition and accident explain as much as conscious decision in their recruitment. For those who make a conscious decision, there is a ready-made proverb—*Allez en avant et la foi vous viendra*. The faith that follows assumes emotional commitment. Where an assignment is given and a commitment has to be cultivated it should begin early and, to bear the most abundant harvest, it needs to be long term. Such a commitment is demanding of energy as well as of time, it will often be inconvenient (especially domestically), and it will call for patience in the face of frustrations.

The days when private incomes were available for investment in scholarly research in other lands have long since passed. Accordingly, there will be financial stresses. Because such work is annually recurrent over a long period, research organizations will generally have limited sympathy for it. There is an impression (though the same does not appear to hold for anthropologists) that research overseas is a matter of a student *wanderjahr* into the territory of the tourist brochure. As a result, it calls at best for personal supplementation of finance, inevitable dependence on the hospitality of the host country (usually on that of its ordinary citizens), and, so far as the British are concerned, adjustment to a more slender purse than that enjoyed by scholars coming from countries which have a greater belief in the value of investment in such pursuits.

The functions of an area specialist are not only to keep abreast of developments in knowledge in a particular territory, to disseminate it, and possibly to add to it. Specialist though he may be on his home ground, he must accept the fact that as a geographer he is likely to become increasingly a generalist. He will have to discipline himself against becoming myopic and developing a one-track mind. He must beware 'the soft charms and subtle comfortable chains' (Butterfield 1951) of foreign governments which would have him write that which is politically and socially acceptable. He must be prepared to disturb the comfortable affirmation of cherished misconceptions about the area of his interest. He must learn that, though he may have the satisfaction of regular absorption in real places and real people, when he attempts to communicate his experiences about them, they will often 'withhold their secrets and dissolve away into sets of characteristics floating on an ether of abstract intellect' (McMurray 1972). He must also be tolerant of those who lack his training and background but whose heightened perceptions frequently enable them to detect a truth which he has overlooked. It is not necessarily the specialist who sees first the nature of the emperor's new clothes. In the final place, he may have to accept the consolation prizes; though among those who share his dedication, he will find a powerful kinship (Mead 1963).

There has, of course, been limited and sometimes imaginative recognition of the need for area specialists. Among the six recommendations for priorities in geographical research identified in the American *Behavioural and Social Science Survey* (ed. Edward J. Taaffe 1970) was the expansion of support programmes for foreign area study. The Social Science Research Council has always been more hesitant in its assistance. Here, however, the context is different, because an alternate source of funding has been available for research in certain foreign areas. One of the really imaginative post-war arrangements was the provision through the University Grants Committee of funds earmarked to enable area specialists to pay regular visits to their fields of research. The facilities that this source of finance have provided have been indispensable to those whose work lies in Eastern Europe, Africa, and the Orient (UGC 1961). Latin Americanists have been similarly assisted (UGC 1965). The current problem is that funding does not keep pace with increasing costs and that there is no parallel assistance to provide the complementary library resources.

## ENTENTE CORDIALE

As with other social scientists, geographers are consciously cultivating a language sense and a sign sense. At the one extreme are those who are dissatisfied with the imprecision of language, move among mathematical models, reduce their arguments to signs and symbols, and embrace regional science. At the other, are the humanistic geographers who acknowledge the fact of subjectivity and who believe that a fuller understanding of perception will lead to a deeper understanding of geography. While the regional scientists are concerned with exactness, the humanistic geographers argue that exactness (the achievement of which remains a matter of degree) is not enough. At a certain stage in investigation (and in education) exactness may be necessary; but the humanistic geographer would concurr with the sentiment of Freya Stark that 'exactness of knowledge is the anteroom; beyond the classroom threshold, the presence chambers shine in clear gold' (1940). Such a point of view is less a pendulum-swing from numerology to Karl Popper's 'almost poetic intuition' than a realization that the same problem may call for different methodological procedures at different stages in its solution. In the field of geography, between those who work with signs and those who labour with words no precise boundary can be drawn (Steiner 1966).

The arithmomorphic and the non-arithmomorphic have not parted company in the last decade—any more than the physical and the human geographers. In the interim, a range of compromise positions has been adopted, which at least facilitates an *entente cordiale*. Some see 'the diligent and able quantitative analysts' as automatically embracing the geographical tradition when they 'search for character and process in the form of spatial interconnections'—interconnections which are likely to be the core of a functional region (Dickenson 1977). Some see the traditional regionalist and the regional scientist as sharing 'a common perspective of knowledge'—an 'outside' view in contrast to the 'inside' view of those whose social and economic activities they investigate. The analysis of at least some 'concrete life situations' favours the joint employment of both approaches (Buttimer 1979). There remain some sets of circumstances that produce 'an acute sensitivity to place and community', that yield 'a parochial humanism', and that are 'locked into the absolute spaces generated by the regional concept' (Harvey 1974). Ultimately, perhaps, it is only 'a work of art' which can succeed in capturing the essence of such a geographical situation (Yi Fu Tuan 1976).

Whether it be the intuitive regionalist, the geographer turned regional scientist, or, indeed, the regional geographer turned area specialist, all must accept that they are working in a discipline which is anything but autonomous. Geography remains a synthetic subject, the weaknesses of which are often compounded for those who work in a regional context. If some regionalists sometimes feel that they acquire a reputation as Jacks-of-all-trades, they may find consolation in the fact that they not infrequently serve the same positive integrating function that a department of geography tends to perform in a modern university. Furthermore, in so far as geography can still be acceptably defined as 'a regional synthesis with a time depth' (Hägerstrand 1977), the regionalists—for all their shortcomings—probably approach nearer to the heart of the subject than most.

# REFERENCES

BERRY, B. J. L. 1964. Approaches to regional analysis: a synthesis. *Ann. Assoc. Amer. Geog.* 54, 2–11.

—— and MARBLE, D. F. 1968. *Spacial analysis.* Englewood Cliffs.

—— and WRÖBEL, A. (eds.) 1969. Economic regionalisation and geographical methods. *Geographica Polonica,* 15. Warsaw.

BLACKSELL, M. 1977. *Post-war Europe.* Folkestone.

BLAKEMORE, H. and SMITH, C. T. 1971. *Latin American perspectives.* London.

BRAUDEL, Fernand. 1972, 1973. *The Mediterranean and the Mediterranean world in the age of Philip II,* 2 vols. London.

BUTTERFIELD, H. 1951. *History and human relations.* London.

BUTTIMER, Anne. 1976. Grasping the dynamism of the life world. *Ann. Assoc. Amer. Geog.* 66: 227–42.

—— 1978. Home, reach and the sense of place, in *Regional identitet och förändring i den regional samverkans samhälle, Acta universitatis upsaliensis symposia,* 13–39. Uppsala.

—— 1979. 'Insiders', 'Outsiders' and the geography of regional life, in Kuklinski, Antoni, and Kultalahti, Olli (ed.), *Regional dynamics and socio-economic change.* Tampere.

CARTER, F. W. 1977. *Historical geography of the Balkans.* London.

CHORLEY, R. J. 1974. *Directions in geography.* London.

CLARKE, J. I. and PINCHEMEL, PHILIPPE 1976. *Human geography in Britain and France.* London, SSRC.

CLOUT, H. D. 1972. *The geography of post-war France.* London.

—— (ed.) 1975. *Regional developments in Western Europe.* London.

—— 1976. *The regional problem in western Europe.* London.

COMMISSION OF THE EUROPEAN COMMUNITIES 1973. *Report of the regional problem in the enlarged community.* Brussels.

COMPTON, P. 1976. *Regional development and planning, Britain and Hungary, Case Studies.* Budapest.

DICKENSON, J. R. 1978. *Brazil.* London.

DICKENSON, R. E. 1970. *Regional ecology, the study of man's environment.* London.

—— 1976. *The regional concept.* London.

DUHAMEL, G. 1926. *La géographie cordiale de l'Europe.* Paris.

ESTALL, R. C. 1977. Economic geography and regional geography. *Geography,* 62: 4.

EVANS, E. E. 1973. *The personality of Ireland: habitat, heritage and history.* Cambridge.

FARMER, B. H. 1973. Geography, area studies and the study of area. *Trans. Inst. Brit. Geog.* 60: 1–15.

FIFER, V. J. 1972. *Bolivia: land, location and politics since 1825.* Cambridge.

FISHER, W. B. 1971. *The Middle East.* London.

—— (ed.) 1968. *The Cambridge History of Iran.* Cambridge.

FISHER, C. 1979. *Three times guest.* London.

GALE, S. 1976. A resolution of the regionalisation problem and its implications for political geography and social justice. *Geogr. Ann.* Series B. 58, 1–16.

GEOGRESCU-ROEGEN, N. 1972. *The entropy law and the economic process.* Cambridge, Mass.

GILBERT, A. 1974. *Latin American development. A geographical perspective.* London.

GRANÖ, O. 1977.  Geography and the problem of the development of science. *Terra*, 89: 1–9.

HAMILTON, F. E. I. 1969.  *Regional economic analysis in Britain and the Commonwealth, a bibliography.* London.

HARRISON-CHURCH, R. J. (*et al.*) 1973.  *North-west Europe.* London. (*Mea culpa.* There is no such region as north-west Europe, but it is an attractive area about which to write.)

HARVEY, D. 1969.  *Explanation in geography.* London.

HAVET, J. (ed.) 1978.  *Main trends in the social and human sciences.* Paris.

HEMMING, J. 1978.  *Red gold. The conquest of the Brazilian Indians.* London.

HOFFMAN, G. 1971.  *Eastern Europe, essays in geographical problems.* London.

HÄGERSTRAND, T. 1970.  Regional forecasting and social engineering, in Chisholm M. (ed.), *Regional forecasting.* Cambridge.

—— 1976.  The geographer's contribution to regional policy, the case of Sweden, in Coppock, J. T. and Sewell, W. R. D. (ed.), *Spatial dimensions in public policy.* New York.

—— 1978.  in Carlstein, Tommy, Parkes, Don, and Thrift, Nigel (ed.) *Timing space and spacing time, II.* London.

ISARD, W. 1960.  *Methods of regional analysis.* Massachusetts, M.I.T. Press.

—— 1975.  *Introduction to regional science.* London, Englewood Cliffs.

JANIK, A. and TOULMIN, S. 1973.  *Wittgenstein's Vienna.* New York.

JOHN, B. *The Scandinavian World* (in press). London.

JONES, M. 1977.  *Finland, a daughter of the sea.* Folkestone.

JORDAN, T. G. 1973.  *The European cultural area.* London.

KIRK, W. 1978.  The road from Mandalay. *Trans. Inst. Brit. Geog.* 3, 4: 382–94.

LAFONTE, R. 1967.  *La révolution régionaliste.* Paris.

LAMBERT, A. M. 1971.  *The making of the Dutch Landscape.* London.

LEE, R. and OGDEN, P. E. (ed.) 1976.  *Economy and society in the E.E.C.* London.

LÉVI-STRAUSS, O. 1973.  *Tristes tropiques.* New York.

LUKERMANN, F. 1965.  Geography *de facto* or de *jure. J. of the Minnesota Academy of Science* 3: 188–96.

MCMURRAY, U. 1972.  *Reason and emotion.* London.

MEAD, W. R. 1963.  The adoption of other lands. *Geography*, 48: 241–54.

MELLOR, R. E. H. 1975.  *Eastern Europe and the economy of the Comecon countries.* London.

MINSHULL, G. N. 1978.  *The New Europe.* London.

MUTTON, A. F. M. 1970.  *Central Europe.* London.

ODELL, P. R. and PRESTON, D. A. 1973.  *Economies and societies in Latin America.* London.

OLSSON, G., and GALE, S. 1979.  *Philosophy in Geography.* London.

PATTISON, J. H. 1974.  Writing regional geography, problems and progress in the Anglo-American realm. *Progress in geography*, 6. London.

SCARGILL, I. D. 1973.  *Problem regions of Europe.* Oxford (in continuation).

SPENCE, N. and TAYLOR, P. 1970.  Quantitative methods in regional taxonomy. *Progress in Geography*, 2.

STARK, F. 1940.  *Perseus in the Wind.* London.

STEINER, G. 1969.  *Language and Silence.* London.

THOMPSON, I. B. 1970.  *Modern France.* London.

TURNOCK, D. 1978.  *Eastern Europe.* Folkestone.

—— 1974.  *An economic geography of Romania.* London.

TUAN, Yi Fu 1976.  Humanistic geography. *Ann. Assoc. Amer. Geogr.* 66. 2: 266–76.

UNIVERSITY GRANTS COMMITTEE 1961.   *Report of the Sub-Committee on Oriental, Slavonic, East European and African Studies.*

—— (The 'Hayter' Report) 1965.   *Report of the Committee on Latin American Studies* (the 'Parry' report).

WARD, P. 1978.   Self help housing in Mexico City. *Town Planning Review*, 49: 38–50.

WEAVER, C. 1978.   Regional theory and regionalism. Towards rethinking the regional question. *Geoforum*, 397–413.

WHEATLEY, P. 1966.   *The golden Kersonese.* Kuala Lumpur.

WHITTLESEY, J. D. 1954.   The regional concept and the regional method, in James, Preston E., and Jones, C. F. (ed.), *American Geography: inventory and prospect.* New York.

WRIGLEY, E. A. 1965.   Changes in the philosophy of geography, in Chorley, R. G., and Haggett, P. (ed.), *Frontiers in geographical teaching.* London, 3–20.